SIGNAL PROCESSING

Using Analog and Digital Techniques

K. G. BEAUCHAMP

C.G.I.A., C.Eng., F.I.E.R.E., F.B.C.S.
Director, Computing Centre, Cranfield Institute of Technology

With a Foreword by

Dr. I. MADDOCK

C.B., O.B.E., F.R.S.
Chief Scientist, Department of Trade and Industry

LONDON · GEORGE ALLEN & UNWIN LTD
Ruskin House   Museum Street

First published in 1973

ISBN: 0 04 621019 9

Printed in Great Britain
in 10pt. IBM Press Roman
by William Clowes & Sons Ltd
London, Colchester and Beccles

SIGNAL PROCESSING

FOREWORD

by I. MADDOCK, C.B., O.B.E., F.R.S.
Chief Scientist
Department of Trade and Industry

There are a number of technologies–the intermediate technologies–which do not have the glamour and panache of the 'high' technologies like aerospace, nuclear energy or communications, but which have a pervasive quality spreading them across technological boundaries and making them vital elements in many technological structures, including those of the giants themselves. Signal processing is one of these.

Like most of these 'intermediate' technologies the literature on signal processing also spreads into the literature of other technologies; often it is found as part of some subject such as control engineering, communications, weapon guidance, seismology or biology. Individual papers on topics in signal processing are often both deep and narrow in their specialisation and anyone setting out to study the subject in a broad way will find both his time and patience heavily taxed, even when he is able properly to select his reading.

I believe that in this book the author has provided a useful service in bringing together and sifting most of the theory on which signal processing depends. He does not pretend to provide solutions to the reader's own problems but offers him the tools with which he can work. Approached in this way the book provides a real service to the practising engineer or research worker who becomes involved in signal processing or wishes to broaden his knowledge of its theoretical basis. For the graduate who is entering this field of work and who needs a fairly self-contained theoretical view of the subject it will be useful also.

The author has been specially interested in this subject for many years; as an engineer, as a researcher and as a lecturer, and in my direct knowledge and experience has shown great expertise in the field. It is of particular value therefore, that he should have distilled this background and insight into this book.

I. MADDOCK

PREFACE

A study of random processes now forms an indispensable part of many practical scientific investigations and is often regarded as part of the design procedure in an engineering environment. Originally confined to mechanical and electrical systems, analysis techniques are currently applied throughout the physical sciences extending to such diverse fields as economics and biomedical work.

As a consequence the understanding of such techniques is no longer restricted to the more mathematically orientated engineers but has become the province of a much wider range of technicians, research workers and scientists who have a need for concise information directly relevant to their field of interest.

The problems of such a study can be considered as falling into three catagories:

1. acquisition of information,
2. processing of information,
3. interpretation of processed results.

This book will offer little help in dealing with step (3). It is hoped however, that it will provide a practical base from which to carry out steps (1) and (2) in a manner designed to avoid the many pitfalls in data acquisition and processing which can occur.

Interpretation is another matter. This needs detailed knowledge of the process under study and, although close liaison between groups responsible for these three activities will obviously create a favourable climate for useful conclusions, correct interpretation will rarely be obtained solely from analysis results.

Processing and analysis procedures are capable of implementation in a number of ways and a secondary objective of this book is to offer some guidance concerning the relative advantages of these alternatives. Often the choice lies between continuous or discrete methods and, in some cases, there is merit in considering some combination of the two. For this reason the relative value of analog, digital and hybrid techniques is constantly discussed throughout the text. The last decade has been extraordinarily productive of improvements in digital equipment and techniques but it cannot be said that analog methods have been entirely supplanted. Much initial processing, including data capture, is continuous by process definition and the economics of signal analysis using a combination of analog and digital techniques continues to exert its influence on the selection of method, particularly for real-time work.

The central core of the book is concerned with signal analysis. This in realistic terms means data reduction. Any practical measurement in a dynamic environment will involve acquisition of a mass of data and the prime purpose of analysis is to reduce this to manageable parameters describing some relevant feature or features of the physical process. A difficulty for the research worker is often a lack of awareness of the techniques that are available and how they can be applied to enable useful parameters to be derived and stated in units that have meaning to his problem. All too often the techniques he requires are available in mathematical terms and expressed in a formalism which is not directly applicable to machine implementation. One of the aims of the analysis chapters is to describe the range of techniques now available and to indicate where and when they can be applied.

Definitions and concepts are used heuristically. Rigorous proofs are not generally included and adequate references are provided for those who desire this information. The intention is rather to state the form of the analysis, its relationships with other methods, and to show the limitations present and their effect on the resultant accuracy.

Chapter 1 is introductory in concept. It discusses signal characteristics and gives a broad outline of analysis possibilities described in detail in later chapters.

Chapter 2 deals with the mechanics of data capture and storage. Transducer selection and use is covered in the first part of this chapter which concludes with a description of the methods and problems associated with the bulk storage of signal information.

Chapter 3 considers those processing operations necessary between the acquisition of raw data and its subsequent analysis. Pre-processing is required for economic and practical reasons associated with the imperfect and uncalibrated nature of the initial signal. Since this is likely to be continuous in form and to require frequency filtering a treatment of analog filters is included.

Chapter 4 considers a particular form of pre-procesisng, namely that of analog-to-digital conversion, and in particular the various forms of conversion error likely to arise.

Chapter 5 gives a brief introduction to statistical terminology which is followed by a discussion of amplitude domain analysis.

Chapter 6 provides a basis for analysis in the frequency domain using Fourier methods with particular reference to the fast Fourier transform.

Chapter 7 is concerned with discrete data filtering and emphasises the use of the z-transform as a tool for digital filter design.

Chapter 8 continues the discussion of analysis methods in the frequency domain, commenced in chapter 6, and describes a number of techniques of spectral analysis. This type of analysis when applied to random data is, at best, an estimation and consideration is given to this and to the value that can be attached to such estimates.

Chapter 9 describes measurements in the time domain and in particular with correlation estimates, their relationship to spectral estimates, and use in signal-to-noise ratio enhancement.

Chapters 10 and 11 consider signals which have transient or time-varying frequency characteristics. Many of the data sequences presented for analysis are unique in the sense of representing unrepeatable transient conditions and may, in addition, contain a frequency spectrum which is not constant over the length of the sequence. Chapter 10 is mainly concerned with transient signals derived from mechanical structures subject to shock and the spectral characteristics so obtained. Chapter 11 presents a review of the current methods of non-stationary analysis and concludes with some notes on the applicability of such methods.

The examples given in this book have been drawn from a wide variety of sources. I wish to thank the Science Research Council for permission to include details of the satellite data processing which is included in chapter 3 and also the United Kingdom Atomic Energy Authority for the many references to the processing of seismic data which appear throughout this work. I also wish to acknowledge the contributions made by many colleagues during the discussions on techniques of analysis which forms the basis for much of the text. In particular thanks are due to Miss S. E. Pittem, Miss M. E. Williamson, Mr P. Hurst, Mr G. E. Johnson, and Mr G. Disbrey for their careful reading of various sections of the text. For the typing of the manuscript I extend to Mrs A. Reynolds my grateful appreciation.

K. G. BEAUCHAMP
Cranfield

CONTENTS

LIST OF SYMBOLS

a	constant coefficient; acceleration
a_0	Fourier coefficient (constant term)
a_k	Fourier coefficient
a.c.	alternating current
A	amplitude; area; constant coefficient
A_k	complex Fourier coefficient
A.D.C.	analog-to-digital converter
b	constant coefficient
b_k	Fourier coefficient
B	bandwidth; bit; magnetic flux density
B.C.D.	binary-coded decimal
B.S.I.	British Standard Interface
c	damping coefficient
C	capacitance; degrees centigrade
C_c	critical damping
C_x	autocovariance
C_{xy}	cross-covariance
$C_{xy}(\tau)$	convolution function
$C_{xy}(f)$	single-sided co-spectral density function
C.R.T.	cathode-ray tube
d.b.	decibel
d.c.	direct current (zero frequency)
D	damping factor
D.A.C.	digital-to-analog converter
D.F.T.	discrete Fourier transform
e	error; strain
e e.m.f.	electromotive force
$\overline{e^2}$	mean square error
$e(t)$	excitation time history
exp	2·71828 base of Naperian logarithms
E	expected value; Chebychev coefficient
$E(\omega)$	Fourier transform of $e(t)$
E.E.G.	Electro-encephalograph
E.M.P.	exponentially mapped past
f	frequency
f_c	cut-off frequency

f_0	fundamental frequency
f_N	Nyquist frequency
f_s	sampling frequency
F	force; statistical F variable
F( )	filter function, Fourier transform of ( )
F.F.T.	fast Fourier transform
F.M.	frequency modulation
g	acceleration
G	acceleration due to gravity (=9·81 m/sec²)
$G(f)$	single-sided spectral density function (auto-spectrum)
$\hat{G}(f)$	estimated value of $G(f)$
$G_x(\omega)$	single-sided spectral density function (angular frequency)
$G_x(k)$	discrete single-sided spectral density function
$G_x(t, \omega)$	generalized spectral density function
$G_{xy}(k)$	discrete single-sided cross spectral density function (cross-spectrum)
$G_{xy}(\omega)$	single-sided cross spectral density function (angular frequency)
h	sampling interval
h_k	filter weight series
H	system gain factor; height
$H(\omega)$	transfer function; frequency response function
$H(s)$	Laplace transfer function
$H(j\omega)$, H(p)	complex angular frequency response function
$H(z)$	z-transfer function
Hz	Hertz (cycles/sec)
i	index; current
i.c.	initial condition
i.p.s.	inches per second
Im	imaginary part of a complex function
I.D.F.T.	inverse discrete Fourier transform
I.R.I.G.	inter-range instrumentation group
j	$\sqrt{(-1)}$
J_n	Bessel function
k	index value: gauge factor
kg	kilogram
k	constant of proportionality; spring coefficient
K	sensitivity constant
$K_x(\tau)$	Cepstrum function
$K_x(k)$	discrete Cepstrum function
i, L	length
$\log_N$	logarithm to base N
L	inductance
m, m	metre; modulation factor; autocorrelation lag

M	mass
n	index; number of variables; degrees of freedom
$n(t)$	noise function
N	number of discrete samples of a variable
N.R.Z.	non-return to zero
p	Heaviside operator, d/dt
p-p	peak-to-peak
$p(x)$	probability density function
$p(x, y)$	joint probability density function
p.s.i.	pounds per square inch
P	power; potentiometer; probability
$\bar{P}$	average power
$P_n(\omega)$	periodogram
$P(x)$	probability distribution function
$P(x, y)$	joint probability distribution function
P.C.M.	pulse code modulation
P.S.D.	power spectral density
q	quantisation step
Q	electric charge; magnification ($\omega L/R$)
$Q_{xy}(f)$	quadrature spectral density function
r	radius of curvature
$r(t)$	response time history
r_{max}	maximum response value
$r(t, \omega)$	two-sided response
r.m.s.	root mean square
R	scanning rate; resistance
Re	real part of a complex function
$R_x(k)$	discrete autocorrelation function
$R_{xy}(k)$	discrete cross-correlation function
$R_x(0)$	maximum value of autocorrelation function
$R_x(t_1, t_2)$	non-stationary autocorrelation function
$R_x(\tau)$	autocorrelation lag function
$\hat{R}_x(\tau)$	estimated value of $R_x(\tau)$
$R_x^c(\tau)$	circular autocorrelation lag function
$R_x^s(\tau)$	single-bit autocorrelation lag function
$R_x^c(k)$	discrete circular autocorrelation function
$R_{xy}(k)$	discrete cross-correlation function
$R_{xy}(\tau)$	cross-correlation lag function
$R_{xy}^c(\tau)$	circular cross-correlation lag function
$R_{xy}^s(\tau)$	single-bit cross-correlation lag function
R.Z.	return to zero
$S = \sigma_x$	standard deviation
s	Laplace variable; complex variable

(s)	smoothed value
sgn( )	sign of ( )
sec	seconds
$S_x(f)$	double-sided spectral density function
S/H	sample and hold
t	time; thickness
T	period of signal; length of signal sample
T_a	averaging time
T_s	total averaging time
$u(t)$	unit step function
v	velocity
V	volts
Var	variance
V_i	input voltage
V_o	output voltage
W	$\exp(-2\pi j/N)$, watts
$W(f)$	frequency window function
W_k	discrete weighting function
$W(t)$	weighting function
$W(\tau)$	lag window function
$\bar{x}$	average value of x
$\overline{x^2}$	mean square value of x
$x(f)$	continuous function of frequency
$x(f, t)$	continuous function of time and frequency
x_i, x_k	discrete function of time
$x(j\omega)$	complex continuous angular function of frequency
x_n	subscripted variable
$x(t)$	continuous function of time
$x(\omega)$	continuous angular function of frequency
$x(\omega, t)$	continuous function of time and angular frequency
X	positional coordinate
$|X|$	modular value of X
$X(f)$	Fourier transform of $x(t)$
$X_c(f)$	cosine Fourier transform of $x(t)$
$X_s(f)$	sine Fourier transform of $x(t)$
X_n	discrete Fourier transform of x_i, x_k
$y(t)$	continuous function of time
$y_i(t)$	sampled function of time
Y	positional coordinate; conductance
$Y_{xy}^2(f)$	coherence function
z	random variable; z-transform
z^{-1}	unit delay
Z	impedance; positional coordinate; displacement

α	constant
β	constant
γ	constant
Γ	gamma function
δ	deviation
$\delta(t)$	delta function; Dirac impulse function
$\delta(f)$	frequency deviation; increment in frequency
δ_l	increment in l
δ_t	delta function; in time
δ_x	increment in x
$\partial/\partial t$	partial derivative
Δt	increment in time
ϵ	normalised standard error
θ	phase angle; angle of rotation
λ	wavelength
$\mu s, \mu sec$	microstrain; microseconds
ξ	random variable
π	product
σ_x	standard deviation
σ_x^2	variance
Σ	summation
τ	time delay lag
$\tau_{max.}$	maximum value of lag
ϕ	phase angle; magnetic flux
χ_n^2	chi-squared variable having n degrees of freedom
$\omega = 2\pi f$	angular frequency
ω_c	angular cut-off frequency
ω_0	fundamental angular frequency
ω_N	angular Nyquist frequency
ω_s	angular sampling frequency
ω_n	angular natural resonant frequency

Other symbols

*	sampled data; convolution, complex conjugate
!	factorial
∞	infinity
$\propto$	proportional to

Chapter 1

INTRODUCTORY

1-1 WHAT IS A SIGNAL?

We can recognise a large number of sources providing information on the presence or change of a physical phenomena. If arrangements are made to convert this information from its original physical form (e.g. temperature, pressure, vibration etc.), into a directly related electrical quantity (e.g. potential, current, power, etc.), then this electrical quantity may be called a Signal. This signal may merely indicate the presence of the physical phenomena or it may provide one or more related parameters concerning its behaviour. Many signals behave as continuous processes during the period of their acquisition and provide a history of the quantity being measured. This continuing but finite length record of the process is termed a 'Time-History'.

In the material that follows, the most common independent variable is time, and we can express the time-history as either a continuous function of time, $x(t)$, or a discrete sampled function, x_i for a single dependent variable.

The second most common independent variable is frequency for which the series of values obtained may be expressed in terms of a direct function of frequency, $x(f)$, and $x_i(f)$ or as a function of angular frequency, $x(\omega), x_i(\omega)$ or as a complex frequency, $x(j\omega), x_i(j\omega)$.

Many other independent variables for the signal are possible. Examples are found in astronomical and economic data which may be related to spatial position or some arbitrary value for which the time of acquisition is irrelevant.

The mathematical analysis techniques to be described however are perfectly general and the substitution of one of these other independent variables for time is frequently carried out and requires only a re-arrangement of terms.

The acquired signal may not always conveniently be obtained as a function of a single variable, although it may very often be considered as such. A common example is a signal which is both a function of frequency and time and expressed as, $x(f, t)$ or $x(\omega, t)$. Representation of the two-dimensional time-frequency history as a single equivalent independent variable or sequential set of variables is often attempted as part of the analysis procedure.

All signals acquired from contact with a physical environment can be expressed in the form of a time-history even if the independent variable is other than time. Before any attempt can be made to extract quantitative data from this, some information must be obtained concerning the environment from which the signal was derived, together with information concerning the signal itself in order to permit recognition of the type of signal we are dealing with. Without this preliminary classification meaningful analysis is not possible.

1-2   SIGNAL CLASSIFICATION

Where the physical quantity being measured can be described explicitly in terms of mathematical relationships then it is likely that we can regard the derived signal in the same way. This type of signal is classed as Deterministic, and may be identified further as belonging to one of a number of groups discussed below. More generally the signal is likely to be Random, that is, we will not be able to predict precisely its value at any given future instant in time. Analysis must then proceed in terms of statistical values from which it is possible that a deterministic relationship may be obtained. One criteria used to decide whether the signal is deterministic or random is to compare several sets of data obtained under identical conditions over a reasonable period of time. If similar results are obtained then the data is likely to be deterministic.

It is important to distinguish between the required information concerning the physical measurement and the distortion and extraneous noise that may be included with the available signal. A signal may appear to be random and yet yield deterministic data by applying a suitable form of analysis, e.g. auto-correlation. In such a case random analysis methods will be succeeded by deterministic measures later in the analysis procedure. Alternatively random data may itself represent the phenomena under investigation and the signal obtained will contain this random data mixed with periodic data of no interest. Analysis methods arranged to detect and remove the periodic component will then precede the required random signal analysis.

1-2-1   DETERMINISTIC SIGNALS

The mathematical form of the physical quantities represented by the deterministic signal, if known (or assumed), will either continuously repeat at regular intervals or decay to a zero value after a finite length of time. The former are known as Periodic Signals and the latter as Transients.

Periodic signals can be considered as comprising one or more sinusoidal signals having a integral relationship with the period of repetition. This period, T, is defined as the interval of time over which the signal repeats itself. (We are considering only single-dimensional time-histories here.)

The simplest example is the sinusoidal function, see fig. 1-1

$$x(t) = A \sin(\omega_0 t + \theta),\tag{1-1}$$

where A is a constant representing the peak amplitude of the wave-form, ω_0 is the angular frequency $= 2\pi f_0$, f_0 is the cyclical frequency in Hz, and θ is the initial phase angle with respect to the time origin in radians.

This signal repeats itself over one complete cycle of frequency so that the period T, will be related to f_0 by

$$T = 1/f_0.\tag{1-2}$$

A more general relationship for periodic signals is obtained by considering $x(t)$ to comprise harmonically related sinusoids and expressed as a Fourier series, where each harmonic component repeats itself exactly for all values of t, i.e.

$$x(t) = A_1 \sin(\omega_0 t + \theta_1) + A_2 \sin(2\omega_0 t + \theta_2) + \ldots A_n.\sin(n\omega_0 t + \theta_n)\tag{1-3}$$

or

$$x(t) = \sum_{n=1}^{n=\infty} A_n.\sin(n\cdot\omega_0 + \theta_n)\ (n = 1, 2, 3, \ldots)\tag{1-4}$$

where A_n = peak amplitude of the nth harmonic, ω_0 = fundamental angular frequency $= 2\pi f_0$, and θ_n = phase of the nth harmonic with respect to the fundamental. The sinusoidal case, equation (1-1), will be seen to be a special case of equation (1-4) for $n = 1$. Several examples of non-sinusoidal periodic signals are illustrated in fig. 1-2.

The frequency content of such periodic signals may be expressed by 'line spectra' as shown in fig. 1-3. Individual frequencies comprising the signal are discrete and located at precise positions on the frequency axis.

A deterministic signal consisting of the synthesis of several sinusoidal elements, which are not necessarily harmonically related, is frequently found. Such signals are common in communication engineering where heterodyne effects between two or more wave-forms take place producing a multi-frequency resultant having no precise repetition period. This combination may be produced by summing or multiplication of sinusoids and any combination of these. An example of this type of signal would be

$$x(t) = A_1 \sin \omega_1 t + A_2 \cos \omega_2 t + A_3 (\sin \omega_1 t)(\cos \omega_3 t).\tag{1-5}$$

This will not result in a periodic signal but will exhibit similar spectral characteristics.

Transient signals are time-varying signals which reduce to zero value over a finite time interval. Some examples of these are given in fig. 1-4.

Such signals are characterised by continuous spectra rather than line spectra, since in theory an infinite number of frequency components are present in the

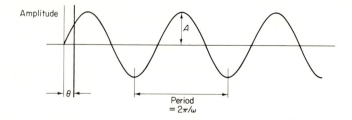

Fig. 1-1. A sinusoidal function.

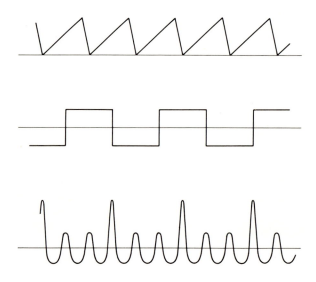

Fig. 1-2. Non-sinusoidal periodic functions.

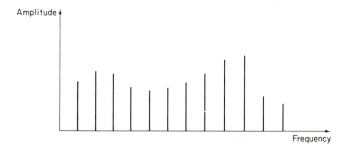

Fig. 1-3. Line spectra.

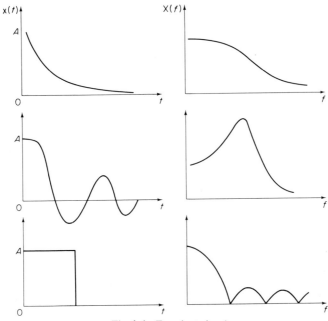

Fig. 1-4. Transient signals.

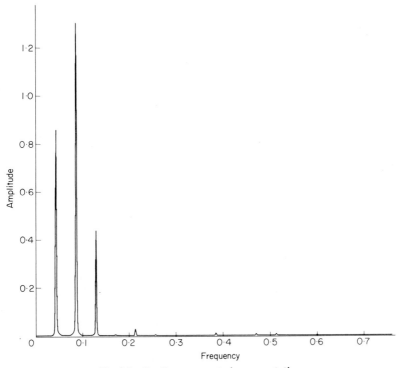

Fig. 1-5. Continuous spectral representation.

signal. Fourier series representation can no longer be used and spectral representation is obtained by means of the Fourier Integral:

$$X(f) = \int_{-\infty}^{\infty} x(t) \exp(-j2\pi ft) \, dt. \tag{1-6}$$

This is a complex quantity and can only be represented completely by an amplitude and phase value for $X(f)$ over the frequency range.

A modulus form $|X(f)|$ is often used to represent the spectra for a transient signal since this form results from several common methods of measurements which do not take into account the phase value of the signal. An example is given in fig. 1-5.

1-2-2 RANDOM SIGNALS

For the deterministic signals discussed above their functions can be obtained from their time-histories, and generally made to repeat exactly under identical conditions. This is not the case with random signals where their functions cannot be determined explicitly and where probabilistic and statistical descriptions will need to be used.

This will be understood from the consideration of a particular random process which we will assume to produce a set of time-histories, known as an Ensemble, each referenced to an identical time instant or commencement time, $t = 0$. See fig. 1.6. This could represent for example, an experiment producing random data, which is repeated N times to give an ensemble of N separate records. Statistical values for this ensemble can be obtained by considering the values of these records taken at specific instants in time.

For example the average value existing at time, t_1 summed over the ensemble is

$$\bar{x}(t_1) = \lim_{N \to \infty} \frac{1}{N} \sum_{k=1}^{N} x_k(t_1). \tag{1-7}$$

Also the average value of the products of two samples taken at two separate times, t_1 and t_2 for each separate record (also known as the Autocorrelation Function) is

$$R(\tau) = \lim_{N \to \infty} \frac{1}{N} \sum_{k=1}^{N} x_k(t_1).x_k(t_2). \quad (\tau = t_2 - t_1) \tag{1-8}$$

The process of arriving at these values is known as Ensemble averaging and may be continued over the entire record length to provide a statistical description of the complete set of records.

A signal is defined as Stationary if the values of $\bar{x}(t_1)$ and $R(\tau)$ are found to remain constant for all possible values of t_1 and τ. In a more realistic and practical case we say that the signal is stationary if $\bar{x}(t_1)$ and $R(\tau)$ are constant

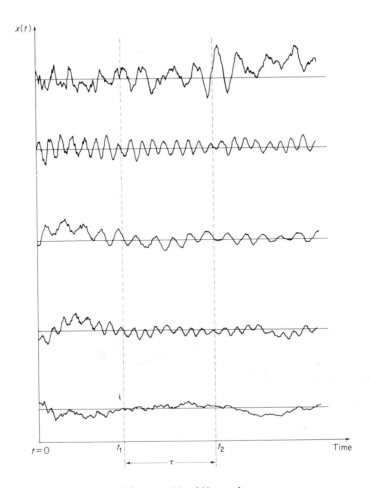

Fig. 1-6. Ensemble of N records.

over the finite record length, T. If the values of $\bar{x}(t_1)$ and $R(\tau)$, vary with time then the signal is defined as being Non-stationary. This is the more general case although for very many practical situations the change with time is so slow as to permit broad classification as a stationary process.

Under certain circumstances we can regard a signal as stationary by considering the statistical characteristics of a single long record. Referring to fig. 1-7 if the record is partitioned into P equal length sections of length T, then the average value of any section, M will be

$$\bar{x}_M = \lim_{T \to \infty} \frac{1}{T} \int_0^T x_M(t).dt \qquad (1\text{-}9)$$

23

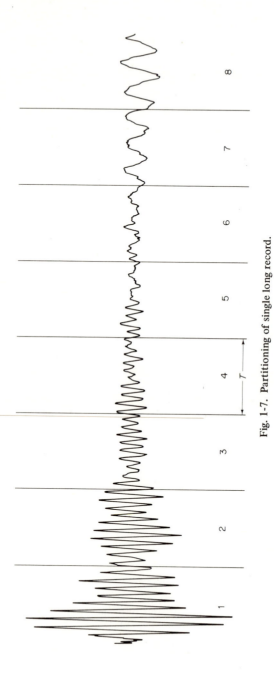

Fig. 1-7. Partitioning of single long record.

and its autocorrelation function taken over any section M will be

$$R_M(\tau) = \lim_{T \to \infty} \frac{1}{T} \int_0^T x_M(t).x.(t + \tau).dt \qquad (1\text{-}10)$$

Now if this single record averaging process is also carried out for a given record taken from the ensemble of fig. 1-6 and we find that

$$\overline{x}_M = \overline{x}$$

and

$$R_M(\tau) = R(\tau) \qquad (1\text{-}11)$$

for all values of t and all sections P, then the process is called an Ergodic Random Process. By definition this must also be a stationary process.

This particular form of a stationary process has the property that all of its statistical quantities can be obtained from measurements carried out on a single record length. Often a process may be assumed to be ergodic. In such cases the property is an important one since many physical processes yield only a single realisation and an ergodic assumption permits more reliance to be placed on the analysis procedures.

Whilst the assumption of stationarity is a necessary one for ergodic processes a stationary random process need not be ergodic. We could for example have a stationary ensemble where each individual record expresses a signal at a different mean level so that the mean level of segments taken from each record would not be equal, although constant over the record length.

As will be considered later, (see chapter 11) in many practical circumstances non-stationarity may be regarded relative to some initial starting point, so that all subsequent records obtained from the same process will be assumed to have similar form. Thus a type of ergodic hypothesis may be applied which allows a single record to be examined and statistical conclusions drawn using only one realisation of the process. This is found to be an acceptable situation for slowly varying time-dependant signals although it is not to be considered as replacing the more rigorous analysis of an ensemble of records.

1-2-3 IDENTIFICATION REQUIREMENTS

Classification is obviously fairly dependent on some prior knowledge concerning the signal itself. To assist signal identification it is necessary to have some idea of its domain limitations, e.g. frequency range, amplitude or power range and time duration. Since this information (other than time) is provided as an electrical analogy of the physical quantity being measured then details of the signal environment are also required so that quantitative conclusions can be drawn. We thus need to study the conversion characteristics of the transducers used, their accuracy and limitations. We will also need to make use of information gained from previous measurements or analytical examination. The signal may not be analysed in real-time and if a temporary storage media is used then its characteristics and time-translation property, (if applicable), will also need to be known.

Finally, when considering the actual results obtained we have to realise that all measurements on random data are estimates and will be obtained at an accuracy dependent on the amount of data made available. Conclusions drawn from a single record should therefore be regarded as tentative and must be accompanied by a statement of the conditions of acquisition and measurement if gross subjective errors are to be avoided.

1-3 THE NATURE OF SIGNAL ANALYSIS

Forms of analysis for random data rely heavily on statistics and probability for reasons stated earlier, and will be considered in detail in the succeeding chapters. The broad areas of analysis are found in the frequency, time, amplitude, power, and response domains. Translation of information from one domain to another is frequently required to facilitate extraction of meaningful characteristics.

One of the earliest examples of analysis in the frequency domain was the use of the periodogram to determine 'hidden periodicities' in a time-history series. The periodogram was suggested by Schuster towards the end of the last century [1], and used to extract information concerning the variable nature of the earth's magnetic field. In its discrete form the periodogram is given by the function

$$P_k(\omega) = \frac{1}{N}\left[\left(\sum_{i=1}^{N} x_i \cdot \cos i\frac{2\pi}{\omega}\right)^2 + \left(\sum_{i=1}^{N} x_i \cdot \sin i\frac{2\pi}{\omega}\right)^2\right] (i, k = 1, 2, \dots, N).$$

$$(1\text{-}12)$$

The application of this function to the discrete data series, x_i, will identify the periodic nature of x_i by giving a peak value of $P_k(\omega)$ at $\omega = \omega_0$ and smaller peaks at $\omega = \omega_0 + (2\omega_0)/N$, where ω_0 is the fundamental angular periodic frequency of the signal x_i.

The essential idea behind the early use of the periodogram for certain time series was that if the periodic values could be determined accurately then sinusoidal wave-forms corresponding to these frequencies could be subtracted in sequence from the original signal leaving ultimately only the random independent series, $e(t)$. This assumed that a model for the data series would take the form

$$x(t) = \sum_{i=1}^{N} A_i \cdot \sin(\omega_i t + B_i) + e(t) \qquad (1\text{-}13)$$

where A_i and B_i are constants.

26

Whilst it has been successfully applied in this form to a few favourable data series [2] it may be shown to be productive of a number of spurious side-peaks which can be misleading [3], particularly as they will be found to extend over a considerable frequency range. Later workers found ways of avoiding this difficulty and under certain circumstances the periodogram can be applied successfully to spectral analysis (see later, chapter 8).

The emphasis in this early work on frequency analysis was directed towards finding a plausible model for the phenomena under investigation. Slutsky [4] suggested moving-average and regressive series models, both of which appeared to represent the behaviour of many physical variables. These particular models have reappeared recently in connection with digital filters of the non-recursive and recursive types but are otherwise little used as spectral representations.

Both of Slutsky's series either individually or in combination played a useful part in spectral representation until the development of the more generalised theories of Wiener [5] and others which reduce the dependence on specific models of the time series. Wiener considered the generation of a series of harmonically related terms

$$x(t) = \sum_{i=1}^{\infty} A_i \exp{(i, \omega_i, t)}. \tag{1-14}$$

which is essentially a Fourier-type representation and appears in one form or another in nearly all the methods of analysis later proposed for the frequency domain. In his paper on generalised harmonic analysis Wiener laid a solid mathematical foundation for the analysis of random processes which has had far reaching consequences in present-day work on communication, control and information theory.

The origins of the present methods of spectral analysis, based on a Fourier representation of a series, are due mainly to the work of Tukey and Bartlett [6, 7]. They realised that a relationship must exist between the easily calculable autocovariance function of the time-history and its power-spectrum.

In fact, the covariance is simply

$$C_x = \overline{(x_t - \overline{x})(x_{t+\tau} - \overline{x}_{t+\tau})} \tag{1-15}$$

and the power spectrum is the cosine Fourier transform of this, i.e.

$$G(\omega) = \frac{1}{2\pi} \int_{-\infty}^{\infty} C_x \cdot \cos{\omega\tau} . d\tau. \tag{1-16}$$

The relationships had been rigorously established earlier by Wiener, but lacked practical application to the problem of power spectral density evaluation.

In its later form the zero-mean autocorrelation replaced the autocovariance and the method of calculating power spectrum from the cosine transform of the mean-lagged products proved the corner-stone of spectral signal processing for a number of years. One reason for the success of this method was the reduction in calculation time. Despite the two-stage method of calculation via the auto-correlation, it proved quicker to do this rather than to calculate all of the Fourier coefficients directly.

A certain measure of success has been obtained with other more direct methods. Of particular interest is the technique of complex demodulation [8] in which the time-history series, $x(t)$ is multiplied by sin $\omega_0 t$ and cos $\omega_0 t$, giving

$$x(t) \sin \omega_0 t, \qquad x(t) \cos \omega_0 t. \tag{1-17}$$

Applying the resultant values through a low-pass filter gives two phase-related terms:

$$a = F(x_t. \sin \omega_0 t), \qquad b = F(x_t. \cos \omega_0 t) \tag{1-18}$$

The complex spectrum can be obtained as an amplitude function

$$A = 2(a^2 + b^2)^{\frac{1}{2}} \tag{1-19}$$

and a phase angle

$$\theta = \tan^{-1}(a/b) \tag{1-20}$$

for a given angular frequency, ω and time, t.

The method is easily mechanised for the analog computer and may also be applied to the analysis of non-stationary signals.

Both the calculation via the mean-lagged product and that of complex demodulation were carried out almost exclusively using analog methods. Integration is both fast and cheap using the operational amplifier and digital computation proved an expensive way to the frequency spectrum.

A considerable change in the methods of calculation for the frequency domain was brought about however by the re-discovery of the fast Fourier transform algorithm in the 1960s [9]. Calculation reverted to the direct method and the digital computer became used extensively for this purpose. To calculate all the Fourier components for N sample values, using the algorithm, requires approximately $N \log_2 N$ complex multiply–add operations compared with N^2 operations using the previous direct transform methods. For a long series this results in a reduction by several hundreds or even thousands of times in computer run time.

The speed of calculation for the fast Fourier transform enabled it to be used not only for the direct calculation of the power spectral density, but also, via the Wiener theorem in the calculation of autocorrelation. (This, it will be noted, is a reversal in the order of two-stage calculations brought about solely by the speed of the fast Fourier transform algorithm.) Its use in convolution and digital filtering is referred to in a number of papers, notably those of Stockham and Helms [10, 11] again for the reduction in computational time which it confers.

28

The spectral analysis of short-term transients associated with the analysis of the particular form of vibrational shock experienced by mechanical structures has led to the development of a special form of spectrum, known as the Shock Spectrum. An interpretation of the effects of shock is required to give information on fatigue and damage to the structure. Consequently only the peak

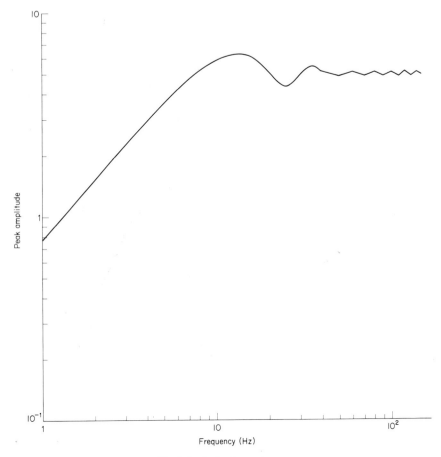

Fig. 1-8. A shock spectrum.

values of the spectra are of importance and, in the shock spectrum, only these are plotted. The times of occurrence for these peaks are of less importance than their magnitude and number so that although the acquired data is time-dependent it can be displayed in a two-dimensional form as a compact graph of peak acceleration or displacement of the structure versus frequency of resonance (see fig. 1-8). A relationship exists between certain shock spectra and the Fourier transform and will be considered in chapter 10.

Whilst spectral methods have been most widely applied and understood, particularly in the engineering field, more recent developments have emphasised the value of correlation and convolution in the time domain. This is due to the value of these functions in the analysis of the behaviour of physical systems and the use of auto- and cross-correlation for signal identification purposes. Correlation is used particularly as a means of improving the signal/noise ratio which permits the extraction of periodic signals immersed in a noisy signal, such as would be obtained, for example, from the reception of long range radar echoes. The technique used is to consider the sum of the cross-products of the signal, $x(t)$ and a delayed replica of itself, $x(t-\tau)$ over a finite time interval. This summation is repeated over a range of delay values with the result that the cyclic similarity between the two signals for the periodic components is preserved whilst the summation tends to zero for the random noise element.

The use of cross-correlation to determine the impulse response of a system is a technique for which the applications appear to be growing. It is now fairly commonplace to inject a perturbation into a system, (mechanical or otherwise), and to determine the system impulse response or transfer function by cross-correlating the system output changes with the input perturbation (noise). The value of such a method of measurement lies in the fact that little disturbance need be made to the system operating normally.

A somewhat similar use of cross-correlation is its use in establishing vibration or shock transmission routes through a structure. From a knowledge of the initial vibration frequency and amplitude at entry point the modification to this signal during its passage can be determined at a suitable exit point and thus provide information regarding the characteristics of the transmission route.

Correlation and its derivatives has also been used extensively in the analysis and study of vocal mechanisms. A particular form of spectrum having close similarities with autocorrelation, known as the Cepstrum, is valuable for this purpose. Whereas the autocorrelation may be defined as the Fourier transform of the power spectrum the Cepstrum is defined as the logarithm of the Fourier transform of the power spectrum, i.e.

$$K_x(\tau) = \left| \int_0^\infty \log_e \left| X(\omega) \right|^2 \cos(\omega\tau).d\omega \right|^2. \qquad (1\text{-}21)$$

This definition permits the unambiguous indication of the rate of repetition for the component parts of periodic signals which are very non-linear in form. Such signals are rich in harmonically related terms and an autocorrelogram would present a complicated indication in the time domain due to the profusion of extracted periodicities. In one sense the Cepstrum techniques can be said to broaden the response at the correlated peaks to improve identification characteristics [12].

The analysis methods referred to above all assume a stationary and often an ergodic signal, so that it is sufficient to consider the signal as a function of time or frequency alone. In many physical situations both must be considered and various definitions of short-term or instantaneous spectra for time-varying signals have been proposed.

An early view was to consider the energy contained in the signal by way of a double integral form of the Wiener theorem:

$$G_x(t, \omega) = \int\limits_{-\infty}^{\infty}\!\!\int R_x(t_1, t_2) \exp\left[j(\omega_1 t_1 - \omega_2 t_2)\right] dt_1 . dt_2 \qquad (1\text{-}22)$$

which has been termed the Generalised Spectral Density Function. A difficulty lies in the interpretation of this result due to the lack of a physical meaning relating it to a stationary form of analysis.

Other forms of analysis, notably those of Priestley [13] are applicable to certain single realisations of a process, and are closer to the single-point idea of spectral density. This estimate is referred to as Evolutionary Spectral Analysis and expressed in terms of energy and frequency density by the function

$$f(t, \omega) = |A(t, \omega)|^2 f(\omega). \qquad (1\text{-}23)$$

The function of time and frequency $A(t, \omega)$ performs the operation of filtering the signal over a limited time and frequency band-width so that the signal may be considered stationary over this period. It is therefore confined to functions that are found to change slowly with time.

The optimum width of the time and frequency windows implied in equation (1-23) is difficult to obtain analytically and the method can result in a large amount of regression analysis being required, using the digital computer to arrive at optimum window values.

Amplitude domain analysis in the form of probability estimates have proved useful in predicting the behaviour of systems to environmental stress. For practical reasons an equivalence between sinusoidal and random inputs for vibration testing is often taken as valid, implying that the process under test behaves in a Gaussian way. This may not always be the case so that initial probability density measurements now form an essential element in the evaluation of physical systems subject to shock and vibration stress analysis.

Other amplitude domain measurements are statistical peak counts and zero-crossing measurements. These are particularly relevant in assessing susceptibility to fatigue damage or establishing boundary conditions. The application of power spectral density methods to fatigue analysis requires certain simplifying assumptions. Generally it is assumed that the stress history may be represented by a stationary Gaussian process which will rarely be the case. Additionally some assumptions concerning the shape of the spectrum will be necessary for a given process [23].

The counting methods are not restricted in this way and permit a linear cumulative damage rule [24] to be applied which considerably simplifies the interpretation of results by the application of Miner's hypothesis [25]. This in simplified form states that if N cycles of a constant amplitude stress cause failure then n cycles of the same stress use up a fraction n/N of the life. This hypothesis and its extensions is at present the only available method of estimating fatigue life having general acceptance.

1-4   BASIC PROPERTIES OF THE SIGNAL

Assuming that the signal is stationary and ergodic then its properties can be defined in probabalistic terms for the amplitude, time or frequency domain. Brief descriptions of these properties will be given here to serve as an introduction to later chapters where they will be considered further and methods of deriving them described in some detail.

1-4-1   AMPLITUDE DOMAIN ESTIMATES
The simplest form of description for a random signal is given by its mean-square value, defined as

$$\overline{x^2} = \lim_{T \to \infty} \frac{1}{T} \int_0^T x^2(t).dt \qquad (1\text{-}24)$$

or the positive square root of this quantity known as the root mean-square value.

These values will give an indication of the amplitude effect of the signal but will not give sufficient information for the detailed understanding of the variable nature of the process. To do this the behaviour of the signal in terms of the probability of its amplitude value exceeding a given level, or lying between specified levels must be determined. Referring to fig. 1-9 the fraction t_n, of the total record time, T, that the signal lies between discrete levels x and $x + \delta x$ over the complete dynamic amplitude excursion X of the signal is given as

$$t_n = \sum_{k=0}^{k=n} t_k (x, x + \delta x, n). \qquad (1\text{-}25)$$

where n indicates the number of levels that the signal amplitude is divided into, and t_k indicates the time period that the signal dwells between these levels.

The probability density function describes the probability that the signal will be found within a given range, x and $x + \delta x$ and is normalised by δx to give a density function

$$p(x) = \lim_{T \to \infty} \frac{t_n}{T \, \delta x}. \qquad (1\text{-}26)$$

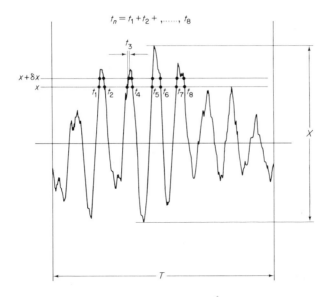

Fig. 1-9. Determination of probability.

The ratio given in equation (1-26) will approach an exact probability value as the record length extends to infinity. If the number of possible discrete values for x is great, then a large number of probability functions are required to describe the function $x(t)$. In such cases it is often more convenient to use the probability distribution function, which describes the probability that the variable will assume a value less than or greater than x. This is given by the integral of the probability density function as

$$P(x) = \int_{-\infty}^{x} p(x).dx. \qquad (1\text{-}27)$$

Examples of these functions are given in fig. 1-10.

These definitions can be extended to include the probability of two associated sets of data being in a given condition at the same time. The joint probability density function defines the probability that $x(t)$ will be found within the range x and $x + \delta x$ whilst $y(t)$ simultaneously assumes a value within the range y and $y + \delta y$.

The time t_n now defines the fraction of the total time that $x(t)$ and $y(t)$ simultaneously fall within these ranges over the record length T, so that

$$p(x,y) = \lim_{T \to \infty} \frac{t_n}{T, \delta x, \delta y} \qquad (1\text{-}28)$$

33

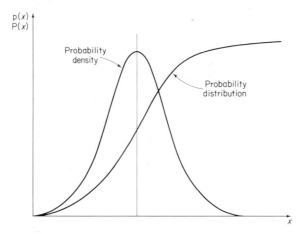

Fig. 1-10. Probability density function.

and the joint probability distribution function is

$$P(x, y) = \int_{-\infty}^{x} \int_{-\infty}^{y} p(x, y) \, dx, \, dy. \tag{1-29}$$

In many joint probabilistic situations it is often easier to use the properties of the cross-correlogram. In others, such as studies of hit probabilities involving conditional probabilities, knowledge of the joint probability density function is required.

1-4-2 TIME DOMAIN ESTIMATES
Simple averages and probability measurements fail to tell us anything about the periodic behaviour of the signal. Statistically this information can be obtained by making measurements of the amplitude of the signal at two times, separated

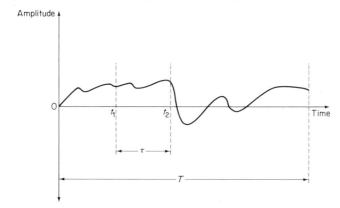

Fig. 1-11. Auto-correlation.

by a delay τ, finding their product, and averaging over the time of the record (see fig. 1-11). This procedure is known as autocorrelation and results in the autocorrelation function

$$R_x(\tau) = \lim_{T \to \infty} \frac{1}{T} \int_0^T x(t).x(t + \tau).d\tau. \qquad (1\text{-}30)$$

It is expressed in graphical form as an autocorrellogram for $R_x(\tau)$ against time or delay (τ).

The main value of the autocorrellogram is its ability to reveal the presence of periodicity in a random signal. For a completely random signal even a slight variation in delay will result in a reduction of the product, $x(t) x(t + \tau)$ to a very small value so that a large value (normalised to unity), will only be obtained at or close to $\tau = 0$. This will not be the case for periodic components, where a

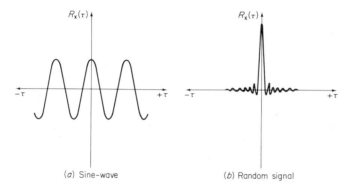

(a) Sine-wave   (b) Random signal

Fig. 1-12. Examples of autocorrelation.

large shift in τ equal to half a period will be required before any substantial change in $R(\tau)$ will be apparent. Examples are shown in fig. 1-12 which indicate clearly the value of the autocorrellogram in differentiating between random and periodic signals.

A similar result is obtained if the measurements are taken from two signals, $x(t)$ and $y(t)$ at times separated by τ. The relationship is expressed as the cross-correlogram

$$R_{xy}(\tau) = \lim_{T \to \infty} \frac{1}{T} \int_0^T x(t).y(t + \tau).d\tau. \qquad (1\text{-}31)$$

This gives an indication of the joint properties which the two signals may share. Examples will be given later to show the value of cross-correlation in determining the relationship between two signals referred to the same energy source.

35

1-4-3 FREQUENCY DOMAIN ESTIMATES

Although information on the spectral decomposition of a signal can be derived from the autocorrellogram a more useful description is obtained by a direct plot of its spectral content. Fourier series analysis can be applied to periodic functions and values of the Fourier coefficients will give directly the peak amplitude of the related harmonics contained within the signal. This form of analysis, is however, inadmissable for random signals which do not necessarily have harmonically related components. Instead a measure of the relative amplitude of the frequency components is obtained via the Fourier transform as

$$X(f) = \int_{-\infty}^{\infty} x(t) \exp(-j\omega t).dt \qquad (\omega = 2\pi f) \qquad (1\text{-}32)$$

The mean-square value is taken to obtain a measure of the instantaneous power at a given frequency over the record length T, viz

$$G_x(\omega) = \lim_{T \to \infty} \frac{1}{T} \int_0^T \frac{x_B(t)^2}{B}.dt \qquad (1\text{-}33)$$

where $x_B(t)^2$ is the instantaneous square of the signal contained within a bandwidth, B. Equation (1-33) gives the Power Spectral Density of the signal obtained by averaging the squared values of the signal over a narrow bandwidth, B. An example of such a spectra is given in fig. 1-13 for a random signal containing several strong periodic components.

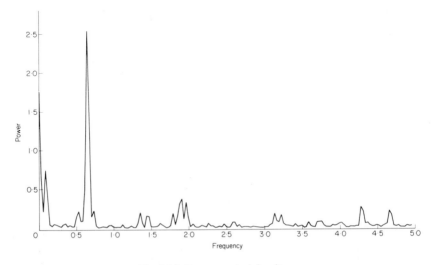

Fig. 1-13. Power spectral density.

The relationship mentioned earlier, between the power spectral density function and the autocorrelation function, is an important one in terms of practical measurement methods. The two functions can be shown to be Fourier transforms of each other, viz

$$S_x(\omega) = \frac{1}{2\pi} \int_{-\infty}^{\infty} R_x(\tau) \exp(-j\omega\tau).d\tau \qquad (1\text{-}34)$$

and

$$R_x(t) = \int_{-\infty}^{\infty} G_x(\omega) \exp(j\omega\tau).d\omega. \qquad (1\text{-}35)$$

These relationships take into account negative values of frequency since the double-sided spectral density function, $S(\omega)$ is used.

In a practical case the relationships for real values at positive frequencies only are given as

$$G_x(\omega) = \frac{2}{\pi} \int_{0}^{\infty} R_x(\tau) \cos \omega\tau.d\tau \qquad (1\text{-}36)$$

$$R_x(\tau) = \int_{0}^{\infty} G_x(\omega) \cos \omega\tau.d\omega \qquad (1\text{-}37)$$

Equation (1-34) to (1-37) are known as the Wiener–Khintchine relationships.

Similar relationships can be derived for the cross-power spectrum for two signals derived from the same physical system. However, since the cross-correlation function is not an even function, its Fourier transformation has the form

$$G_{xy}(\omega) = \text{Re}\ [R_{xy}(\omega)] + \text{Im}\ [R_{xy}(\omega)] \qquad (1\text{-}38)$$

where the real part is given as

$$\text{Re} = \frac{1}{\pi} \int_{-\infty}^{\infty} R_{xy}(\tau).\cos \omega\tau.d\tau \qquad (1\text{-}39)$$

and the imaginary part as

$$\text{Im} = \frac{1}{\pi} \int_{-\infty}^{\infty} R_{xy}(\tau). \sin \omega\tau.d\tau. \qquad (1\text{-}40)$$

The cross-power spectrum can be considered as a plot of the power contained in the product of the two time histories, $x(t)$ and $y(t)$ taken over successive narrow-band frequency intervals.

1-5 PROCEDURE FOR ANALYSIS

The dependence of the method of analysis on the environmental conditions under which the signal becomes available has been mentioned earlier. This applies particularly to the conversion of the physical quantity into an electrical signal. Any set of procedures for analysis of transducer-derived time-histories must include pre-processing and calibration of the signal in order to take this into account.

Pre-processing can play a major role in conditioning the signal into a form suitable for analysis. For example the presence of a large amplitude slowly-varying trend can prevent effective analysis of small rapid changes in the signal by restricting the useable dynamic range to a function of that of the trend. Trend removal, which in this case amounts to simple filtering, can considerably improve this situation. Filtering is also required prior to digitisation or decimation of a discrete time series. Calibration involves or generates detailed information concerning the environment of the transducer, method of storage and the processing system itself. All of this information, with the pre-processing conditions and physical characteristics of the system under study, needs to be acquired to permit interpretation of the processed signal. The importance of an established procedure, and particularly documentation procedure, cannot be stressed too highly, since a time-history alone is valueless unless associated with the necessary information regarding its manner of acquisition.

Following the establishment of a data-acquisition procedure a decision regarding 'on-line' or 'off-line' processing must be made. There is seldom justification for direct on-line processing unless the form of the results can be anticipated well enough to preclude the need for subsequent reprocessing. Even then the process may be expensive and, since operation in real time can result in high data rates, it may be necessary to accept a lower accuracy in processing than is desirable.

It is more usual therefore to process the signal as an 'off-line' operation which implies some form of temporary storage. This can take analog or digital form, with analog storage on magnetic tape preferred in very many practical situations. One reason for this is the extremely flexible characteristics of the analog recording and playback process. The wide dynamic range of recording and favourable frequency characteristics permit the signal acquired from a transducer or other source to be stored and recovered with hardly any modification to the original signal. A second reason is the ability to translate the time-base of the stored signal which has many practical uses.

The analysis process can employ, analog, digital or hybrid means. Analog processes are generally cheaper, particularly if special-purpose equipment is used (e.g. wave-analysers or correlators), although hardware digital fast Fourier transform methods may well be more economical under some circumstances (e.g. long low-frequency signal records). Hybrid processes can also result in a cheap form of analysis with the additional advantage of ease in specification of analysis characteristics under program control.

Digital computer methods are highly repeatable and extremely accurate as regards the calculations involved. Cost is no longer a major limitation with fast Fourier transform algorithms and efficient hardware. A more serious difficulty is that of the initial conversion of the continuous signal into a sampled form, which must inevitably lead to loss of information. This may not necessarily lead to an unacceptable resolution, but does mean that care will have to be taken in processing the discrete time-histories formed. In particular, quantising level and sampling errors can result in a difficulty in identifying the constituent frequency terms of the complex signal. The problem is referred to as Aliasing and is the direct result of Shannon's sampling theorem [14], which states that the sampling frequency must be at least twice as great as the frequency of the highest Fourier component contained in the continuous signal.

The analysis procedure for the signal must follow a preliminary classification as discussed above. Tests for the signal characteristics are therefore required as part of the procedure. Of particular importance is the establishment of a testing procedure to detect the presence of periodic components in an otherwise random signal. The most powerful method is the use of the autocorrelation function as indicated earlier. Measurements taken from the autocorrelogram can be used to define a region for investigation using power spectral density analysis to determine the value of the detected periodic components. A search for these will often reveal sinusoidal peaks by their relationship to the bandwidth of the analysis filter. If the periodic component is truly sinusoidal then the measured peaks will take on the bandwidth characteristics of the filter no matter how narrow this is, although it is important to remember that the peak height will vary to maintain unity area.

It will also be necessary to reject any unwanted data from the signal before analysis commences. This is important since the procedure for data acquisition will almost always result in extraneous information being included, either in the form of an extended time-history which include pre- and post-transient recording, or in extended bandwidth outside the range of interest or both. Time-compression or decimation techniques can result in reduction in the quantity of data to be stored and processed and may be justified in many cases.

The form of analysis is dependent upon the physical requirements of the process and will be discussed in detail in subsequent chapters. The choice of a desirable record length for various forms of analysis is considered in the next section.

1-5-1   REQUIRED LENGTH OF RECORD

Analysis accuracy for the statistical estimates will be dependant basically on the amount of data available. Each method for analysis will define its own minimum sample length for a given estimation accuracy, and these lengths will be different.

It will be shown later that for all the basic properties of the signal discussed in section 1-4 the record length required, T, is inversely proportional to twice the analysis bandwidth multiplied by the square of the estimation error and will also

39

be proportional to a constant, K whose value is dependent on the property being measured [15], i.e.

$$T = \frac{K}{2B\epsilon^2}.$$ (1-41)

For mean-square value and power spectral density estimates a value of $K = 2$ can be assumed (equation (8-53)). Half this record length is required for correllation estimates where measurements take place around the peak value. As the lag (τ) increases then the record length required for equal error increases, i.e.

$$K = 1 + \left(\frac{R_x(0)}{R_x(\tau)}\right)^2.$$ (1-42)

Record length requirement for mean value estimation is dependent on the square of the ratios of the standard deviation to the mean value, i.e.

$$K = \left(\frac{S}{\bar{x}}\right)^2,$$ (1-43)

and therefore depends on the dynamic range of the signal.

Probability estimates require the longest sample of the record since we are equally interested in the probability of very small as very large, values and may need a long record to collect sufficient high amplitude values to form a statistically viable estimate. For probability estimates

$$K = \frac{1}{A.p(x)}$$ (1-44)

where A is the amplitude resolution, and $p(x)$ is the probability value.

1-6 PRESENTATION OF RESULTS

All measurements on real signals are necessarily estimates and the final results of an analysis should include some quantative information concerning the accuracy of these estimates. It may be sufficient to state the practical conditions under which the result was achieved, e.g. calibration accuracy, length of signal, sampling rate, filter bandwidth etc., since these can then be related to measurements obtained under comparable conditions. Where only a single record is available for analysis, the statistical conditions for a more precise statement of accuracy may simply not be present, and the estimates may have to be accepted in these terms. If an ensemble of records is obtained then statistical measures can be applied and figures for variance and mean square error of the estimations obtained. Hypothesis testing on the lines proposed by Bendat and Piersol [15] will enable a realistic expected variance value to be determined for a given record, which can then be compared with the values actually obtained. A method of assigning a confidence level to spectral estimates is described in detail in chapter 8.

Since, with few exceptions, signal processing is primarily concerned with continuous phenomena, the results are invariably required in graphical form. Some information may become available in tabular or histogram form, e.g. mean-square values, standard deviations, probability analysis etc., but it is conceptually difficult to consider for example, power spectral density functions in other than graphical form.

In order to aid interpretation it may be necessary to filter or smooth the data in either the time or frequency domain. Where the data becomes available in digital form a numerical form of smoothing is used. A number of smoothing algorithms are available for this purpose, a well-known example being that due to Hann;

$$y_i(s) = 0 \cdot 25\, y_{i-1} + 0 \cdot 5\, y_i + 0 \cdot 25\, y_{i+1} \quad (i = 1, 2, \ldots, N). \qquad (1\text{-}45)$$

Smoothed data points are obtained by weighting and summing of sets of three adjacent values to form a new series of N data points. This can be shown to be equivalent to low-pass filtering and represents a moving average form of the digital filter. Selective plotting of analog or digital information via a frequency filter is a fairly common requirement, either to enhance the features of interest, or to reduce the amount of processing or plotting required. The subject of filtering for both analog and digital information will be considered in later chapters from the point of view of application rather than rigorous analysis. A similar operation in the time domain is known as convolution, where the signal is modified by the shape of a convolution window, such that the value of units in a time series is reduced to zero outside the window region. By suitable shaping of the window boundaries the signal is reduced gradually to zero at either extreme of the window length. (fig. 1-14). This is a necessary operation with limited length records which avoids injecting high frequency transients into the desired record arising from the abrupt beginning and end to the record.

Annotation of a graphical record is difficult to carry out using analog systems without resorting to manual methods. The position is made easier when a hybrid linkage to a digital computer is available, since a software system of commands from the digital machine can cause integrator ramp control and pen movement to be used in the production of a library of symbols for direct plotting on the analog X/Y recorder.

The use of the digital incremental plotter provides by far the most flexible means of recording of analysis results. A fairly extensive library of software plotting routines is available with most scientific computing installations and which include axis plotting, scaling and comprehensive annotation procedures. This flexibility is particularly useful where special facilities are required such as logarithmic scaling, contour plotting and three-dimensional plotting. The latter is a prerequisite for the representation of non-stationary data where amplitude is plotted against both time and frequency. A further practical advantage of digital plotting is that if the calculation results are stored on magnetic tape or disc then the relatively slow process of actually producing the plot can generally proceed on a time-sharing basis from tape to plotter, involving the computer in very little central-processing activity.

C.R.T. display methods form another method of presentation for graphical results. This is often associated with analog computer methods to display fairly high frequency real-time results, which cannot be plotted using strip-chart recorders or X/Y plotters and which are too fast for digital computation. Photographic means are also used to create a permanent record and this may be cheaper than transfer to a magnetic storage media and the use of slower off-line

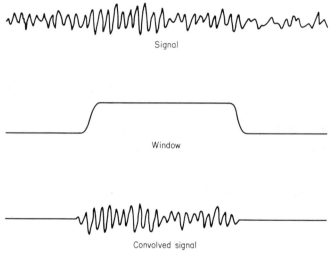

Signal

Window

Convolved signal

Fig. 1-14. Use of a convolution window.

reproduction. Digital computer graphics can provide interactive operation using a light-pen and elaborate software dedicated to the technique of analysis selection. Freedman has described a system whereby choice of analysis method and analysis parameters can be determined by the operator using interactive methods [16]. Applied to experimental data or subjective analysis on non-linear systems the rejection and selection capability of interactive graphics can considerably shorten the investigation period.

BIBLIOGRAPHY
1. SCHUSTER, A. On the investigation of hidden periodicities. *Terr. Mag.* **3**, 13, 1898.
2. SCHUSTER, A. The periodogram of magnetic declination. *Trans. Camb. Phil. Soc.* **18**, 107, 1900.

3. WHITTAKER, E. T. and ROBINSON, G. *The Calculus of Observations*. Blackie & Sons, London, 1924.
4. SLUTSKY, E. The summation of random causes as the source of cyclic processes. *Ecometrica* **5**, 105, 1937.
5. WIENER, N. The extrapolation, interpolation and smoothing of stationary time series. M.I.T. Press, Cambridge and New York, 1949.
6. TUKEY, J. W. *The sampling theory of power-spectrum estimates*. Symposium in application of autocorrelation analysis to physical problems. U.S. Office Naval Research, NAVEXOS-P-735, 1950.
7. BARTLETT, M. S. Periodogram analysis and continuous spectra. *Biometrika* **37**, 1-16, 1950.
8. GODFREY, M. D. An exploratory study of the biospectrum of economic time series. *Appl. Stat. C*, **14**, 1, 48-69, 1965
9. COOLEY, J. W. and TUKEY, J. W. An algorithm for the machine calculation of Fourier series. *Math. Comp.* **19**, 297, 1965.
10. STOCKHAM, T. G. High-speed convolution and correlation. *A.F.I.P.S. Proc.* Spring Joint Comp. Conf. **28**, 229-33, 1966.
11. HELMS, H. Non-recursive digital filter design method for achieving specifications of frequency response. *I.E.E.E. Trans. (Audio and Electroacoust.)* AU-16, 1968.
12. NOLL, A. M. Short-term spectrum and cepstrum techniques for vocal pitch detection. *J. Acoust. Soc. Amer.* **36**, 2, 296-302, 1964.
13. PRIESTLEY, M. B. Evolutionary spectra and non-stationary processes. *Jl. Roy. Stat. Soc.*, B, **27**, 204-37, 1965.
14. SHANNON, C. E. A mathematical theory of communication. *Bell Syst. Tech. J.* **27**, 623-56, 1948.
15. BENDAT, J. S. and PIERSOL, A. G. *Random Data: Analysis and Measurement Procedures*. John Wiley, New York, 1971.
16. FREEDMAN, R. A. *Interactive Signal Processing*. Computer Graphics Symposium, Brunel University, April 1970.

Additional references
17. CRAMER, H. On the theory of stationary random processes. *Ann. Math.* **41**, 215, 1940.
18. BLACKMAN, R. B. and TUKEY, J. W. *The Measurement of Power Spectra*. Dover, New York, 1959.
19. JENKINS, G. M. and WATTS, D. G. *Spectral Analysis*. Holden-Day Inc., San Francisco, 1969.
20. HANNAN, E. J. *Time Series Analysis*. John Wiley, New York, 1960.
21. DAVENPORT, W. B. and ROOT, W. L. *An Introduction to the Theory of Random Signals and Noise*. McGraw-Hill, New York, 1958.
22. BARTLETT, M. S. *An Introduction to Stochastic Processes with Special Reference to Methods and Applications*. Cambridge University Press, London, 1953.
23. LIVESEY, J. *et al. Recording and interpretation of strain measurements in Military bridges Conference: The Recording and Interpretation of Engineering Measurements.* Institute of Marine Engineers, (April 1972).
24. LANGER, B. F. Fatigue failure from stress cycles of varying amplitude *Trans. A.S.M.E.* **59**, A160-2 (1937).
25. MINER, M. A. Cumulative damage in fatigue. *Trans. A.S.M.E.* **67**, 159-4 (1945).

Chapter 2

DATA ACQUISITION AND STORAGE

2-1 INTRODUCTION

This chapter will be concerned with methods of detection, electrical conversion and storage for the physical quantity being measured. Since the purpose of data acquisition is to obtain quantitative information about a physical system, it will be necessary to consider the operation and accuracy of these processes and to determine their limitations related to the characteristics of the system being studied. It is extremely important to understand the mechanics of data capture and storage since the entire validity of the conclusions drawn from a signal can be nullified through ignorance of the environmental conditions under which the measurements were taken. It is an unfortunate fact that all too often insufficient consideration is given to the limitations of the data capture methods through the mistaken belief that compensation for these limitations can be made at the analysis stage.

Devices used for the measurement of physical quantities in terms of an electrical output signal are called Transducers. The classification is a broad one and it is the intention here to consider a fairly limited range of applications which will, it is hoped, cover the main requirements of system measurement as are generally understood in the signal processing environment. Choice of transducer to detect and translate physical energy will depend upon the form of the energy, accuracy requirements, period of operation, environmental conditions, dimensions and cost. These considerations will form the framework of the discussion concerning transducer types which follows.

Real-time processing of transducer outputs is unusual. Some form of storage is generally required to permit subsequent processing and, indeed, to carry out re-processing at a later time when different analysis characteristics are seen to be required. Choice of storage media is dictated by the characteristics of the signal, economy, and convenience for later processing. A cheap form of recording for continuous low-frequency signals is the analog strip-chart recorder and this is widely used. A visible record is obtained from which essential information can readily be extracted by inspection. Difficulties arise, however, when subsequent

44

analysis of the record is required. Conversion of the trace recording to analog or digital form requires specialised equipment and the cost of this may well exceed the cost of recording the data in a suitable form for retention in the first instance. A similar problem arises if cinematograph film is used to capture a light image derived from the system, although under some circumstances the characteristic of the signal may preclude other means of data capture (e.g. where very high transient frequencies or rapid movement are involved).

A method of recording and storage is clearly required that presents no major conversion problems when the recovery of the stored information is required. It must not impose its own characteristics on the signal being stored (e.g. distortion and undue protraction of the time scale), and must be simple and economic to use. Magnetic tape storage has obvious advantages in this respect and is widely used for data storage.

For some purposes where the required data has discrete form (e.g. from a counting process), then direct digital recording using paper tape or an incremental digital magnetic tape recorder may be used and methods of entry for this data into the processing equipment will be considered in later chapters.

Direct digital recording of physical quantities does not find wide favour since most of these processes are continuous by definition. A notable exception to this is the class of counting transducers, using the properties of a radioactive source and a detection device, used to measure displacement, thickness, etc., largely for industrial purposes. Other counting transducers rely on the variation in natural resonant frequency of a high-Q element caused by fluctuations in the measurable variable acting on the element, e.g. pressure and fluid flow. In all these counting techniques a direct binary output signal is obtained which can be entered into a digital computer directly or stored in binary form.

Most transducers give an output in continuous form and conversion to a digitally coded form is required in order to permit entry into a digital computer. This process has a number of attendant dangers.

The most significant of these is that once the digitisation rate has been decided, then so has the upper frequency limit of analysis. For reasons of economy in order to limit the number of samples obtained, the rate selected will be the lowest considered necessary at the time. Later information concerning the process may well indicate that higher frequency components, present in the original analog signal but not recoverable from the digitised data, are important. It will then be too late to extract the required information from the available digitised record. A second consideration which is very relevant for long continuous signals, is the economy of analog *vs.* digital methods of storage. As will be shown later, the density of recording is considerably lower with digital than with analog methods of storage, although the former will permit of much higher accuracy for the recorded data.

Also, in contrast to analog methods of magnetic tape storage, translation of the time-scale upon replay is not possible with conventional blocked digital tape

45

recordings. For these and other reasons, analog magnetic tape recording is the preferred medium for much of the recording of results from physical measurements and its consideration will dominate other forms of storage discussed in this chapter.

As indicated above, the recording media used may not always be the most convenient for later forms of analysis, and the problem of conversion from one storage media to another may have to be considered. A number of conversion techniques will be considered briefly towards the end of this chapter, excluding analog-to-digital conversion which is of sufficient importance to merit a separate chapter to itself.

2-2 TRANSDUCERS

Considered within the context of this book, transducers may be defined as devices which will translate physical input quantities into electrical output signals. We expect them to have an accurately known input/output characteristic under a given set of conditions so that meaningful properties can be ascribed to the physical process being measured.

From the wide range of possible physical measurements, three basic quantities can be recognised. These are:

1. Displacement. Linear and angular measurements representing stress, strain, pressure, thickness, creep, force, angle of incidence etc.

2. Velocity. Linear and angular measurements representing speed, rate, momentum etc.

3. Acceleration. Linear and angular measurements representing vibration impact, inertia, etc.

Other characteristics such as light, heat, and time may also be recorded either as a direct electrical analog or in terms of one of the basic quantities given above.

A variety of physical effects can be utilised in the design of practical transducers and exhaustive treatment of their design and characteristics is not possible within the confines of this chapter. All that can be hoped to achieve is a classification of the main types in use for signal processing purposes, together with sufficient details of their characteristics and use to be of value when assessing the reliability of measurements made using them. A broad classification of the physical effects used can be stated in terms of:

1. Energy conversion transducers–where energy is abstracted from the system under measurement and converted (with some energy loss) into an equivalent electrical form.

2. Passive transducers–in which a measured change in the physical quantity merely causes a corresponding change in some electrical quantity (e.g. resistance, capacitance, inductance etc.), and requires an auxiliary source of energy in order to produce an output signal.

46

3. Feedback transducers–these are characterised by a feed-back loop which effects an equilibrium between the input physical quantity and that of an opposing electrical quantity. The force necessary to achieve this equilibrium gives a measure of the physical quantity being measured.

A sub-classification widely used to define transducer types is by the particular physical effect utilised in their operation. Fig. 2-1 shows the division of these types in terms of energy conversion or passive control action. The third type, that of equilibrium feedback can utilise one or more of any of these effects in its operation.

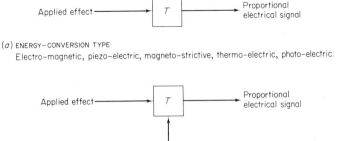

(a) ENERGY–CONVERSION TYPE
   Electro–magnetic, piezo–electric, magneto–strictive, thermo–electric, photo–electric.

(b) PASSIVE–CONTROL TYPE
   Resistive, inductive (reluctance), capacitive, mechanical strain (resistive), Semi–conductive, thermo–resistive (thermocouple), photo–resistive, hall effect, radioactive.

Fig. 2-1. Types of transducers.

2-2-1 DISPLACEMENT TRANSDUCERS

The biggest single group are those used for the measurement of strain, commonly known as strain gauges.

The amount of strain may be measured by its effect on a long thin wire attached to the structure. Referring to fig. 2-2 we can define strain as:

$$\text{strain} = e = \delta l / L \qquad \text{m/m} \qquad (2\text{-}1)$$

from which the stress, defined as force per unit area, can be obtained from the product of e and Young's modulus of elasticity.

The cross-sectional area of this wire is very small (typically 0.0001 mm^2) so that the change in length, δl is reflected in a measurable change in its electrical

47

resistance, δR. For most metals a constant gauge factor can be derived [1], as:

$$k = \frac{\delta R/R}{\delta l/L}.$$ (2-2)

Since k is reasonably constant over a wide range of values for strain a linear relationship between δl and the measured output voltage, $\delta V_o = \delta R.i$ can be obtained. Modern resistive strain gauges use thin foil rather than a circular wire conductor which eases the problems of manufacture and results in a more robust design. The foil is bonded to a thin insulating backing and the bonded strain gauge affixed to the structure. The measurement is obtained by direct

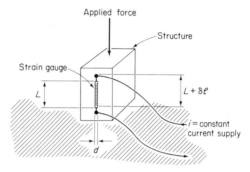

Fig. 2-2. A thin-wire strain gauge.

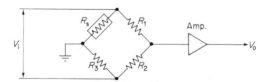

Fig. 2-3. Strain gauge unit–d.c. operation.

transmission of the strain through the backing material to the gauge. This is used incorporated in a bridge configuration as shown in fig. 2-3. Changes in resistivity, caused by the alteration in length of the gauge under the applied strain, will unbalance the bridge and cause a potential to be developed across it which is linearly proportional to strain. The sensitivity of foil gauges is fairly low and for this reason other strain gauge transducers can take the place of the bridge resistors shown, so that the output signal may be multiplied by the number of gauges used.

Although the auxiliary energy source shown in fig. 2-3 is shown as a d.c. supply, the use of an alternating current source generally leads to a more reliable system. The output from the strain gauge bridge will be amplitude-modulated by the strain signal and require demodulation and filtering to obtain an output

48

proportional to strain. A typical arrangement is shown in fig. 2-4 to include the phase-sensitive detector, used for detection.

Due to the risk of fracture the dynamic range of the bonded strain gauge is limited and the permissable current density (and hence V_i) must be kept small to minimise the resistance changes due to temperature variations. Compensation for these variations is generally included by means of a thermally sensitive resistor shunting the transducer terminals. An alternative method of compensation is to mount the strain gauges in pairs. One gauge is used for the actual measurement whilst the other acts as a dummy gauge affixed in the direction of minimum strain. The measurement gauge and the dummy gauge are connected in series and situated in opposite arms of the bridge so that the resistance changes due to temperature effects will tend to cancel.

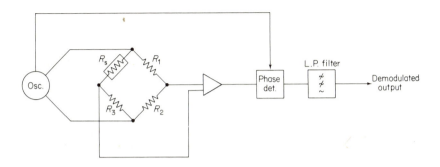

Fig. 2-4. Strain gauge unit–a.c. operation.

The magnitude and direction of strain can be determined from a group of similar bonded gauges, often three in number, and referred to as strain-gauge rosettes. The constituent gauges are accurately manufactured on a common base to lie at convenient and accurately known angles to each other. Thus the principle strains and stresses and their angular orientation with respect to the gauge angles can be computed from the three measured strains allowing a small correction for transverse sensitivity.

An unbonded version of a strain gauge finds a use for certain applications, particularly where high temperature operation is required. Here the strain gauge sensing element is wound between insulated pins as shown in fig. 2-5, so that a large change in resistivity can be obtained within a given space. The gauge is, however, more sensitive to overload and operates at a lower current density than the bonded type. This design is not suitable for confined locations where the bonded strain gauge would be used.

The low sensitivity of the wire resistance gauge had led to the development of a form of resistance gauge known as the thin film strain gauge. A ceramic film is vacuum-deposited on to the sensing element and acts as an insulating base. Upon this is deposited, again by vacuum methods, a film of resistive or semi-conductive material through a mask to give a bridge form having the correct strain orientation. A high resistance gauge is obtained which can withstand high applied voltage and hence produce a higher output for a given power dissipation.

A considerable improvement in sensitivity is obtained by the use of semi-conductor filaments in place of thin resistance wire or foil. The gauge factor, k for silicon or germanium filaments is of the order 100 to 200 compared with about 2 for copper–nickel alloys. To a large extent the advantage in terms of sensitivity of the semi-conductor gauge is now eroded by the availability of cheap integrated circuit amplifiers which can be used with the resistance strain gauge. The disadvantages of the semi-conductor gauge are its poor linearity and

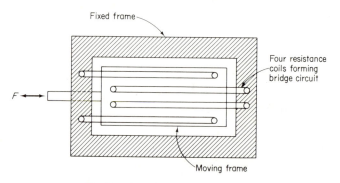

Fig. 2-5. An unbonded strain gauge.

its noise characteristics and also sensitivity to temperature changes. In contrast with the good linearity of the wire resistance strain gauge the semi-conductor gauge is distinctly non-linear and having to form

$$\delta R/R = ae + be^2 \qquad (2\text{-}3)$$

where a and b are constants dependent on the semi-conductor material used, and e is a measure of strain. This gauge is more temperature sensitive than those previously discussed. Compensation can be obtained by balancing pairs of gauges or the use of various circuit artifacts [2]. For extremely slow changes in strain a more accurate measurement can be made by balancing the bridge for each measurement taken. This would be the case for mechanical creep measurements where automatically balancing bridge control arrangements are made. The results of such measurements can be read in highly accurate digital form and output as a punched paper tape sequence at suitable time intervals [3].

A resistive displacement transducer which is considerably more robust than the resistance strain gauge is the potentiometric transducer shown in fig. 2-6.

This couples the motion caused by pressure directly to the arm of a recti-linear potentiometer through which a constant current is passed, thus affecting the position of the wiper and hence the amplitude of the voltage developed between this and one end of the potentiometer. When connected to a high-input impedance amplifier the arrangement is highly linear and suitable for measuring large pressure differentials. It is not suitable for the measurement of very small displacements, such as strain, however, due the mechanical limitation of the linkage arrangement.

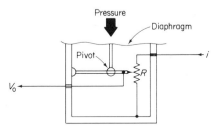

Fig. 2-6. A potentiometric pressure transducer.

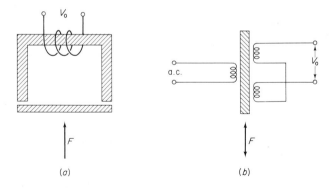

Fig. 2-7. Variable inductance transducers. (a) Self-inductive type. (b) Differential type.

Other passive methods of displacement measurement utilise a change in inductance or capacitance value. There are several forms of variable inductance transducer, two of which are shown in fig. 2-7. The variable self-inductive type (a) gives an inductance change proportional to pressure over a limited range. The resolution is good and the transducer can be used for the measurement of both static and fluctuating pressures.

The second method (b) uses the change in mutual coupling between two coils which occurs due to movement of their common core. This is known as the differential transformer transducer [4]. Whilst the sensitivity of this arrangement is high, a linear relationship between displacement and the output voltage can only be obtained over a very small part of the total core movement ($\simeq 5\%$). An alternating current energy source is required and an amplitude-modulated output signal is obtained.

Recent developments in variable capacitance transducers have resulted in extremely accurate methods of measuring displacement. The principle used is that of fringe capacitance measurement which is dependent only on area and separation of the two electrodes and the dielectric between them (fig. 2-8). The relationship between the capacitance and distance separating the fixed electrodes shown from an earthed member (x in fig. 2-8), is a fairly linear one, even for

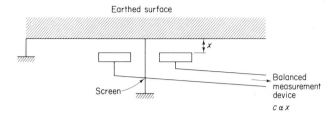

Fig. 2-8. A variable capacitance transducer.

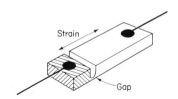

Fig. 2-9. A piezo-electric strain gauge.

large values of x. Transducers employing this technique have been designed to measure strain, pressure, and displacement over a wide dynamic range and at extremely high temperatures [5]. The small changes in capacitance values involved can be carried out using a balanced impedance bridge. A Blumlein bridge has been used for this purpose but suffers from the effects of cable capacitance and load impedance. A differential charge amplifier system has been successfully developed to avoid these difficulties and permits accurate displacement measurements over a range of 0·001 to 12 inches [6].

The transducers considered above are all passive devices. An energy-conversion transducer which is in wide use is the piezo-electric strain gauge (fig. 2.9). The piezo-electric material is subject to mechanical stress and generates a surface charge due to deformation of the crystal lattice. This charge can be made to leak away through an external resistance thus producing a voltage proportional to the imposed stress. The duration of this voltage however is finite, so that it can only be used for dynamic measurements. The output voltage produced is given by

$$V = het \text{ volts} \qquad (2\text{-}4)$$

where, h is the piezo-electric strain coefficient (V/m), e isthe strain (m/m), and t is the active material thickness (m).

The active material used in early makes of piezo-electric transducers was Rochelle salt or barium titanate. These materials have now been replaced by lead zirconate titanate due to its high sensitivity. The linearity of the piezo-electric transducer is found to be excellent and the sensitivity high. It finds wide use for the detection of short-term shock strains. Unlike many other displacement transducers the piezo-electric strain gauge is unaffected by temperature over a

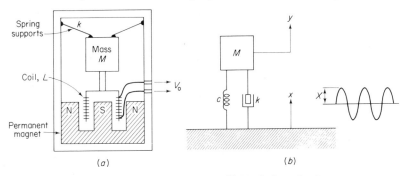

Fig. 2-10. (a) A seismometer. (b) Equivalent circuit.

very wide range (typically −200 to +450 deg. C) and can be used under fairly rigorous environmental conditions.

An important energy conversion method of measuring small displacement is the seismic transducer. It is a dynamic instrument and will also measure velocity and acceleration. One design for a seisomometer, used for detecting small earth movement, is shown in fig. 2-10a. A mass, M is suspended by a spring k and is subject to a single degree of freedom movement. A coil L is attached to the mass and moves with it across the field of a permanent magnet. This movement generates a small e.m.f. in the coil, which is a measure of the mass velocity (see equation 2-13). The process of measurement will impose a damping component on the spring-mass system so that the equivalent circuit will take the form shown in fig. 2-10b.

The equation of motion for this system is given as

$$Mp^2z + cpz + kz = -Mp^2x \qquad (2\text{-}5)$$

where z = relative movement, $y-x$ and p = operator d/dt.

53

We can study the response of this system to random signal inputs (base movement at x) by either harmonic or impulse methods. It is convenient to take the former and let

$$x = X \sin \omega t. \tag{2-6}$$

where X is the peak displacement of x.

Substituting this in equation (2-5) and taking the steady-state case, the solution is

$$z = Z \sin (\omega t - \phi) \tag{2-7}$$

for the relative motion between the mass and the base. Here Z is the peak amplitude of the relative motion and ϕ is the phase shift between x and z. Substitution of the values for x and z in equation (2-5) enables the extent of energy transfer between the base movement and the free mass to be determined in terms of the natural resonant frequency of the system [7]:

$$\omega_n = \sqrt{\left(\frac{k}{M}\right)} \tag{2-8}$$

and a damping factor, $\quad D = \dfrac{c}{2\omega_n . M} = c/2\sqrt{(Mk)}$

as $\quad Z = X \dfrac{(\omega/\omega_n)^2}{\sqrt{[[1 - (\omega/\omega_n)^2]^2 + (2D.\omega/\omega_n)^2]}}. \tag{2-9}$

From this we see that if $\omega \gg \omega_n$ then the relative displacement of the mass is almost identical to the initial base (earth) movement. Hence, if the mass is made very large and the damping coefficient made at or near to critical value ($D = 0.7$), then the mass displacement will give a very accurate measurement of earth displacement down to very low frequencies. This is the basic principle of seismic measurement. Early seismic instruments used extremely large masses, many tons in weight, to determine earth movements having very long periods.

An adaptation of this principle for the measurement of small base displacement is the potentiometric transducer. The mass is mechanically connected to the arm of a linear potentiometer, so that the movement of the mass results in a potential, developed between the arm and the base, proportional to displacement. Damping of this movement is obtained at about 60–70% of critical by containing the mechanism within a silicone fluid having a high viscosity.

2-2-2 MEASUREMENT OF VELOCITY AND ACCELERATION

Using the simple relationships $v = ds/dt$ and, $a = dv/dt$, we can obtain any desired quantity of motion (displacement s, velocity v, or acceleration a), by differentiating or integrating the signal obtained. Calibration may present some difficulties, particularly if a reference point is not available (e.g. measurement of velocity and acceleration of an artificial earth satellite).

Bonded strain gauges are widely used to measure the velocity of a moving body. If we consider, for example, the angular acceleration of the rotating mass structure shown in fig. 2-11 this will be given as

$$a_n = r\omega^2 \text{ (m/sec}^2\text{)} \tag{2-10}$$

where ω is the angular velocity of the rotating body, and r is the radius of curvature.

a_n is related to normal force by Newton's Second Law:

$$F_n = Ma_n \tag{2-11}$$

where a_n is produced by a force F_n acting on a mass, M kg, hence

$$\omega^2 = F_n/Mr. \tag{2-12}$$

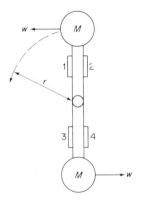

Fig. 2-11. Cantilever velocity transducer.

Four strain gauge transducers are shown in fig. 2-11. Two of these are arranged on opposite sides of the cantilever arm connecting the mass to the centre of the structure. By connecting these in the sequentially numbered order, shown in fig. 2-11, then cancellation of the tangential forces is obtained.

By far the most common velocity transducers are those relying on the generator effect for their realisation. This is shown in fig. 2-12 where the peak value of the induced e.m.f. across the loop rotating within a magnetic field, B is given by

$$E = AnBv \text{ volts peak,} \tag{2-13}$$

where A represents the area of the loop, n is the number of turns, and v is the loop angular velocity.

55

An example has already been given of this principle as used in the seismometer linear displacement system.

The measurement of angular displacement velocity is made by means of a tachometer, which is a compact form of a dynamo generator having the position of the magnet and coil reversed, i.e. the magnet is caused to rotate within a fixed coil. Several windings are included spaced about an axially rotating permanent magnet. The induced e.m.f. in the windings provides a rectangular output wave-form having a value proportional to the product of the total number of turns in the stator, n, and the rate of change of flux, and hence velocity, viz.

$$e = -n \frac{d\phi}{dt}.$$
(2-14)

Another form of angular velocity measurement utilises a series of small magnets

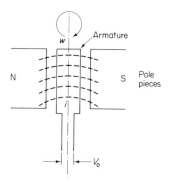

Fig. 2-12. Dynamo velocity transducer.

attached to the rotating body. These cause a small e.m.f. to be induced in an fixed loop adjacent to the rotating shaft with each rotation of the shaft. A train of pulses will be produced, the frequency of which will be proportional to shaft speed. By using a pair of such loops accurate measurements of axial and radial displacement of the shaft can also be obtained.

2-2-3 ACCELERATION TRANSDUCERS

Acceleration value is usually obtained indirectly by measuring the force necessary to produce the observed acceleration in a moving structure. Referring to equation (2-9) if ω/ω_n is made very small then the relative displacement is given approximately by

$$Z \simeq \frac{\omega^2 X}{\omega_n^2}$$
(2-15)

and, since for sinusoidal excitation $x = X \sin \omega t$,

$$p^2 x = - \omega^2 X \sin \omega t,$$

so that the peak value of acceleration is $-\omega^2 X$ and we can write

$$Z \simeq \frac{A_{\max.}}{\omega_n^2}. \qquad (2\text{-}16)$$

This is the principle of the potentiometric accelerometer, which is identical in form to that of the displacement potentiometric transducer shown in fig. 2-6 except that the pressure sensing element is replaced by flexural springs supporting a seismic mass. When subject to acceleration forces the position of the mass remains constant whilst the body housing moves. The relative displacement between them is sensed by the movement of the potentiometer

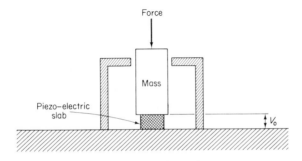

Fig. 2-13. Piezo-electric accelerometer.

arm which causes a potential voltage to be developed providing a direct measure of acceleration. Considerable care has to be taken in the design of the transducer to ensure that the natural resonant frequency of the mass and the damping force chosen permit a linear relationship to be obtained over the desired acceleration range.

Acceleration measurement using strain gauges involves four gauges affixed to the moving mass such that two of these are orientated to give maximum output, whilst the other two act as temperature compensation gauges, mounted for minimum acceleration. It can be shown [2] that the measured acceleration in this case is also proportional to strain. The mass is generally made fairly large to obtain a reasonable output voltage and this results in the measurement being confined to fairly small low-frequency variations in acceleration.

The piezo-electric accelerometer differs from the strain gauge version described in section 2-2-2 by the presence of a mass, M, which provides a pressure on the face of the crystal, proportional to the acceleration magnitude (fig. 2-13). The voltage developed across the crystal is dependant on the generated charge, Q, and the shunt capacitance of the crystal, C_p plus the cable capacitance, and is given by:

$$V = Q/C_p \text{ volts.} \tag{2-17}$$

The sensitivity is high and the unit can be designed for a very small size, but, as with the pressure transducer, the operation is confined to high frequencies.

Due to its favourable sensitivity:mass ratio this accelerometer is widely used for impact testing. The problem with this form of testing is the very high acceleration values experienced, which can rise to as high as 100 000 G in a few microseconds. As a consequence the effective weight of other forms of transducer can reach values of several tons. With miniature piezo-electric accelerometers the increase is limited to a few hundred pounds and, also due to the small mass, the resonant frequency is generally much higher than most structural elements, upon which they are mounted. For similar reasons such transducers are ideal for the measurement of high velocity shock transients.

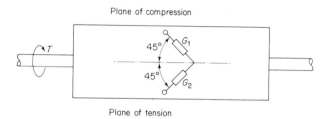

Plane of compression

Plane of tension

Fig. 2-14. Measurement of angular velocity.

Under these conditions the poor long-term zero stability, and sensitivity to temperature variations are of secondary importance.

Piezo-resistive accelerometers utilise the change in resistivity obtained when the element is subject to a bending stress. As with wire-resistive gauges they are used in pairs mounted in a bridge configuration.

Recent forms of piezo accelerometers incorporate a solid-state impedance converter within the body of the transducer, resulting in output impedances as low as 100 Ω. This obviates the need for a charge amplifier (see section 2.2.7) and permits the use of a long cable connection to the recording device.

The cantilever type of accelerometer was referred to in the previous section. This method is also used to measure the torque of a rotating shaft. Here two strain gauges are affixed to the shaft at +45° with respect to the shaft axis (fig. 2-14). Reduction of noise generated by a slip ring contact is made by incorporating the entire bridge on the rotating shaft, so that only the supply and output signals will be subject to contact error, and will not affect the measurement accuracy. Measurement of multi-directional force movements can be obtained using pairs of strain gauges or rosettes attached to the structure at

TABLE 2-1 Comparative transducer characteristics

Type	Range	Sensitivity	Frequency response	Resonant frequency	Accuracy	Temperature range (deg C)	Function
Bonded strain gauge	0–10 000 μS	2·0–3·0 (k)	0–50 kHz	5–100 kHz	± 1%	−40 + 180	
Semi-conductor strain gauge	0–2000 μS	100–300 (k)	0–50 kHz	5–100 kHz	± 1%	−50 + 100	
Foil strain gauge	0–200 000 μS	2·0–3·0 (k)	0–50 kHz	5–100 kHz	± 0·5%	−30 + 100	Displacement
Seismometer	0–100 μS	1 mV/μS	0–2 Hz	1–5 Hz	± 2%	+40 + 80	
Inductance	0·0025–0·1 in	+20 kHz/f.s.	0–3 kHz	>10 kHz	± 0·1%	−40 + 100	
Capacitance	0·001–10 in	0.01 pF/mm	20–2000 Hz	100 Hz	± 1%	up to 650 °C	
Potentiometric	0–1000 p.s.i.	–	20–2000 Hz	100 Hz	± 1%	−50 + 100	
Resistive strain gauge	0–50 000 p.s.i.	0·5–5 mV/V	0–4 kHz	4–20 kHz	± 0·5%	−40 + 100	
Foil gauge	0–500 p.s.i.	0·5–5 mV/V	0–4 kHz	4–20 kHz	± 1%	−50 + 100	Pressure
Piezo-electric	0–150 000 p.s.i.	5 pC/p.s.i.	2–40 kHz	50–200 kHz	± 1%	−150 + 250	
Capacitance	0–2000 p.s.i.	1–20 pF/f.s.	1–50 kHz	2–100 kHz	± 0·1%	−50 + 500	
Inductance	0–30 000 p.s.i.	40 kHz/f.s.	1–20 kHz	0·5–10 kHz	± 1%	−50 + 400	
Strain gauge accelerometer	20–200 G	3–40 mV/G	0–100 Hz	100–1000 Hz	± 2%	−40 + 200	
Potentiometric	5–200 G		0–20 Hz	100 Hz	± 1%	−50 + 100	
Piezo-resistive	50–300 G	1–5 mV/G	0–2 kHz	5–50 kHz	± 1%	−50 + 100	Acceleration
Piezo-electric	0–20 000 G	10–100 mV/G	2–50 kHz	20–120 kHz	± 2%	−200 + 400	
Piezo-electric (miniature accelerometer)	100 000 G	0.01 mV/G	0·1–50 kHz	60–250 kHz	± 1% per 10 kg	−50 + 250	
Feedback accelerometer	0–15 G	250 V/G	0–500 Hz	1000 Hz	± 0·1%	−50 + 100	Velocity
Permanent magnet generator	1–50 G	20–100 mV/G	20 kHz–2 kHz	10–20 kHz	± 1%	−40 + 300	

k = gauge factor; μS = microstrain; p.s.i. = pounds per square inch.

suitable angles and locations. An interesting application of a combined gauge assembly for this purpose is the sting balance [7], used to measure the six degrees of freedom possible for a solid body suspended in space, e.g. the aerodynamical behaviour of an aircraft model in a wind tunnel. The strain gauges are disposed about a specially shaped block of material to which the model is attached. Movements of the model in terms of pitch, yaw, roll, axial and side forces are mechanically conveyed to the sting balance which provides multi-channel electrical outputs corresponding to their value.

The method of defining sensitivity for accelerometers is generally stated in terms of the output potential generated in volts for an acceleration due to gravity $(G = 9\cdot81\text{m/sec}^2)$. Some typical figures are given in table 2-1 for a number of different transducer types. The detection of structure-borne vibrations demands care in accelerometer mounting to avoid resonance effects. For example if the mounting surface is irregular then a good transmission of energy to the transducer will not be possible. As a result the accelerometer will be decoupled from the vibrating motion of the structure. The mounting structure itself will have its own resonance characteristics which can affect the detected signal. Various forms of damped mountings have been used to prevent this two degree of freedom resonance but these can lead to a degradation in the higher frequency response and a downward shift of a detected resonance peak by as much as a decade in frequency [8]. Similar considerations apply to the bonding of strain gauges to a structure. A major source of error in detection of motion using strain gauges is directly applicable to the methods of bonding used. An excellent discussion of bonding techniques is given in ref [9].

2-2-4 FEEDBACK TRANSDUCERS

Feedback techniques permit the design of transducers which have good linearity and a wide range of application. These techniques were introduced in section 2-2. In order to achieve an equilibrium between the applied force and an opposing electrical voltage some form of sensing device is included to control the magnitude of the opposing voltage. This can be a subsidiary feedback loop detecting velocity, rather than linear movement, to permit a rapid equilibrium, and hence higher frequency response.

The advantages of feedback transducers over other methods are increased accuracy, linearity, sensitivity and a closer control over natural frequency and damping [10]. A good example of a feedback accelerometer is the Endevco design illustrated in fig. 2-15. This consists of a small pendulum-pivoted mass constructed of quartz, which is free to move within very small gaps maintained between the mass and the fixed side reference plates of a capacitive pick-off device. The mass also carries with it a coil moving within a permanent magnetic field. Movement of the mass is sensed by means of the change in capacitance at the base of the pendulum. This causes a small change in current which may be applied through a servo-amplifier circuit to cause a restraining force to be given by the coil. Equilibrium is quickly reached and the current maintained through the coil is then a measure of the acceleration obtained. Due to the large amount of loop gain employed, the actual motion required for the mass is extremely small so that its weight and size can be quite small.

Another example of a feedback transducer is the seismic transducer shown in fig. 2-16. This differs from the seismometer described in section 2-2-1 in that the coil is no longer a generator, but is provided with a driving current related to the

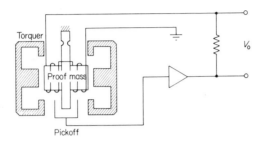

Fig. 2-15. The Endevco feedback accelerometer.

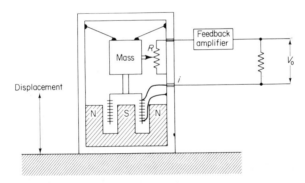

Fig. 2-16. A seismic transducer.

potentiometric displacement value, and hence position of the mass. The current is in such a direction as to oppose any movement of the mass by means of a counter-force acting on the coil. The equilibrium current flowing in the feedback path is thus a measure of the original displacement of the mass.

Feedback systems used to measure pressure may employ a combination of a pressure-sensing diaphragm linked to a differential transducer or a capacitance transducer. The signal produced by the pressure variation is amplified by a servo-system and used to restore the diaphragm to its original position. As with the seismic transducer, the value of the restoration force is a direct measure of the input quantity (in this case pressure).

61

2-2-5   SIGNAL CONDITIONING

It will have become apparent from the preceding discussion that transducers need a given electrical environment in order to function correctly. Passive transducers, such as the resistance strain gauge, require a stable constant current or voltage supply and additional resistors to complete a null bridge arrangement. Some energy conversion types require to be coupled to a high input impedance amplifier if their sensitivity is not to be seriously reduced. Additionally many transducers require calibration before use and the necessary calibrating components will need to be made available.

Signal conditioning may be considered as comprising all those operations which are ancillary to the functioning of the transducer itself, and which are necessary in order to extract a signal from it related to the physical quantity being measured. A signal conditioner is the name given to the unit which carries out the process of signal conditioning between the transducer and the recording or analysis equipment (fig. 2-17).

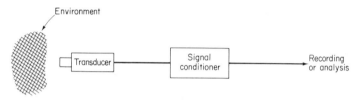

Fig. 2-17. Signal conditioning.

Two essential features of a signal conditioner are a stable operational amplifier having a defined gain, high input, and low output impedance, and a stable constant voltage (or current) power supply.

Where piezo-electric transducers are used the voltage amplifier is replaced by a charge amplifier which permits a reduction in the effects of cable capacitance on the transducer gain and frequency response. A charge amplifier is essentially an analog integrator which maintains its input at nearly zero voltage so that the output is a function of the integral of the input current, i.e.

$$V_o = \int i. \, \mathrm{d}t = Q \qquad (2\text{-}18)$$

The reflected dynamic capacitance component at the input to the integrator will be large compared with the stray capacitance, so that the effects of varying capacitance at the input will be negligible. The signal conditioning unit is designed in modular form such that additional features can be added to suit the particular transducer to which it is coupled. Examples are, a low-pass filter to reject the high frequency components generated by the transient excitation of an accelerometer, and the bridge completing and balancing resistors of a strain gauge transducer.

Variable reluctance and other transducers will require an energising alternating current source. This is modulated by the action of the transducer and detection of the modulated signal is carried out using a phase-sensitive detector followed by a low-pass filter in order to remove the carrier and intermodulation harmonics. These operations can also form part of the signal conditioning process. Finally, certain calibration components may be included to facilitate setting-up of the transducer for use. Details of these will be considered in the next section.

2-2-6   CALIBRATION OF TRANSDUCERS

Wire resistance strain gauges are manufactured in batches and their characteristics will vary from batch to batch. It is necessary, therefore, to calibrate a number of sample gauges in each batch to define the usable gauge factor. For small strain values a simple bending beam jig is used. The gauge is mounted on a flat beam which is subject to a known arc bending moment. From the dimensions of the jig and the measured deflection the strain is determined and plotted against change in resistive value.

A relative calibration method which is easy to use under operational conditions is to shunt the gauge element R_g of the bridge with a calibration resistor R_c and the output indication V_o is observed. The equivalent indicated strain can be shown to be

$$e_{\text{equiv.}} = \frac{-R_g}{k(R_g + R_c)} \qquad (2\text{-}19)$$

where k is the gauge factor.

A number of values of R_c are shunted across the gauge element in turn and values for $e_{\text{equiv.}}$ calculated. From the ratios of

$$\frac{V_{o_n}}{e_{\text{equiv.}_n}}$$

that are obtained the sensitivity of the system, and also linearity over the calibrated range, can be derived. A similar method may be used with piezo-electric strain gauges by shunting the gauge with fixed capacitators of known values.

Calibration of accelerometers must be carried out under dynamic conditions so that the effects of frequency on amplitude sensitivity and phase shift become important. The frequency of calibration should be chosen to be similar to that of the phenomenon under investigation. Three methods are in general use for the determination of amplitude sensitivity. These are:

1. Constant acceleration,
2. Sinusoidal motion,
3. Transient motion.

An example of the constant acceleration method is to subject the accelerometer to a normal acceleration by mounting it on the end of a centrifuge. The acceleration produced will be

$$a_n = r\omega^2 \tag{2-20}$$

where r is the distance of the accelerometer from the centre of rotation, and ω is the angular velocity in rads sec^{-1}. Since r and ω can be accurately measured a_n can be determined in terms of G over a wide acceleration range.

Constant acceleration methods will not give any information concerning frequency performance and for this sinusoidal or transient methods are used. A sinusoidal motion can be produced by means of a 'shake-table' which consists of a flat plate mechanically connected to the moving coil of an electro-magnetic transducer, very similar to that of a moving-coil loud speaker. Sinusoidal movement of the plate can be obtained by applying a driving current wave-form to the moving coil. The transducer to be calibrated will be mounted on the table

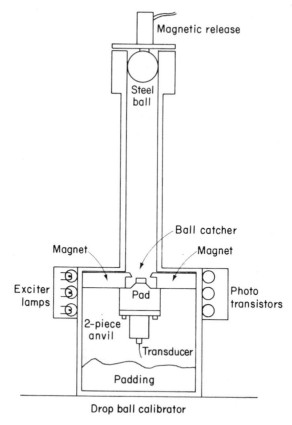

Drop ball calibrator

Fig. 2-18. Calibration using a drop ball calibrator.

and excited at a given amplitude of vibration over a range of sinusoidal frequencies. Measurement of peak displacement places a limit on the accuracy of this method. Since acceleration is proportional to ω^2 then the displacement becomes very small and hence difficult to determine. Strain gauge [11] or optical methods [12] are used to calibrate this small displacement.

Transient methods involve some form of ballistic device producing a known and repeatable impact. Ballistic pendulums have been used for this purpose but more reliable results are obtained from a drop-test machine. An example is the Endevco drop-ball calibrator shown in fig. 2-18. The transducer under test is fixed to the underside of the anvil. A steel ball is dropped down the tube and on contact drives the anvil away from the retaining magnets. Photo-electrical methods of measuring anvil speed are used and since the velocity of the ball is known, the energy given to the anvil can be calculated. A range of shock pulses can be obtained covering the range 100 G, 3 μsec duration to 100 000 G, 1 μsec duration. Through the use of a magnetic ball release [35] a high consistency in results can be obtained.

2-3 ANALOG MAGNETIC TAPE RECORDING AND STORAGE

Development of instrumentation systems over the last decade have resulted in techniques of information storage which are accurate, convenient to use, and, as a consequence of their widespread use, have attained a high degree of standardisation between manufacturers. These techniques use magnetic tape as a storage media and analog methods of recording. They are widely used for signal processing operations due to their low cost and high performance when compared with alternative forms of storage. Analog tape replay systems permit linear changes in the duration and frequency of the recorded information during replay in a way not easily obtained by other media.

A further advantage concerns the overload characteristics of the analog record. These are less severe with analog recording than other media (e.g. film) so that useful information can be obtained under conditions of poor scaling. Problems associated with sampled data are not present, so that the difficulties experienced with digital methods such as aliasing, quantisation, and sampling rate do not arise. Finally, due to the early standardisation of such methods of recording, the interchange of data and the ability to replay information collected in the field using different equipment is valuable in the economic processing of signals, recorded in this way.

2-3-1 BASIC ELEMENTS OF A RECORDING SYSTEM

For the purpose of description we can consider a magnetic tape recording system to be comprised of three basic elements:

1. The electronic coding system, which is responsible for the recording process, that is, the encoding of signal information into a suitable form for

recording, and for decoding the signal on playback to its original form. This will include signal amplification, automatic gain control, and correction for differing recording speeds as well as the actual encoding electronics.

2. The magnetic head, which converts the recorded signal into an electrical signal identical in form to the original recording. Most of the non-linear characteristics of an analog recording system may be attributable to this element, so that a clear understanding of the conversion process is essential if advantage is to be taken of the inherent low-distortion capability of the system.

3. The tape transport mechanism, whose function is to move the tape smoothly across the magnetic heads at a constant speed. Malfunction in this area can give rise to a number of distortions of the signal which are peculiar to the process of magnetic tape recording.

Each of these elements will be discussed in the following sections, commencing with the recording process. Many distinct types of coding systems have been evolved for various purposes. We will consider first an amplitude recording technique, known as the direct recording process. Although this method is of little value for many signal processing purposes a study of its characteristics will show clearly the boundary limitations of magnetic tape used as a storage media.

2-3-2 THE DIRECT RECORDING PROCESS

This is a method familiar in the design of domestic tape recorders used for the recording of speech and music. Its chief advantages are the availability of simple techniques of recording and playback together with an extremely wide recording bandwidth. For the recording of very short transients containing a high frequency spectrum, it will be superior to any of the methods to be discussed later. It does possess, however, a number of disadvantages for signal processing work and to understand these the method of direct recording will be considered in some detail.

The signal to be recorded in the form of a varying electrical current is passed through the windings of a recording head as shown in fig. 2-19. This consists of a

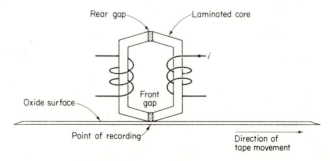

Fig. 2-19. The recording process.

magnetic core in the form of a closed ring, having two narrow non-magnetic gaps inserted at opposite points, as shown in the diagram. The rear gap does not play any significant part in the recording process and its purpose will be described later.

The magnetic path in the head core is completed by the ferrous nature of the coating on the magnetic tape, which is shunted across the front gap formed at the recording head. The signal current flowing through the pair of windings located on each limb of the head core produces a varying magnetic flux across

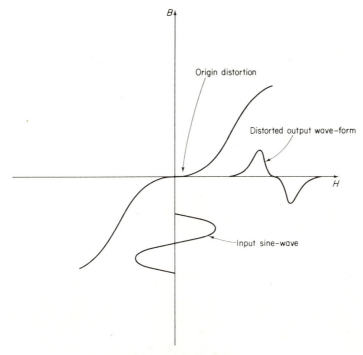

Fig. 2-20. Origin distortion.

the gap. Due to the non-linearity of the *B–H* curve for the magnetic material used in the core, the variation in flux will not be proportional to the variation in the recording current so that considerable distortion will take place about the origin, as shown in fig. 2-20.

A simple way of linearising this operation lies in the addition of a d.c. bias current to the signal current, so that only the linear portion of the transfer characteristic is traversed. This method, whilst successful in overcoming the fundamental problem of non-linearity has the effect of deteriorating the signal/noise ratio and has been replaced by an a.c. bias technique which does not suffer from this disadvantage. The method of a.c. bias is shown in fig. 2-21. An alternating current having a frequency of some three to five times the highest signal frequency, is added to (not modulated by) the signal frequency. During its passage past the gap the tape is subject to a number of complete hysteresis loops of magnetisation. If there is no recording current then the mean magnetisation on the tape, as it leaves the gap, is zero. The presence of a recording current displaces this mean value about a positive or negative value, determined by the signal current, so that the final value of magnetisation is

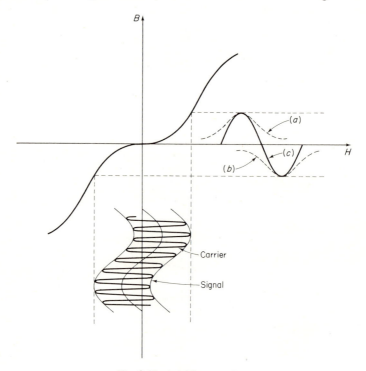

Fig. 2-21. h.f. bias recording.

linearly proportional to the recording current. This will be understood if we trace the upper and lower envelopes of the biased input waveform across the B–H characteristic. The output flux at any instant will be the difference between curves (a) and (b) resulting in a change of flux (c) having a linear relationship to the input signal current. It is important to note that the combination of the a.c. bias and the recording signal is accomplished with no new sum and difference frequencies being introduced. Consequently the bias frequency does not enter into the recording or subsequent playback processes.

The magnitude of magnetic flux established across the recording head gap is proportional to the alternating current through the windings. The actual change in magnetic domains only takes place as the tape leaves the gap area, i.e. on the trailing edge of the gap.

It will be apparent that if the recording current is sinusoidal, so that $i = I \sin \omega t$, then the wavelength of the signal to be recorded can be expressed in terms of distance along the tape enabling a relationship between tape velocity, v and frequency f, to be expressed as

$$\lambda = v/f \text{ in.} \tag{2-21}$$

The remanence flux will be given by

$$\phi_r = kI \sin (2\pi v/\lambda). \tag{2-22}$$

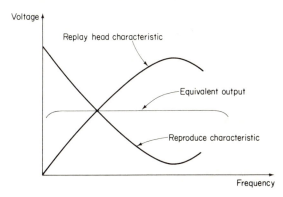

Fig. 2-22. Replay characteristics and equalisation.

To reproduce the original electrical signal from these variations in remanence flux, ϕ_r, the magnetised surface of the tape is drawn past the gap of a reproduce head, similar in construction to the record head. At any given instant a section of the tape is shunted across the gap and some of the flux surrounding the aligned elemental domains will be conveyed around the head core. The voltage induced in the reproduce head will be proportional only to those flux lines which emerge from the oxide surface, and not to the total flux present in the tape.

It is given by the rate of change of this flux, i.e. from equation (2-22):

$$e = KI \omega \cos (\omega t) \tag{2-23}$$

where K is a constant of proportionality and is $\ll k$.

From this we see that the reproduced voltage is proportional to frequency and for constant-current recording the output voltage will have the relationship shown in fig. 2-22. In order to obtain an overall flat frequency response characteristic from the reproduce system then this trend must be removed by

69

incorporating an inverse frequency characteristic as shown in the diagram. This technique is known as 'play-back equalisation'. It is also apparent from fig. 2-22 that the signal recoverable from the head decreases with frequency and results in a very poor signal/noise ratio at very low frequencies where equalisation becomes ineffective. This illustrates a major disadvantage of the direct recording process, preventing storage of signals having very low or zero frequency.

2-3-3 LIMITATIONS OF THE DIRECT RECORDING PROCESS

Some further limitations of this process will be considered below. An important limitation affecting all recording processes is known as the 'gap effect'. It is well known that as the recorded wavelength approaches the effective gap length then the ratio of the flux linking the core to the flux in the tape decreases. It can be shown [13] that the rate of decrease is described approximately by the periodic function

$$f(x) = \frac{\sin x}{x} \tag{2-24}$$

which reduces to zero when the wavelength equals the gap length. This effect is shown in figs. 2-23a and b. The first figure shows the reproduction of a relatively long wavelength of recorded signal on tape. Since the average value of magnetic flux across the gap is changing smoothly an output voltage results in accordance with equation (2-23).

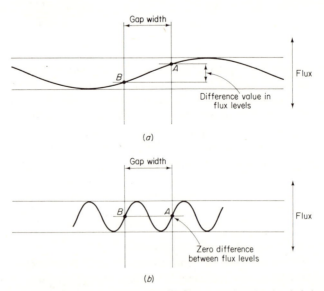

(a)

(b)

Fig. 2-23. (a) Long wavelength recorded signal. (b) Short wavelength recorded signal equal to gap width.

In the second figure a much shorter recorded wavelength is shown, equal in length to the dimension of the gap itself. Under this condition the average value of magnetic flux across the gap will be zero and will not change as the tape traverses the gap. The proportional response of the reproduce head, the physical limitations of the gap size, together with other factors discussed below, set bounds to the range of frequencies that can be reproduced by the direct

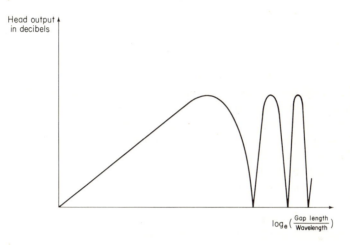

Fig. 2-24. Reproduce characteristics of the replay head.

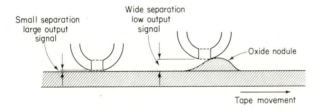

Fig. 2-25. Tape surface imperfections.

recording process. This is illustrated by fig. 2-24 for one particular tape speed. A second factor affecting reproduction of high frequencies is the sensitivity of the direct recording process to tape surface irregularities. These cause instantaneous attentuation of the reproduced signal, which are termed 'drop-outs'. The effect of these irregularities can be seen from fig. 2-25, which shows the variation in spacing that occurs between the uniform level of the tape oxide coating and the tape head as the tape is drawn past the head. The same effect, of course, can occur due to dust particles adhering to the tape surface.

It has been shown by Wallace [14] that a space, d, between the oxide coating and recording head will produce a wavelength-dependant loss given by

$$\text{loss} = 54 \cdot 6 \, d/\lambda \, (\text{db}) \qquad (2\text{-}25)$$

Other limitations of the direct recording process are

1. Self-demagnetisation–the process of recording effectively causes directional alignment of the magnetic domains contained within the tape material, like poles of which are adjacent. For very small wavelengths the domain strings are short and demagnetisation, due to the proximity of like poles, is considerably increased. This effect decreases as the recorded wavelength becomes longer since the like poles of the effective domain string are further separated.

2. Gap alignment–azimuth tilt can also cause attenuation of the higher frequencies, since for small wavelengths, the induced flux during playback varies along the length of the gap. This has the effect of reducing the peak flux value and increasing the minimum value so that the actual flux separation is reduced, the reduction being 100% (i.e. zero induced flux), when the tilt reaches a complete wavelength. This is similar to the effect obtained with a finite gap length and becomes less important as the wavelength increases.

3. Penetration loss–this refers to oxide coat thickness and produces effects which are similar to spacing loss, i.e. shorter excursions of lines of flux from the tape surface occur as the wavelength decreases. With finite tape thickness, as the wavelength is reduced a decreasing proportion of the coating, starting from the inside, fails to make any appreciable contribution to the external field.

4. Recording demagnetisation–the maximum susceptibility of the oxide coating of the tape occurs at a given bias current. The location of this point will, however, vary in depth from the surface of the oxide coating in a manner proportional to frequency, with the result that a reduction in sensitivity occurs as the frequency is increased.

These effects are present with other forms of magnetic recording but due to the high sensitivity of direct recording to changes in recording wavelength, are particularly serious with this method of storage.

2-3-4 FREQUENCY-MODULATION RECORDING

It has been noted previously that the amplitude performance of magnetic tape systems is subject to a number of limitations, resulting in a degradation of the accuracy for the overall system. In particular the frequency sensitive characteristics of direct recording places a low limit on the minimum frequency that can be satisfactorily recovered from the tape and precludes the recording of d.c. or slowly varying signals completely. Carrier techniques provide a solution to the problem.

Using these techniques the recorded information is contained in the change and rate of change of the carrier frequency. Consequently, saturation recording is often used to obtain the maximum signal-to-noise ratio at the reproduce head, without affecting the accuracy of the information obtained. However it is not essential to use saturation recording, and there are advantages in recording the carrier frequency using an additive bias level, as with direct recording, particularly for wide-band operation. This is particularly the case where a high packing density is desirable. Using saturation recording the packing density is limited by the thickness of the oxide coating, whereas no such limitation is applicable to non-saturation recording.

Many techniques of carrier modulation are in use, examples are: pulse-code modulation, phase modulation, frequency modulation and, of course, the various forms of digital recording. In practical recording systems there is little to choose between the use of phase or frequency modulation, and only the latter will be considered, although much of the treatment is also relevant to phase modulation. Various forms of pulse code modulation have been devised for analog recording and recent developments have indicated considerable advantages over frequency modulation, particularly with regard to packing density [15]. The method will be discussed briefly later in this chapter.

Frequency modulation carrier systems (F.M. systems) are used extensively in instrumentation and signal processing work to achieve a very high standard of performance. Here a high carrier frequency is employed which is modulated in frequency by the signal to be recorded (unlike direct recording where an additive carrier is used.)

If we express this signal as a voltage, $e = A \cos \omega t$, the instantaneous frequency of the F.M. signal will be given as

$$f_i = f_0 + m\delta f \cos \omega t \qquad (2\text{-}26)$$

where f_0 is the unmodulated carrier frequency, m is a modulation factor $\leqslant 1$ and represents a normalised form of the amplitude of the signal voltage (i.e. $m = A/A_{\max}$.), δf is the frequency deviation and represents the extreme value of the instantaneous frequency of the carrier.

In this form, information regarding amplitude and frequency of the modulating signal is conveyed by the characteristics of the carrier frequency. such that the amplitude of the signal determines the shift of the carrier from its mean value, and the frequency of the signal determines the rate of change of this shift in carrier frequency.

If we represent the F.M. carrier by $e_c = B \cos \omega_i t$, then from equation (2-26)

$$e_c = B \cos (\omega_0 + m\delta\omega . \cos \omega t) t$$

$$= B \left\{ \cos \omega_0 . \cos \left(\frac{m\delta\omega}{\omega} . \sin \omega t \right) - \sin \omega_0 . \sin \left(\frac{m\delta\omega}{\omega} . \sin \omega t \right) \right\}. \qquad (2\text{-}27)$$

Representing $(m\delta\omega)/\omega$ by $\delta\theta$ then it can be shown [16], that the modulating terms; $\cos(\delta\theta.\sin\omega t)$ and $\sin(\delta\theta.\sin\omega t)$ can be expanded as a series of Bessel functions:

$$\cos(\delta\theta.\sin\omega t) = J_0(\delta\theta) + J_2(\delta\theta).\cos 2\omega t\ldots,$$

$$+ 2J_{2n}(\delta\theta).\cos(2n\omega t) \qquad (2\text{-}28)$$

and

$$\sin(\delta\theta.\sin\omega t) = J_1(\delta\theta).\sin\omega t + 2J_3(\delta\theta).\sin 3\omega t\ldots,$$

$$+ 2J_{2n+1}(\delta\theta)\sin((2n+1)\omega t) \qquad (2\text{-}29)$$

so that the substitution in equation (2-27) will give

$$e = B[J_0(\delta\theta).\cos\omega_0 t + J_1(\delta\theta).\cos(\omega_0+\omega_1)t - J_1(\delta\theta).\cos(\omega_0-\omega_1)t$$
$$+ J_2(\delta\theta).\cos(\omega_0+2\omega_1)t + J_2(\delta\theta).\cos(\omega_0-2\omega_1)t$$
$$+ J_3(\delta\theta).\cos(\omega_0+3\omega_1)t - J_3(\delta\theta).\cos(\omega_0-3\omega_1)t\ldots] \qquad (2\text{-}30)$$

where

$$\left.\begin{array}{l}\cos\omega_0.\cos n\omega = \\ \sin\omega_0.\cos n\omega = \end{array}\right\}\cos(\omega_0+n\omega)\pm\cos(\omega_0-n\omega).$$

Reference to a table of Bessel functions will show that the bandwidth of a frequency modulated system will not be contained within the deviation range (as obtained with amplitude modulation, where the bandwidth equals twice the modulation frequency) but will depend on the deviation ratio

$$= \frac{\text{carrier frequency deviation}}{\text{modulation frequency}} = \frac{\delta\omega}{\omega}. \qquad (2\text{-}31)$$

This parameter determines to a large degree the susceptibility of the system to noise. For frequency modulation recording this ratio generally lies between 1 and 2.

A second defining parameter for a frequency modulated system is the percentage deviation

$$= 100.\delta\omega/\omega_0 \qquad (2\text{-}32)$$

and, in order to achieve a high degree of linearity with the system, the percentage deviation is large and of the order 20 to 80%.

A frequency modulation record/replay system is shown in fig. 2-26. The input signal is applied to a voltage/controlled oscillator to which is connected the recording head. The signal recovered on playback is usually quite small (see fig. 2-27), and must be amplified before being presented to the demodulation circuit. A low-pass filter follows the demodulator in order to remove the carrier frequency and side band frequencies generated during the recording process.

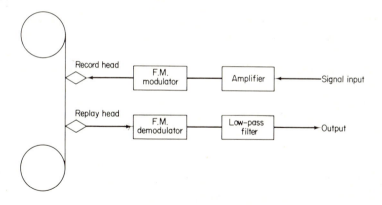

Fig. 2-26. A frequency modulation record/replay system.

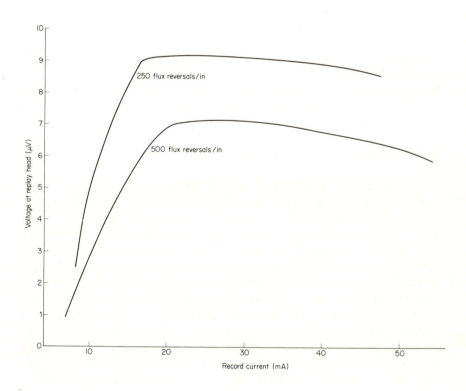

Fig. 2-27. Input–output characteristics of the record/reproduce head.

75

This will have a cut-off frequency approximately one fifth that of the carrier frequency permitting the recording of signals from 20kHz down to d.c. at a tape speed of 60 i.p.s. (inches per second). These are typical figures and assume a carrier frequency of 1800 Hz/in.sec which, as we shall see later, is an internationally agreed standard. Recording at a low speed, using a proportionately reduced frequency range, will increase the recording time available from a given magnetic tape length. If the centre frequency is scaled down in the same way, keeping the deviation at ± 40% as before, then the wavelength recorded in the tape will be the same. This makes it possible to record at one tape speed and reproduce at an entirely different tape speed. The change in signal time base obtained in this way represents one of the most important characteristics of the F.M. recording process.

Accuracy of performance for a F.M. recording/playback process is dependent to a large extent on the characteristics of the tape transport mechanism. The coding and decoding electronics will generally have a performance considerably in excess of the minimum requirements for the overall system. The demands on the ability of the tape transport to move the tape across the heads at a precisely uniform speed are indeed quite stringent. Any speed variations introduced into the tape movement at its point of contact with the heads will cause an unwanted modulation of the carrier frequency and result in additional periodic or random components to the wanted signal. It is important to note that the effect of these unwanted speed variations, commonly referred to as flutter, will introduce a background noise level which is directly proportional to the deviation obtained. Hence it is important to reduce these to a minimum, particularly with wide deviation systems, such as sub-carrier F.M. systems. We can regard these effects as limiting factors which control the dynamic range and accuracy and will merit further consideration in a later section.

2-3-5 MAGNETIC RECORDING AND REPLAY HEADS

It will have now become apparent that many of the limitations of the recording process are intimately related to specific characteristics of the magnetic heads. The construction of a typical magnetic head was shown in fig. 2-19. The two identical core halves are constructed from thin laminations of a material, having both high permeability and low electrical resistance. The latter is required to minimise the effects of eddy-current losses induced in the core, and which increase with frequency. Both halves of the core carry identical windings which are connected in series. Two non-magnetic gaps are shown in the diagram but only the one in contact with the tape enters into the recording process. The purpose of the rear gap is to increase the reluctance of the magnetic circuit such that the heads do not become magnetised easily. This could occur accidentally (due to, for example, undesired surges in power supplies), and result in permanent magnetisation which will reduce the efficiency of the recording and playback process. Since the efficiency of the magnetic head is determined by the ratio of the front gap to the rear gap, the latter is made very small. The gap size of a record head is a compromise between the need to achieve deep flux penetration and high frequency performance. A common value of gap width is 0·005 in. A much smaller gap is used for the replay head.

Multiple track heads allow several channels of information to be recorded simultaneously. The number of tracks varies from 2 to 96 and are accommodated in tapes of varying width from 0·25 to 2 in. Such a head consists of several cores stacked one above the other (fig. 2-31). To obtain a good signal–noise ratio each track should be made as wide as possible. This conflicts with the need to maintain adequate track spacing and so avoid inter-track coupling or cross-talk. A compromise design inevitably results. Cross-talk can be improved by using two recording heads and arranging that the even-numbered tracks are recorded or played back by one head and the odd-numbered tracks by the other. Close track-to-track spacing on the tape is thus achieved with wider separation between adjacent channels.

Extremely close mechanical tolerances are crucial to obtain optimum performance using interleaved heads. One of the most vital of these dimensions is that between the gap centre lines of the two inter-leaved stacks. The two head-stacks are required to be precisely positioned one relative to the other, so that the distance between the two lines passing through the centre lines of the gaps is exactly 1·5 in. within the tolerance of ± 0·001 in. (an international standard). This establishes the relative timing accuracy between information channels recorded on separate stacks. It should be noted that this accuracy is subject to error resulting from tape stretch or shrinkage, which can occur with changes in temperature or humidity. Two other vital mechanical factors are gap scatter and azimuth. Gap scatter denotes alignment of track gap centre lines within the stack. Deviation from the linear alignment is held to less than 100 micro inches.

Gap azimuth refers to the perpendicularity of gap centre line with the head-mounting plate surface, and is maintained within plus or minus a single minute of arc.

2-3-6 MAGNETIC TAPE TRANSPORT

The tape transport performs the function of moving the magnetic tape at a constant linear velocity across the magnetic heads. It must do this without disturbing the fixed relationship between the position of the recording tracks on the tape and the magnetic head record or replay gaps relative to a fixed datum (e.g. the base plate of the machine).

As mentioned earlier, F.M. recording is extremely susceptible to variations in transport speed, resulting in a form of signal distortion known as flutter. Two methods of speed control are in current use. The majority of instrumentation systems use a system of capstan speed control in which the output of a tachometer, driven by the capstan, is compared with a signal derived from a stable crystal oscillator. The error signal produced is used to control the speed of the capstan motor so as to maintain this at a precise synchronous speed. To change the capstan speed it is only necessary to alter a frequency division network, interposed between the reference oscillator and the frequency comparator.

This method of control will not necessarily produce a constant tape speed however, due to slip between the capstan and the pinch rollers, and also due to the dynamic distortion of the tape. A second method of control is therefore used and referred to as tape speed control. Here a reference signal is derived from the crystal oscillator and recorded on one track of the tape. During playback it is this signal which is compared with the crystal oscillator instead of the tachometer output. With this method of control the capstan speed is permitted to vary although the actual tape movement is maintained constant to a high degree of accuracy. Both methods of control are used for example, in Sangamo magnetic tape systems, where the replay tape speed is maintained to within 0·001% of the speed at which the tape was originally recorded.

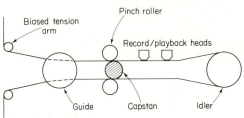

Fig. 2-28. A closed loop transport system.

Transport mechanisms are complex and it will not be possible to describe adequately details of their design here. The general principles involved are, firstly to ensure a constant speed of movement for the tape across the record/replay heads and secondly to avoid any long unsupported length of tape in this area which could give rise to high frequency fluctuations of tape speed. This is generally achieved by using a closed loop head configuration, as shown in fig. 2-28. Control of tape tension is obtained by means of one or more biased tension arms supporting the tape which are free to move, whilst maintaining tape tension substantially constant.

An alternative to the tension arms shown in fig. 2-28, is a vacuum-bin storage system as shown in fig. 2-29. The photo-sensing devices shown in the diagram detect the position of the two loops of tape contained within the bin. They cause signal levels to be supplied to the drive and take-up motors via a servo-mechanism, which is arranged to maintain the required tension on the tape, for which the extent of the loop is a measure. This method is used particularly for high-speed systems and is found in digital recording systems since it also permits very rapid braking of the moving tape.

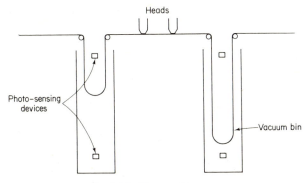

Fig. 2-29. Vacuum bin system.

2-3-7 CONTROL FACILITIES

Operating modes for an instrumentation tape transport can include:

1. Forward movement under capstan control at one of a number of selected speeds.

2. Fast forward movement (tape preferably lifted clear from the heads).

3. Forward movement under variable manual control speed (known as the search mode).

4. Reverse movement under capstan control.

5. Fast rewind.

6. Manual halt.

7. Automatic halt from sensing strip on the tape.

8. Loading mode.

9. Shuttle operation.

10. Loop operation.

All of these operational functions are not, of course, found in every tape transport. Some means of remote control of some or all of these modes is desirable to facilitate automated analysis procedures. The shuttle facility enables a specified section of tape, identified by means of sensing marker strips at either end, to be repeatedly replayed at a selected speed, and automatically re-spooled to the commencement of the section between each play-back period. This facility is known as 'shuttle' and is of value in iterative operations, such as, for example, the resolution of the power spectral density of data recorded on the tape by analog methods.

Facilities for recording and reproduction from endless tape loops is also required for analysis purposes and for the introduction of time delays. Small loops can generally be accommodated within the tape transport region. Longer loops of tape are accommodated by winding the tape between a parallel series of idler rollers, or preferably the use of a tape bin. In the latter case difficulty may be experienced from static charges accumulating on the tape causing folds of tape within the bin to adhere together.

2-3-8 TAPE MOTION ERRORS

Ideally, the tape should traverse the record reply heads at a uniform velocity at a precisely known speed. In practice a number of departures from this ideal are found and are attributable to very many sources, not all of which are amenable to design correction.

A shift in the average velocity can be compensated by velocity control of the capstan motor from its servo-system, so that precision in tape speed can be achieved fairly easily. Short term speed variations are another matter and the control techniques described in section 2-3-6 can be applied to reduce these. Flutter can take two forms; those in which the flutter variations are identical across the width of the tape, and those which are not. It may or may not be possible to distinguish between the two by measurement of performance. Flutter may be caused by eccentricity in the rotating parts, irregularity due to tension variations, or friction resulting from the rough texture of the tape itself.

Tape traversal speed is thus subject to a combination of effects which can be regarded collectively as a number of individual cyclic variations superimposed on a fairly wide-band random base. The positon is still further complicated by differing characteristics of the recording and playback equipment which may not share a common transport. When a signal has been recorded and is later reproduced, the flux circulating in the playback head will be frequency modulated by the flutter obtained during the recording process, as well as with that present during playback. Since the output from the playback head also produces a voltage proportional to the rate of change of flux then an amplitude modulation component will also be present. An analysis of the overall effects of flutter during the record-replay process has been given by Davies [13], who shows that the flutter is accentuated by the value of the ratio between the carrier frequency and frequency deviation, so that a large deviation is found desirable for a frequency modulation system. Some measure of separation in the recording and playback effects can be made by recording at a high speed (when the effects of flutter are least) and playing back at a low speed. The measured flutter is then almost solely due to that of the replay equipment.

Measurement of flutter for the direct recording process is generally carried out from a r.m.s. sum of the magnitudes of the deviation for all frequencies within a narrow band, up to about 200 Hz, and expressed as a total percentage figure [17]. This approach is not adequate to express the performance of wide-band frequency modulation systems. Instead, a flutter density measurement is made of the percentage r.m.s. flutter obtained over a narrow bandwidth, repeated at continuous intervals over the frequency range of interest. This corresponds closely to the method of specifying the power spectrum for a random phenomenon (see chapter 8). An alternative method is commonly used in which a cumulative peak-to-peak curve of the flutter density spectrum is plotted by adding the contribution of successively selected higher frequencies (fig. 2-30).

Related to flutter is time-displacement error. This is defined as the error between the separation in time of two events recorded on the same track at the time of recording, and the measured separation time during playback. It has been demonstrated [13] that this particular type of error is cyclic and will attain a number of zero values when the time between recorded events becomes equal to integer multiples of the flutter period. The error will be small for small flutter values and for large time intervals.

A form of timing error occurring between adjacent tracks of the tape is known as skew or yaw error. Here the time or phase displacement takes place

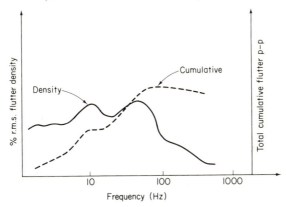

Fig. 2-30. Flutter characteristics.

across the width of the tape and can have a fixed value or can change with the motion of the tape past the heads. In the former case it may be due to errors in head stack alignment, incorrect traversal path of the tape across the heads, or phase-shift errors introduced by the reproducing electronics. Variable skew produced by tape motion irregularities is rather more difficult to correct. The actual direction of tape traversal across the head can vary in a non-uniform way and produce a time or phase displacement between tracks which is linearly proportional to the spacing between them. This can be caused by a tension gradient across the tape, which may not be uniform with time. Due to the tensional strains experienced by frequent stop-start operation, the effect is more noticeable with digital transports. The timing errors produced by skew are accentuated when a tape is recorded on one machine and played back on another. If the same machine is used for recording and replay then some of the linear forms of skew can reasonably be expected to cancel out. Skov [18] discusses a number of causes for skew and describes suitable measurement techniques that can be applied.

2-3-9   ERROR COMPENSATION

The introduction of a first-order noise component as a result of flutter is one of the major disadvantages of carrier recording systems. It is, however, one which can fairly easily be corrected by electrical means. Correction for second-order effects such as time displacement and skew becomes more complex and both mechanical and electronic methods have been attempted [19,20]. A simple form of correction for the effects of flutter assumes the generation of coherent errors on all recording channels. Due to this identity in time the noise produced by the flutter, in an unmodulated track, can be sign-reversed and used to cancel the effect produced on other modulated tracks. One track of the tape is allocated for flutter correction and an unmodulated carrier signal recorded. No signal should be recoverable from this track during replay, but due to variations in the speed of the tape during the recording process, the reproduced carrier frequency will vary and produce a reference signal proportional to the flutter variations of the tape. Correction is obtained by sign inversion of this reference and adding it to each of the recorded signals during replay. Under certain conditions this technique has been successfully applied to earlier types of transport which used a synchronous motor and relied on a heavy flywheel to affect speed stabilisation. The phase differences between the reference signal and the signal being compensated prohibits the simple subtraction of wide-band signals, and, due to the skew effect, this correction is not so effective for tracks geographically removed from the reference track. The reference track must be located on the same head stack as the compensated track so that a separate reference track is required for each recording head. The developments in servo-control, outlined in section 2-3-6 have reduced the flutter in tape transport speed to such a low level that electronic compensation methods are now not nearly so effective. This is because the effects of phase differences prevent complete cancellation of the small amounts of flutter remaining, particularly at high tape speeds.

Relative timing errors between tracks caused by skew can be corrected by tilting the playback head about an axis normal to the plain of the tape and passing through the centre of the head gap line. A mechanical method of correction for dynamic skew errors has been suggested by Maxwell and Bartley [20] involving a servo loop actuating a head-tilting mechanism and which is effective for low skew frequencies. Electronic methods of correcting skew involve inserting delays into each demodulating signal from the head stack. These delays can be adjusted fairly readily for static head skew but their adjustment for dynamic head skew is not readily obtained.

2-3-10   I.R.I.G. STANDARDS FOR F.M. SYSTEMS

In order to allow maximum interchange of information recorded on magnetic tape, using frequency modulating systems, an international standard has been agreed and found fairly universal acceptance. This is known as the I.R.I.G.

standard (inter-range instrumentation group). [21]. A condensed form of this standard is reproduced below for frequency modulation record/reproduce systems only.

Mechanical characteristics

Magnetic tape.

Tape width either 0·5 or 1 in.

Track geometry. This is shown in fig. 2-31.

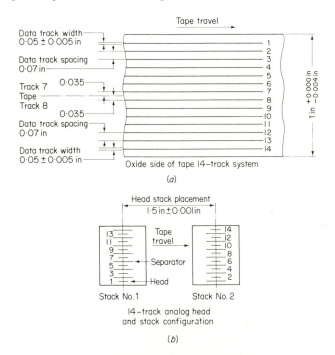

Fig. 2-31. I.R.I.G. track and head geometry.

Track width is defined as 0·05 ± 0·005 in.

Track spacing is defined as 0·070 in centre-to-centre.

Track numbering shall be consecutive starting with track one.

Numbering is taken from top to bottom of the tape when viewing the oxide-coated side of the tape with the earlier portion of the recorded signal to the observers right.

Head. Two heads are specified for recording and reproduction with the even numbered tracks on one head and the odd numbered tracks on the other (see fig. 2-31). The centre lines through the head gaps are to be parallel and spaced

1·500 ± 0·001 in apart. Head stack tilt is defined by stating that the plane tangent to the front surface of the head stack at the centre line of the head gaps shall be perpendicular to the mounting plate to within plus and minus 3 minutes of arc. Gap scatter shall be 0·0001 in or less. Mean gap azimuth alignment shall be perpendicular to the mounting plate to within plus and minus 3 minutes of arc. Head location shall be positioned within plus and minus 0·002 in of the nominal position required to match the track location shown in the diagram.

Transport. The standard tape speeds are

$$\tfrac{15}{16}, 1\tfrac{7}{8}, 3\tfrac{3}{4}, 7\tfrac{1}{2}, 15, 30, 60 \text{ and } 120 \text{ i.p.s.}$$

Record reproduce parameters
  Bandwidth. Three bandwidths are designated as follows:
  (a) Low band: 0 to 10kHz at 60 i.p.s.
  (b) Intermediate band: 0 to 20 kHz at 60 i.p.s.
  (c) Wide band: (Group 1) 0 to 80 kHz at 120 i.p.s. (Group 2) 0 to 400 kHz at 120 i.p.s.

Record characteristics. Input voltage of 1·0 to 10·0 volts peak-to-peak shall be adjustable to produce full frequency deviation. Deviation direction should be such that increasing positive voltage will give increasing frequency.

Reproduce parameters. Output level shall be a minimum of 2 volts peak-to-peak with increasing input frequency giving a positive going output voltage. Deviation index should be plus and minus 40% for full deviation. The tape speeds and related carrier frequencies are detailed in the table 2-2.

TABLE 2-2 I.R.I.G. tape speeds and related carrier frequencies

Carrier centre frequency (kHz)	Tape speeds (i.p.s.)		
	Low	Intermediate	High
1·688	$1\tfrac{7}{8}$	$\tfrac{15}{16}$	—
3·375	$3\tfrac{3}{4}$	$1\tfrac{7}{8}$	$\tfrac{15}{16}$
13·500	$7\tfrac{1}{2}$	$3\tfrac{3}{4}$	$1\tfrac{7}{8}$
27·000	15	$7\tfrac{1}{2}$	$3\tfrac{3}{4}$
54·000	30	15	$7\tfrac{1}{2}$
108·000	60	30	15
216·000	120	60	30
432·000	—	120	60

I.R.I.G. performance testing. The reader is referred to the document quoted earlier for accepted methods of testing for the performance parameters given above.

2-3-11 P.C.M. RECORDING

Other forms of saturation recording define the characteristics of the modulating signal by varying the width of the carrier signal (pulse duration modulation, or P.D.M.), or by recognition of two levels of recording corresponding to a digital 0 or 1, and so recording the amplitude of the signal in digital coding terms (pulse code modulation, or P.C.M.).

An advantage of P.D.M. over F.M. recording is that it is relatively unaffected by tape speed variations. A serious disadvantage is that a very much wider bandwidth is required to contain a given frequency of the modulated signal at a given tape speed than the corresponding F.M. system. For this reason it is not widely used as a continuous carrier system, although it has certain advantages for multi-channel telemetry systems.

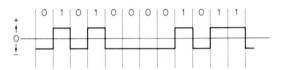

Fig. 2-32. N.R.Z. coding.

A number of alternative coding sequences are in use for P.C.M. A method widely employed for digital computer purposes is the N.R.Z. (non-return to zero) code shown in fig. 2-32. A change in the level occurs for each transition from a 0 to a 1 or a 1 to a 0. Two major limitations are apparent with this coding sequence when used for magnetic tape recording. A response down to d.c. will be required and the system will be susceptible to time-base errors [22]. This latter is a consequence of the nature of the code which permits long strings of ones or zeros to exist during which time no changes in the signal can occur. This precludes accurate extraction of the clock period from the recorded signal.

An improvement in time-base jitter is obtained if a transition is obtained for each bit period, using a form of coding known as bi-phase modulation. The necessity for a bandwidth extending down to zero frequency is removed although the total bandwidth required will be double that of N.R.Z. recording. Other methods of coding have been suggested which result in some improvements in jitter performance without incurring a bandwidth penalty [15]. It remains to be seen however whether the additional complication of digital encoding will result in a major improvement in performance or cost when compared with conventional F.M. systems in use for continuous recording.

2-4   DIGITAL RECORDING AND STORAGE

An example of digital recording was given in the previous section when P.C.M. recording was discussed. The initial signal was however considered as being continuous in form, and digital means used to overcome some of the deficiencies of analog methods of recording. There are circumstances where digital means of temporary storage are necessary for other reasons.

Occasionally the acquired data will assume a limited number of discrete values and may become available only in digital form. An example of this is that of data derived from solar X-ray observations carried out in the upper atmosphere, [23], in which the nuclear counters used for detection quantise the data into a small number of finite levels. A second need for digital data storage arises where the transducers have a capability of measuring to a degree of accuracy outside the dynamic range of analog storage methods. Digitisation of the signal may then be carried out and the data stored in this form on digital magnetic tape. A third possibility is where it may be found necessary to convert the analog signals, possibly after some analog processing and data reduction, to digital form in order to carry out digital computer operations on the signal. Subsequent storage of the processed data will then be required from data available in digital form. For reasons of speed, convenience, and economy this storage may need to be carried out using magnetic tape so that a treatment of some of the problems involved in digital magnetic tape recording is relevant to this chapter.

Digital tape recording is well-developed for digital computing equipment and has attained a high degree of standardisation, permitting data exchange between widely scattered installations. The application of this technique is far less critical in performance than analog tape recording and it is not so essential that the user understand the detailed mechanics of its operation.

Some characteristics of digital recording are:

1. Considerably reduced dynamic range (only 2 levels of storage required).

2. The data may be read out under synchronous conditions thus reducing the importance of timing errors.

3. Digital recording is relatively insensitive to tape transport speed variations.

4. The digital process is capable of an extremely high order of accuracy.

5. Record/replay takes place at a high transport speed so that at no time does the magnitude of the signal approach that of the noise level of the process.

6. A lower packing density is employed.

7. Simplified read/write electronics are used giving high reliability.

Digital information is written onto magnetic tape as a series of discrete areas of magnetisation, one for each track of the tape, having a width slightly wider than that of the recording head gap (see fig. 2-33). Each pulse of magnetisation can take one of two saturation states corresponding to a binary 1 or 0. A set of such pulses across the tape width is referred to as a 'stripe', and may be interpreted as a unique coded character consisting of 6 or 8 binary bits. Unlike analog information the characters are not written or read as a continuous sequence on the tape, but are broken up into groups of characters, referred to as 'blocks' where each block length is separated by a 0·75 in inter-record-gap in which no information is recorded. Block lengths vary between 128 and 4096 bits and, since the inter-record-gap is fixed, the tape utilisation deteriorates as the block length is reduced. This is indicated in the following expression for total real storage capability:

$$\text{total no. of bits} = \frac{\text{length of tape} \times \text{length of record}}{(\text{record length/packing density}) + \text{inter-record gap}} \qquad (2\text{-}33)$$

The record length is given in characters, and gap and tape length given in inches. The packing density for digital tape varies between 200 and 1600 b.p.i.

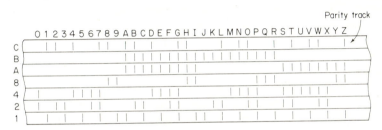

Fig. 2-33. Digital magnetic tape storage.

This compares with the ability of analog tape to store up to 16 000 cycles of information per inch of the tape, so that from the point of view of information storage digital tape is considerably inferior.

The main problems in the use of digital magnetic tape storage in computer compatible format are caused by the intermittent start/stop operation, consequent upon this block-by-block record/replay process, and the high dependence on tape quality for satisfactory performance. The former results in skew error which is corrected in the design of the recorder by mechanical or electronic means as discussed earlier. The limitations in packing density resulting from this error have been considered by Ziman [22] who has shown the advantages of a clocking system in minimising this.

Digital recording systems are very sensitive to drop-outs as will be apparent if we consider the method of information storage. Since all the information is contained in the presence or absence of pulses upon playback then the loss of signal or generation of spurious signal by tape inhomogenities cannot be tolerated. This is a result of the discrete and uncorrelated nature of the recording

87

digital signal. Each signal is accepted as a unique number and not one of a related series as is the case with frequency modulation analog recording, so that no inherent 'smoothing' is present in the process. We have seen earlier, when discussing direct recording, that drop-outs are most critical at short wave-lengths approaching the size of the replay head gap. This sets a minimum duration for the recorded digital pulse and limits the maximum packing density to a value considerably lower than that obtained with analog tape. Due to the often catastrophic effects of drop-outs in digital systems and the inability to completely remove magnetic tape surface imperfections, various methods of parity checking are used. This is shown in fig. 2-33 for a 7-track digital recording system. Six tracks are used to record a digital word corresponding to an alpha-numerical character across the tape. An additional binary 1 is recorded on the seventh track, when a 6-bit character containing an even number of 1's is recorded. During the read-out of the character the 1's in all the 7 tracks are summed so that the loss or addition of one pulse can be detected. This arrangement would, of course, fail to detect the loss or addition of an even number of pulses so that some small measure of uncertainty exists. A similar check can be carried out along the tape so that a parity bit can be written in an additional stripe added to the character block (see chapter 4). The use of both of these parity check methods, will in nearly all cases, ensure accurate detection of bit losses.

For signal processing purposes data is frequently stored in binary-coded-decimal form as a 3- or 4-character word. This is a useful form from the point of view of subsequent analysis and will be considered later in chapter 4 when the digitisation of analog records is discussed.

The process of digital storage on magnetic tape, although not continuous in the same sense as analog storage, is a continuous process within the blocks of data. It is necessary to provide the data in the form of complete blocks of information at a rate compatable with the speed of transport for a single block. This may, or may not, match the rate of availability of data to be stored, which will, in any case, become available as a continuous character string. Some form of buffer storage designed to hold a complete block is therefore necessary to act as a reservoir to the block-by-block writing sequence. Where the rate of arrival of the data in character form is extremely slow (e.g. recording of pulses from a slowly decaying radio-active source), then the conventional means of magnetic tape recording becomes inefficient. If the amount of data is small then the information can be recorded on punched paper tape for later digital analysis. For larger quantities of information then a special form of digital tape recorder, known as an incremental recorder may be used. Here the tape transport is arranged to move in incremented steps of one stripe at a time. Each input digital character consisting of 6 bits is written as a stripe across the tape together with a parity bit, pausing after each character has been written until the next character becomes available and assembled ready to be written. The pulse packing density is necessarily smaller than with continuous block-organised tape transports but the available total storage is considerably greater than would be available from punched paper tape and entry into the digital computer for subsequent analysis is consequently faster. The incremental recorder generally includes the control logic necessary to insert an inter-record gap at appropriate intervals, so that the resultant tape is compatible with conventional digital forms of storage.

2-5 GRAPHICAL RECORDS

Data acquisition in the form of continuous chart recording has a value for immediate visual inspection of the signal. The basic unit is a galvanometer element arm carrying a pen which records the signal variations on a continuously moving paper chart. The main disadvantage is the low frequency response (about 50 Hz), which is dependent on the dynamic properties of the galvanometer and moving arm assembly. A slightly faster movement can be obtained if a hot stylus is employed using heat-sensitive recording paper. A considerable improvement is obtained if the arm is replaced by a beam of light directed on to photo-sensitive paper. Such mirror-galvanometer combinations will permit recording up to several kHz and also permit the crossing of tracks. The penalty to be paid with this arrangement is the delay associated with chemical processing of the results.

None of these methods will present the acquired data in a form which will permit direct entry into a computer for further processing. Methods of conversion for such graphical records will be considered in the next section.

2-5-1 GRAPHICAL–DIGITAL CONVERSION

Conversion of graphical data to a digital computer compatible medium can be by semi-automatic or automatic means. The former involves an operator following a recorded line with crossed-hairs or a servo-tracked pencil-follower (see fig. 2-34). The reading device, shown as a hand-guided pencil indicator in fig. 2-34 carries a coil energised with an alternating current signal. This is detected by means of a second coil situated on a moving gantry beneath the plotting surface. The gantry is capable of movement in the X and Y directions in response to signals from servo amplifiers arranged to drive the controlling motors in such a direction that the follower coil beneath the table surface always centres itself on the reading device. In one design of follower the servo motors also rotate shaft position encoders which provide two sets of binary signals indicating the position in X-Y coordinates referenced to a specified position on the plotting table. After code-conversion the digital coordinates are fed to a paper-tape punch, card punch or magnetic tape for output data retention.

Automatic equipment requires more complex methods in which the servo-controlled follower is locked to the recording trace by means of a balanced signal derived from a pair of photo-cells situated on the moving gantry. Some analog methods of follower exist where the trace to be followed is prepared

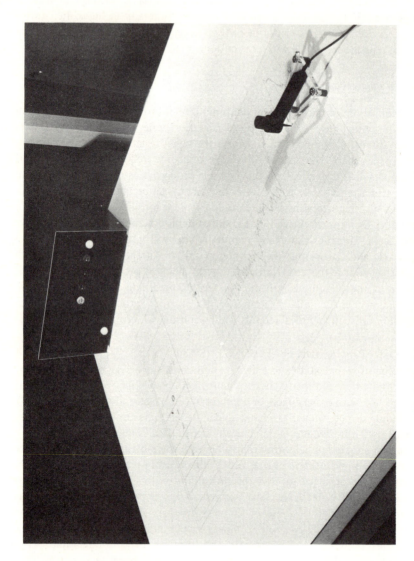

Fig. 2-34. A servo pencil-follower graphical converter.

using conducting ink which carries the energising signal. This is detected by the gantry assembly situated beneath the plotting surface so that by applying a continuous ramp signal to the X amplifier the gantry is enabled to follow the curve at a linear rate in the X direction and to produce a signal at the Y servo amplifier output proportional to the trace waveform [24]. The speed of this type of curve follower is limited to about 1 sec/in and is only practical for short waveforms.

2-5-2 FILM READERS

Much faster reading and the ability to follow fast-recorded transients is obtained using an automatic film reader. This is a flying spot cathode-ray tube system, based on the measurement of the time interval between interruption of the scan by a reference line and the recorded trace. Tube deflection errors are eliminated by referencing the deflection to a fixed precision graticule. In a design due to Noyes [25] the flying-spot scans the film record and graticule simultaneously. The graticule pulses are counted and the number of pulses reached at the trace position gives a measure of the Y coordinate. The X coordinate is obtained in a similar way with reference to a second graticule. By associating such a design with a memory element and a digital magnetic tape transport a completely automatic device can be obtained which will transfer film recorded signals rapidly into annotated and blocked data suitable for direct entry into the digital computer.

BIBLIOGRAPHY

1. NEUBERT, H. K. P. *Strain Gauges.* Macmillan, 1967.
2. NEUBERT, H. K. P. *Instrument Transducers.* Clarendon Press, Oxford, 1963.
3. WOLFENDALE, P. C. F. Precise capacitative displacement transducers and a.c. instrumentation methods. I.E.E. Colloq. on transducers, London, Feb. 1971.
4. ATKINSON, P. D. and HYNDS, R. W. Analysis of a linear differential transformer. *Elliot J.* 2, London, 1954.
5. NOLTINGK, R. The measurement of clearance using fringe capacitance. I.E.E. Colloq. on transducers, London, Feb. 1971.
6. WALTON, H. A high temperature transducer system based upon a differential charge amplifier. I.E.E. Colloq. on transducers, London, Feb. 1971.
7. DOVE, R. C. and ADAMS, P. H. *Experimental stress Analysis and Motion Measurement.* C. E. Merrill Ltd, 1964.
8. BROWNSEY, C. M. Data acquisition and processing – the measurement system. Inst. Mech. E. Conf. on acoustics as a diagnostic tool, London, Oct. 1970.
9. HARRIS, C. M. and CREDE, C. E. *Shock and Vibration Handbook,* Vol. 1. McGraw-Hill, New York, 1961.
10. JACOBS, E. D. *New Developments in Servo Accelerometers.* J. Env. Sci., St Louis, Missouri, April 1968.

11. YATES, W. A. and DAVIDSON, M. Wide range calibration for vibration pick-ups. *Electronics,* **26**, 2, 183-5, Sept. 1953.
12. TYZZO, F. G. and HARDY, H. C. Accelerometer calibration techniques. *J. Acoust. Soc. Amer.* **22**, 454, 1950.
13. DAVIES, G. L. *Magnetic Tape Instrumentation.* McGraw-Hill, New York, 1961.
14. WALLACE, R. L. The reproduction of magnetically recorded signals. *Bell Syst. Tech. J.* **11**, 30, 4, 1195, 1951.
15. The Miller coding in direct PCM recording and reproducing. Ampex Ltd Engineering evaluation report, 1969.
16. WARREN, A. G. *Mathematics Applied to Electrical Engineering.* Chapman & Hall, London, 1939.
17. I.R.E. Standard, 1953. Method of determining flutter content. *Proc. I.R.E.* **42**, 3, 1954.
18. SKOV, R. A. Pulse time displacement in high-density magnetic tape. *I.B.M. J. Res. Dev.* **2**, 2, 1958.
19. PESHAL, R. L. The application of wow and flutter compensation technique to F.M. magnetic tape recording systems. *I.R.E. Nat. Conv. Rec.* **5**, 7, 1957.
20. MAXWELL, D. E. and BARTLEY, W. P. Synchronisation of multiplexed systems for recording video signals on magnetic tape. *I.R.E. Nat. Conv. Rec.* **3**, 7, 1955.
21. Telemetry Standards I.R.I.G. Document 106-66 March 1966, Secretariat Range Commanders Council White Sands Missile Range, New Mexico 88002.
22. ZIMAN, G. C. Maximum pulse-packing densities on magnetic tape. *Elec. Eng.* **34**, 414, 1962.
23. ACKROYD, J. *et al.* X-ray spectrometer for the scout satellite. *J. Br. I.R.E.* **22**, 3, 1961.
24. KEMSHALL, C. D., BEAUCHAMP K. G. and BENJAMIN, P. W. The attenuation of proportional counter pulses by R.C. amplifier time constants. *Nuclear Instr. Methods* **68**, 153-6, 1969.
25. NOYES, J. G. An automatic film reader to prepare data for digital computation on the design of data processing systems. U.K.A.C. Conf., Edinburgh, April, 1964.

Additional references

26. LION, K. S. *Instrumentation in Scientific Research – Electrical Input Transducers.* McGraw-Hill, New York, 1959.
27. ZIENKIEWICZ, O. C. and HOLISTER, G. S. *Stress Analysis; Recent Developments in Numerical and Experimental Methods.* John Wiley, New York, 1965.
28. DEAN, M. and DOUGLAS, R. D. *Semiconductor and Conventional Strain Gauges.* Academic Press, New York, 1962.
29. REID, D. H. Transducers: detecting methods used in industrial systems. *Eng. Mat. Des.* **13**, 5, 619-26, 1970.
30. THOMSON, J. Instrument transducers *J. Sci. I. (London)* **34**, 217, 1957.
31. BAKER, W. E. and DOVE, R. C. Transient calibration of piezo-electric accelerometer *J. Env. Sci.* **14**, 10, 1962.
32. STEWART, W. E. *Magnetic Recording Techniques.* McGraw-Hill, New York, 1958.

33. HEIDE, A. R. More bits/inch. *Datamation,* **16**, 66-71, July, 1970.
34. AXON, P. E. Instrumentation magnetic recording. *J. Br. I.R.E.* **20**, 10, 1960.
35. RAPLEY, E. J. Evaluation of the performance of shock transducers for the measurement of high environment. *Seventh International Aerospace Instrumentation Symposium* Cranfield, March 1972.

Chapter 3

PRE-PROCESSING

3-1 INTRODUCTION

Pre-processing forms a part of any signal processing operation and must be considered carefully in conjunction with known information about the raw input data and the methods of analysis required. Signal information is rarely obtained in precisely the right form to suit the analysis methods available. We find this situation for analog signals where pre-processing is required prior to recording, analysis, or conversion to digital form, and also for digital data which often requires modification before it can be accepted as input to a digital analysis program.

The need for pre-processing may be seen if we consider the practical difficulties inherent in, for example, the 'capturing' of a transient effect occurring at an imprecise time but within a given time interval. This time interval may be long and as a result the useful record period may be unduly extended. Apart from the errors that this will introduce in the analysis estimates, obtained from processing the summation of background noise plus signal, the analysis process will become unnecessarily protracted and hence become more expensive. This is particularly the case for digitised data, where the analysis methods are likely to be time-consuming, whether applied to the background or the transient signal. Arrangements need to be made, therefore, to delay the initiation of the analysis program until slightly before the onset of the recorded transient, using control or elimination methods, and to terminate analysis following the completion of the transient so that only the transient itself will be analysed.

Similarly if the required signal contains noise or an unwanted discrete frequency component then the time history can be represented as:

$$x(t) = s(t) + n(t) \tag{3-1}$$

where $s(t)$ represents the desired signal, and $n(t)$ represents the noise signal, Equation (3-1) may be transformed to give a similar equation in the frequency domain, i.e.,

$$x(\omega) = s(\omega) + n(\omega) \tag{3-2}$$

If $n(\omega)$ is sufficiently remote from $s(\omega)$ along the frequency spectrum, it is possible to remove this by filtering in the frequency domain and to carry out an inverse transform to recover the value $x(t) = s(t)$, representing the wanted signal.

A particular form of filtering, known as pre-whitening, is of value when power spectral density estimates are required, in order to make best use of the dynamic range available with the analysis method. More severe methods of rejection filtration are necessary to deal with the problem of large localised peaks in the spectrum. Two analyses are required, one on the unfiltered signal to assess the importance of the peaks, and one on the filtered signal to analyse the remainder of the spectrum.

Sampling a continuous signal will also demand filtering to precede the operation, as indicated later, if the effects of aliasing are to be avoided. Continuous filtering techniques, therefore, play an important role in the pre-processing operation and will be fully dealt with later in this chapter.

Other constraints necessary in order to extract meaningful data from the raw signal may be found in the amplitude or power domain. We may, for instance, wish to reject the signal if its standard deviation falls outside a permitted value. This implies continuous monitoring of the raw signal to detect this characteristic, and a control action to modify the processing operation. Sometimes the unwanted information added to the required signal can represent a modulation of the signal by the characteristics of the measuring device, such as a decrease in sensitivity as a function of temperature or time. The removal of this function is termed Trend Removal.

Finally, the refined data for analysis may be defined or represented by a combination of signals recorded on a number of separate channels. To achieve a signal–noise enhancement, ensemble averaging by parallel addition of these signals may be all that is required. The more complex method of cross-correlation is needed for small-signal event discrimination and will be considered in some detail in chapter 9.

The object of most of these pre-processing operations is to obtain data reduction or compression such as to reduce the computing time required in later analysis. Other operations which may properly be considered during this stage are concerned with calibration and identification of the acquired data.

The raw signal or data consists essentially of a sequence (continuous or discrete) of numbers representing a measured parameter (e.g. pressure, displacement, temperature, velocity, etc.). In order to carry out calculations on these numbers conversion into the appropriate units is needed, requiring a knowledge of the conversion factor, i.e. calibration is required. This may form part of the recorded ensemble and given as a separately recorded sequence. The process of calibration may be obtained within the computer analysis program, but it is often convenient to carry out the necessary arithmetic operations in the pre-processing stage.

Handling time following data acquisition can be reduced considerably if a proper identification routine is used. This is carried out by including run and task identification during the recording or pre-processing operation. Where analog recording is employed, a separate channel can be coded to indicate this essential housekeeping information. The first few blocks of a digital tape can be similarly allocated.

The range of techniques discussed briefly above and required in the pre-processing stage is best illustrated by a discussion of typical acquisition and pre-processing systems, and the next section will be concerned with a general definition of such systems. A number of specific examples will then be described and the chapter will conclude with a discussion of the individual techniques used. Particular attention will be paid to analog filtering. Smoothing of a digital data series and digital filtering will be deferred to a later chapter.

3-2 DATA ACQUISITION SYSTEMS

Such a system is defined as that existing between the source of a random or other process, which generates the time-sequential signals, and a computer or other device used to extract the characteristics of the process. The data acquisition system thus acts both as a buffer and a converter between the data source and analysis operation.

A generalised system is described below, which shows all the features to be found in practical systems, although a given system will not necessarily include them all. This system is only assumed to operate in real time up to the time of actual data capture so that the data of interest is obtained and stored during a test run and analysed at a later time. This is nearly always true of signal processing operations, although the point of actual data recording may follow a considerable amount of real-time electronics (e.g. the acquisition of satellite information). Re-processing and data reduction operations may then also form part of the total analysis of the acquired data.

Apart from the actual real-time recording of the information, the processing and analysis can be carried out at a rate greater or smaller than that of real time. This change of time scale can be of enormous advantage in reducing the time of analysis or in some situations permitting analysis of data originally captured at a rate far greater than could be accommodated with practical computing equipment.

A schematic diagram of a generalised data acquisition system is given in fig. 3-1. For the purposes of description, the system can be defined under six headings, namely:

1. Environment.
2. Transmission.
3. Reception.
4. Recording.
5. Ancillary.
6. Playback.

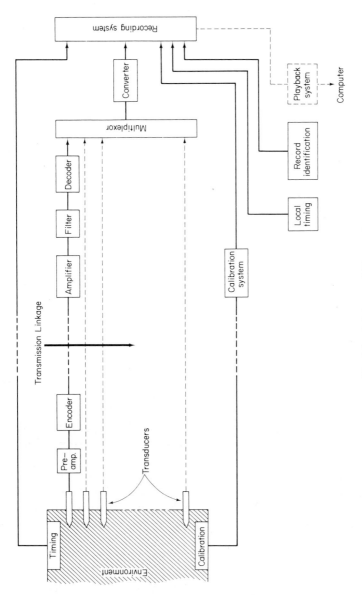

Fig. 3-1. A data-acquisition system.

The playback system may be simply the converse of the recording system and included in the computer or analysis facility (e.g. analog or digital magnetic tape replayed on the appropriate computer peripheral), or it may include further complex pre-processing equipment prior to computer analysis. Cases in which this latter possibility occur are generally specialised and applied to a particular kind of data or analysis method, and as such are not described here. One important exception is where the final recorded signal is stored in analog form on magnetic tape.

Analysis methods may demand a sampled and digitised version of the signal to be available. A full discussion of the process of digitisation will be given in chapter 4.

Referring to fig. 3-1, let us consider first the environment in which the phenomena of interest is expected to occur. In intimate contact with the environment will be the transducers described in chapter 2. Electrical currents or

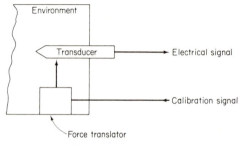

Fig. 3-2. Calibration within the environment.

potentials, having a known relationship to the values of the phenomena, are generated at the transducers and are available for transmission. We would also expect to find some means of calibration of the transducers within their environment location. This generally implies a reversal of the role of transducer in which an e.m.f. or current is translated into a known displacement, or other physical quantity, capable of acting on the transducer and thus providing a means of calibration within the real environment (see fig. 3-2). Where timing is important in relation to the phenomenon being measured a relative or absolute timing signal may be included within the environmental area of interest and transmitted, together with the data.

If the electrical signal derived from the transducers is small (which is very likely to be the case) a low-noise pre-amplifier will be situated close to the transducer in order to improve the signal–noise ratio to a level permitting valid transmission to the reception equipment.

Transmission arrangements vary from a simple cable to complex telemetry equipment. Cable transmission may be a single-ended or balanced-wire line. Both are shown in fig. 3-3. The latter is valuable as a means of reducing interfering signals picked up by the transmission line. Where any considerable distance is involved, the signal is imposed on a higher frequency carrier by means of a process of modulation. At the receiving end the signal is de-modulated back into its original form and amplified to a level suitable for entry to the next processing stage. A detailed discussion of the various methods of modulation used falls outside the scope of this book, and the interested reader is referred to the sources given at the end of this chapter for further information. Quite apart from this carrier modulation process we may find that the signal itself may be subject to an encoded process prior to transmission. A decoding process may or may not be included at the receiving end. If the encoding takes the form of digitisation of a continuous signal then it may be convenient to retain the data in this form for all subsequent analysis.

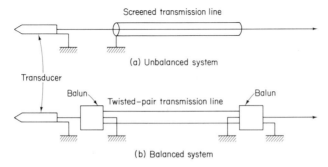

Fig. 3-3. Transmission arrangements.

Multichannel analog signals can be applied to a time-division multiplexing device prior to recording. The assumption is made here that the number of channels being transmitted is greater than can be simultaneously stored by the recording device. It is necessary to make some arrangement for preserving the channel identity during this process of recording. The general method is to program the multiplexor in a sequential fashion and then identify one particular channel (e.g. a zero level channel) as a reference marker for adjacent channels, numbered in known sequential order. More complex methods involve the generation of channel identification, within the multiplexor, as a digital word. A problem associated with the multiplexing process is that of skewing. The values, taken sequentially from the random time series existing at each sample channel, will not be obtained simultaneously. A timing error is present which will increase progressively through the complete sampling cycle so that comparison between channels (e.g. for the purposes of correlation), can introduce a large error which varies with the channel chosen and with the rate of sampling. Where such an error is unacceptable a simultaneous sampling of the channel signals can be carried out. A sample/hold amplifier is used, preceding each multiplexor input channel. The amplifiers are placed in the hold state simultaneously and held at the output values reached at this time for a sufficiently long period to allow the multiplexor to sample all the channel values sequentially, thus achieving the same effect as if it had sampled all the channels at the same time. The amplifiers are then reverted to their sampling mode in order to obtain the next set of samples of the signals taken simultaneously. This is illustrated in fig. 3-4.

In its practical form, the pre-processing operation (and for that matter later analysis operations) must take cognisance of the essential information about the signal that accompanies the record. This often takes the form of a log-sheet or pencilled notes, both of which have the severe disadvantage that they are

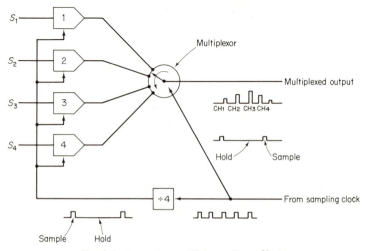

Fig. 3-4. Removing multiplexor skew effect.

capable of being lost or separated from the relevant data. A preferable scheme is to record this information on the recording media that carries the data. Separation or loss is then less likely and the information is already in the correct form for abstraction and action by the analysis program. This applies particularly to calibration constants which will be required for use in a scaling operation. Methods for carrying this out will be discussed later. The calibration information can be introduced locally or derived from the transducer environment. Both are indicated in fig. 3-1. Timing and identification information will usually be generally locally and recorded, together with the data and book-keeping information, on the recording media.

This concludes an outline of general pre-processing operations. In order to further illustrate the characteristics of a practical system, and to indicate the range of requirements necessary for such physical systems, several examples are described below. The methods and equipment described refer to the acquisition

100

of physical data whose parameters are varying at a fairly rapid rate. The acquisition and pre-processing methods used for data-logging of industrial and other processes will not be discussed here.

3-2-1 A SEISMIC DATA ACQUISITION SYSTEM

The following description of a seismic data acquisition and pre-processing equipment is included with permission from the U.K. Atomic Energy Authority. This system [1] has been developed primarily for research into the detection and identification of distant underground nuclear explosions, and is representative of a class of recording systems used for the acquisition of data from shock and vibration experiments. The purpose of this system is to detect the minute earth movements resulting from distant seismic events and to record these in

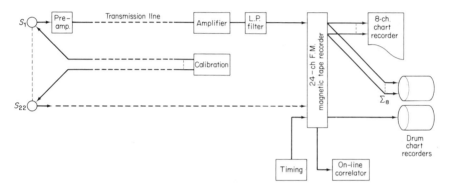

Fig. 3-5. A seismic data acquisition system.

such a way that data analysis can yield significant information about the origins of these events. These earth movements are very small and, in order to improve the signal/ambient seismic noise ratio a number of seismometers (transducers) are arranged, grouped in a given geometric array, covering an area several kilometres in extent. For various reasons, the signals resulting from the seismometers need to be recorded individually, so that this system can be regarded as a multi-channel recording system.

A simplified schematic diagram is given in fig. 3-5. Each seismometer has associated with it a pre-amplifier in order to raise the level of the signal prior to transmission over a land-line to the main recording amplifier. Incoming signals are amplified and filtered before being recorded by a 24-channel F.M. tape recorder. Due to the very slow rates of change involved in seismic events, a bandwidth of 15 Hz is adequate to record all the information of interest. This narrow bandwith permits magnetic tape recording of the signal using very slow-speed tape transports. A continuous recording period of several days is, therefore, possible using only one reel of magnetic tape.

Monitoring of a number of selected channels is carried out by demodulation of the signal immediately following its recording on the F.M. tape, and applying these outputs to a multi-channel strip chart recorder. This provides a check that the recording process has been successfully carried out, as well as providing a 'quick look' copy of the received signal time-history. The acquisition stations are operated continuously so that the monitoring method, using a strip chart recorder, is only suitable for checking over short intervals of time. In order to provide a continuous visual examination of the data recorded on the tape some sacrifice of resolution is accepted and a given channel is selected and recorded using a drum chart-recorder. This incorporates a continuously tracking pen recorder monitoring the seismic signal and recording on chart paper wrapped around a slowly rotating drum. The pen assembly is arranged to move slowly

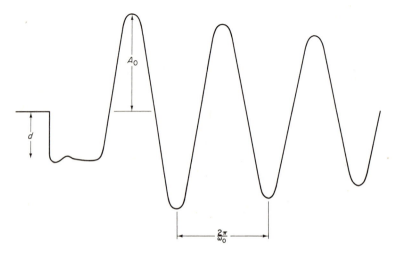

Fig. 3-6. Seismometer calibration wave-form.

across the width of the paper providing a helical trace covering a period of up to 24 hours. A further 'quick look' facility is also shown in fig. 3-5. This provides a cross-correlation output on chart paper for selective events and is mentioned briefly in chapter 9.

The calibration system is of interest and employs a method of direct calibration within the transducer environment [2]. Separate lines are conveyed directly from the remote calibration system to each individual seismometer. A system of remote relay switches enables a known step function of current to be passed through the transducer coil and to deflect the mass by a given known amount. After equilibrium has been reached this steady current is removed and the mass then undergoes a first order oscillatory movement, shown in fig. 3-6. The electrical signal resulting from this movement is transmitted along the signal lines and results in the trace recording shown being produced at the recording station. The ratio A_0/d and the natural period $2\pi/\omega_0$ of the seismometer are measured from the record and, from these values and a knowledge of the velocity transducer coil resistance, R_s, and seismometer mass, M, the sensitivity constant, K_s of the seismometer can be calculated from the expression:

$$K_s = \frac{R_d}{(R_d + R_s)} \cdot 10^{-7}(\omega_0 M R_s A_0/d)^{\frac{1}{2}} \quad V/\text{cm sec} \quad (3\text{-}3)$$

where R_d is a damping resistance shunted across the seismometer output terminals.

Absolute timing of events forms a very significant part of the subsequent analysis of the seismic signals. One channel of the F.M. tape recorder is therefore allocated to locally generated timing signals derived from a highly stable crystal controlled chronometer, checked periodically against standard time broadcasts.

Due to the vast amount of information produced in this way, data compression and reduction processes carried out at the pre-processing stage play a large part in subsequent computer operations. One technique used is to combine the signals with appropriate phasing delays and to monitor the standard deviation of the combined signal in order to detect when significant events are present. Only these events are then subject to further processing, so that a large reduction in the volume of data being handled is obtained. These selected records are subject to phasing, normalisation, filtering, summation and cross-correlation. The main purpose of the processing is to improve the signal-noise ratio to a level where significant seismic events may be studied. Information regarding the direction of approach and characteristics of the events can then be derived using frequency analysis and correlation techniques. Digitisation of pre-processed analog signals is carried out for selected events to permit storage in a form accessible for later digital analysis, using some of the fast digital techniques described later in this book.

3-2-2 A SPACE SATELLITE DATA ACQUISITION AND PRE-PROCESSING SYSTEM

One of the most complex of pre-processing operations is that associated with data telemetered from a scientific research satellite in orbit. The following description of the operations carried out on the data derived from the British satellite 'Ariel 1' has been included with permission of the Science Research Council.†

The primary objectives of this satellite were to study the characteristics of the ionosphere, to monitor solar ultraviolet and X-radiation, and to measure the primary cosmic ray energy spectrum as a function of latitude and longitude [3].

† It should be noted that in this operation the pre-processing operations were carried out by NASA in the United States resulting in a series of digital tapes which were later processed for the experimentors in the United Kingdom.

The satellite carried a number of transducers designed to produce an electrical signal corresponding to the level of various physical states occurring in the environment through which the satellite passed, including some concerned with

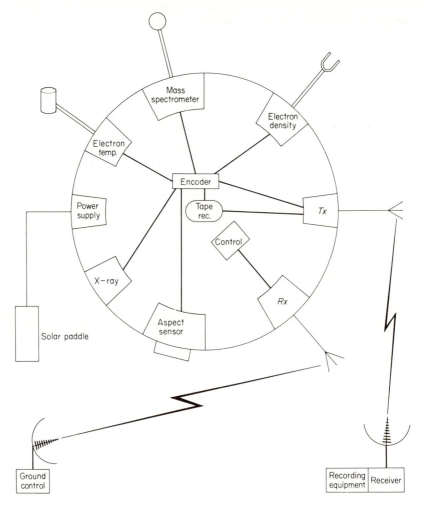

Fig. 3-7. Satellite data acquisition and transmission system.

the detection of extra-terrestrial phenomena (e.g. cosmic rays), and other transducers concerned with the calculation of solar aspect.

These signals were presented in continuous form to encoding circuits, carried within the satellite, and conveyed to the ground receiving station via a V.H.F. telemetry link. Figure 3-7 presents a simplified schematic diagram of the overall acquisition and transmission system.

The analog signals were coded in the form of a pulse sequence at a rate dependant on the amplitude of the signal. Sampling of the pulsed frequency-modulated signal occurred at two rates. A high rate of 50 samples/sec was used for direct transmission of data, which was recoverable only when the satellite was within range of a ground receiving station. A much lower rate of sampling was used for certain selected data, recorded during each orbit of the satellite, and stored on a magnetic tape system carried within the satellite.

The transmitted information was time-division multiplexed in frames of sixteen samples, the first sample of each frame being distinguished by a 50 per cent greater length than that of the other channels. It should be noted here that a frame is defined as a group of sixteen samples and does not represent the recurring group of all samples, which is designated as a sequence. This is made clear in fig. 3-8 which illustrates the composition of two complete data sample sequences, one for high-speed data and the second for low-speed data.

The high-speed data was transmitted as a sequence of sixteen frames. Odd-numbered frames have their first wider channel used for synchronisation by transmission of a frequency outside the data modulation band. Even-numbered frames use this position for frame identification by transmission of a series of digital numbers representing the decimal digits 0, 1 to 7.

Some further multiplexing of the first three data channels was carried out for transmission of calibration and reference information, e.g. state of the solar batteries and satellite aspect position. Low-speed data was transmitted as a sequence of two frames of sixteen channels each. A calibration frequency occupied a space between each channel and thus enabled a direct measure of the speed of the recording tape transport to be obtained.

Reception of data, transmitted from the satellite, was carried out by a network of ground stations. After de-modulation the data, still in the form of pulse-frequency modulated signals, was recorded directly on analog magnetic tape together with locally generated timing information. Four tracks of this tape record were concerned with time measurement and included reference frequencies and a broadcast time indication from a ground transmitting station.

The pre-processing equipment accepting this recorded information, produced as its output a blocked tape suitable for direct entry to a digital computer. A simplified block diagram of this equipment [4] is shown in fig. 3-9. Before digitisation the signal data, expressed in the form of pulsed frequency modulation, is first passed through a comb filter designed to improve the signal–noise ratio as well as carrying out the necessary quantising process. The time information is also coded and multiplexed with the data. A programmer unit controls the sequence of events and conveys the multiplexed signal into the computer format control buffer. Synchronisation of these sequences is controlled by means of a flywheel synchronisation system designed to operate in the presence of momentary fading of the telemetry signal.

The comb filter indicated in fig. 3-9 is a particularly successful demodulation

105

HIGH-SPEED DATA

Frame No.	Ch. 0.	Sub-multiplexed data channels			Other data channels				
0	Sync.	1	2	3	4	5	6	— —	15
1	Ref. 0.	1	2	3	4	5	6	— —	15
2	Sync.	1	2	3	4	5	6	— —	15
3	Ref. 1.	1	2	3	4	5	6	— —	15
4	Sync.	1	2	3	4	5	6	— —	15
.	.	.	.	.	.	.	.	— —	15
.	.	.	.	.	.	.	.	— —	15
.	.	.	.	.	.	.	.	— —	15
.	.	.	.	.	.	.	.	— —	15
.	.	.	.	.	.	.	.	— —	15
.	.	.	.	.	.	.	.	— —	15
.	.	.	.	.	.	.	.	— —	15
.	.	.	.	.	.	.	.	— —	15
.	.	.	.	.	.	.	.	— —	15
15	Ref. 7.	1	2	3	4	5	6	— —	15

LOW-SPEED DATA

Frame No.	Ch. 0.	Sub-multiplexed data channels			Data channels	
1	Sync.	1	2	3	— — — — — — — — — — — — — — — — — — —	15
2	Blank	1	2	3	— — — — — — — — — — — — — — — — — — —	15

Fig. 3-8. Composition of telemetered data sequences.

and quantising device, having the merit of increasing quite considerably the signal–noise ratio of the pulsed frequency modulated signal detected by the system. It comprises 128 narrow band-pass filters whose centre frequencies are equally spaced across the used band. The filters are arranged so that their $-3db$ points are shared by neighbouring filters, as shown in fig. 3-10. When a signal is fed simultaneously to all these filter inputs a logical operation decides which filter is receiving the greatest amplitude of signal. The logical circuits then operates to remove the signal from all other filters, thus allowing the comb filter to act as a frequency-determining device. The output of the comb filter is not used directly. Instead it is made to initiate a pre-coded output to give an integral number in the

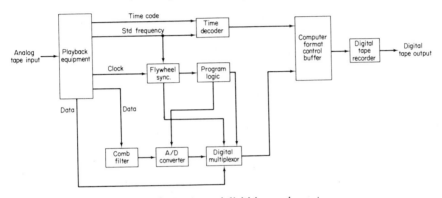

Fig. 3-9. Replay and digitising equipment.

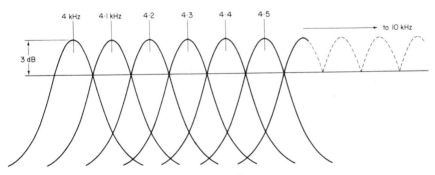

Fig. 3-10. Comb filter response.

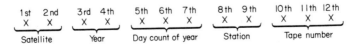

Fig. 3-11. Identification label for satellite data.

107

Speed Identification → 999900

I.D. →

Synch.	Time	Ch. 1	Ch. 2	Ch. 3	Ch. 4	Ch. 5	Ch. 6	Ch. 7	Ch. 8	Ch. 9	Ch. 10	Ch. 11	Ch. 12	Ch. 13	Ch. 14	Ch. 15
I.D.		62	116	08	002	51	62	116	08	002						
000	.06233812869	000	400	000	104	104	102	104	030	075	047	066	060	081	103	104
005	.06233813190	400	100	100	104	104	102	104	029	077	047	065	060	081	103	104
100	.06233813510	044	021	104	104	104	103	104	033	076	048	067	059	081	103	104
005	.06233813830	067	069	102	103	103	102	104	028	076	047	067	059	081	102	104
200	.06233814149	047	084	104	104	104	102	104	032	075	047	066	060	081	103	104
005	.06233814470	070	079	028	104	103	103	104	031	075	047	066	065	081	103	104
300	.06233814789	037	051	085	104	104	103	105	028	074	047	066	064	079	103	104
005	.06233815110	051	050	032	104	104	102	104	030	079	047	068	064	081	102	104
400	.06233815429	051	500	000	104	103	102	104	033	078	048	066	065	081	102	104
005	.06233815749	500	100	100	104	103	102	104	028	077	048	068	064	081	103	104
500	.06233816069	400	020	105	105	104	103	104	035	074	047	066	064	081	103	104
005	.06233816389	047	038	102	105	104	103	104	028	077	048	065	064	081	102	104
600	.06233816709	067	063	105	104	104	102	104	033	058	048	064	064	081	103	104
005	.06233817029	051	041	028	104	104	102	104	029	067	047	065	064	081	102	104
700	.06233817349	028	500	100	104	104	102	104	034	076	047	066	064	081	103	104
005	.06233817669	030	700	700	104	103	102	104	026	078	047	067	064	081	103	104
I.D. 999900 51		62	116	08	002	51	62	116	08	002						
000	.06233823108	700	600	000	104	103	103	104	075	077	050	066	064	081	102	104
005	.06233823429	400	700	100	104	103	102	104	066	071	046	067	064	081	102	104
100	.06233823748	056	020	104	104	103	102	104	071	067	051	065	064	081	103	104
005	.06233824068	073	033	102	104	104	102	104	075	073	056	060	064	080	103	104
200	.06233824388	060	083	105	104	104	103	104	075	079	057	061	064	081	103	104
005	.06233824708	075	079	028	104	104	102	104	074	078	067	058	064	081	103	104
300	.06233825028	030	041	085	104	104	102	104	071	076	068	057	064	081	102	104
005	.06233825348	050	051	032	104	104	102	104	072	074	058	057	064	081	103	104
400	.06233825668	050	700	000	104	104	102	104	074	077	065	057	064	081	103	104
005	.06233825988	600	700	100	104	104	102	104	075	079	068	059	064	081	103	104
500	.06233826308	400	021	105	104	104	103	104	074	079	068	063	064	081	103	104
005	.06233826628	059	079	102	104	103	103	104	074	077	068	067	064	081	102	104

Fig. 3-12. Sample listing of high-speed data.

range 1 to 128. This coded number represents the value of the frequency from the comb filter output and hence the amplitude of the decoded signal. The improvement in signal–noise ratio has been increased by effectively narrowing the signal bandwidth at the centre frequency of the responding filter to a fraction of that of the signal bandwidth.

The format control buffer shown in fig. 3-9 serves to multiplex and code-convert the data into a computer-compatible from. It also establishes the block length, record length and inter-record gaps. Horizontal and vertical parity bits are also inserted at the stage.

An identification label shown in fig. 3-11 is entered into the digitised data by means of manual switches and is arranged to identify the satellite, receiving station, date and the particular recorded tape.

Two types of magnetic tape data records are prepared; one for high-speed and one for low-speed data. Each tape contains a number of 'files' of information. For low-speed records a file consists of data acquired during a complete orbit of the satellite. For high-speed records a file consists of data received during a single pass of the satellite over the ground receiving station. A fairly complex format is necessary to identify the separate pieces of information transmitted. As mentioned earlier the encoded data is recorded in the form of sequences of data, each sequence consisting of a number of telemetry frames, the number of frames in each sequence depending upon the type of data (high- or low-speed), Identification records are included in the edit tape to indicate sequence separation. The high-speed frames contain sixteen three-character data words corresponding to sixteen of the experimental information samples. The low-speed frames contain thirty-two data words corresponding to sixteen data samples plus sixteen calibration signals. Although most of the data telemetered

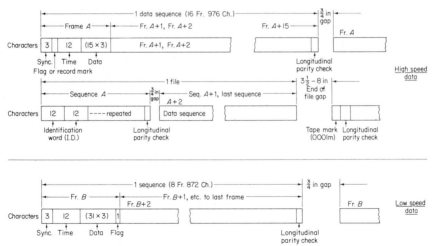

Fig. 3-13. Format of digital output tape.

represents continuous analog signals and requires three digits for its representation, some of the data is derived directly in digital form (e.g. X-ray count), and is reproduced in the digital tape as a three-digit number in which only the first digit has significance, e.g. 200, 500, 700, 000 etc. The complete format is shown diagrammatically in figs. 3-12 and 3-13. The digital magnetic tape represents the final output of the satellite data pre-processing system.

Post-processing analysis was carried out using a digital computer to extract calibrated information on the physical processes and to add other relevant information, e.g. orbital position [5]. Some of the techniques used in these operations will be referred to in a later chapter.

3-3 CALIBRATION AND IDENTIFICATION

The actual process of recording information from a physical system need not be very complex. The real problem comes in attempting to extract meaningful data from the mass of recorded information that can be acquired. With a properly planned acquisition operation this is a matter of data reduction involving scaling or curve fitting to reduce the raw data to significant values. Where no thought has been given to the relationship between the acquired values and the physical units involved, i.e. a previously considered calibration procedure, then real values can never be ascribed to the data and the best that one can hope for is to obtain relative information between data ensembles obtained at the same time under the same conditions.

Of even greater importance is the need to associate each record with sufficient information in order to identify the record itself, to distinguish it from other records taken at other times in other conditions. It is also necessary to include all relevant characteristics of the measurement itself, e.g. type of transducer, its sensitivity and frequency characteristic, amplifier gain, filter characteristics, sampling rate, etc. Where the data is multiplexed, the order of multiplexing and method of channel identification is required. This is in addition to information on the source and measurement environment, such as source constants, physical characteristics, timing, etc., which establish the conditions under which the measurements were taken. All too often the data is separated from its essential identification, rendering processing impossible without repeating the whole sequence of measurements.

Calibration considerations begin when the experiment is being planned and, in many cases, calibration hardware may form a major part of the source acquisition equipment, particularly if an automatic calibration procedure is necessary during the course of data acquisition.

Calibration will be required to:

1. Relate the characteristic of the transducer to the parameter being measured, e.g. pressure in pounds per square inch/volts output.

110

2. Check the linearity, gain, dynamic range, frequency response, etc., of the system following the transducer.

3. Check the changes that may have occurred during the actual period of recording, e.g. amplifier drift, change of transducer characteristic, etc.

We can distinguish three periods when calibration information can be made available:

1. Prior to data recording (pre-experiment calibration).

2. During the recording of data, the calibration information either multi-plexed with the signal data or recorded on a parallel channel (real-time calibration).

3. After the data has been recorded (post-experimental calibration).

In addition, we can recognise the two forms of calibration:

(*a*) Direct calibration on the transducer within its environment.

(*b*) Substitution methods in which the transducer is replaced by an equivalent source generator.

Pre and post-experiment calibration is frequently carried out using the substitution method. A typical example is where a transducer is replaced by a highly accurate resistor through which a measured source current is passed, (which can be a.c. or d.c.). The resultant known voltage is recorded, both before and after the desired signal measurement. The absolute value and difference values between the two readings will give information on the fixed gain of the system and its change with time. In a practical case a series of resistor values would be used to check the dynamic range of the system.

It is generally assumed that the changes occurring during the record are linearly related to the two values taken prior to and after the run. This is not necessarily always the case and where a non-linear response is suspected then a 'dry run' with the pre-calibration input voltage is also carried out for a duration equal to the record. The continuous record obtained can then be analysed by a least-squares or polynomial method to derive the shape of the system response. The pre- and post-calibration values can then be fitted to the curve and the raw data modified in accordance with the calibration curve so obtained. If the post-calibration procedure is not carried out, it must be assumed that the constants of the acquisition system remain unchanged for the duration of the record.

In general, these calibration methods can only provide information on the system itself. The characteristics of the transducer need to be known either by means of a separate calibration curve, obtained on the unit itself, or dynam-ically as part of the acquisition system. The latter is preferred since transducers are sensitive to environment and a calibration made in the laboratory under laboratory conditions may differ considerably from a similar calibration repeated under totally different environmental conditions.

111

One such method was described in section 3-2-1, for the dynamic calibration of a seismometer. The general principle is, either to operate the transducer as a reverse generator (as has been described for the seismometer), so that it may be driven by an electrical signal and calibrated in terms of its physical change, or to activate the transducer with a known physical change and observe the output voltage or current produced. An example of this latter technique will now be described.

This concerns the calibration of the detection arrangements used for a micro-meteorite investigation carried out in the upper atmosphere using a scientific satellite. The method of detection used is shown in fig. 3-14.

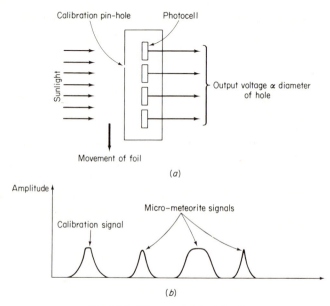

Fig. 3-14. Micro-meteoric detection.

A very thin metallic foil is exposed to the environment through which the micro-meteorite particles are passing. Particles above a certain size are capable of puncturing the foil and leaving a pin-hole, slightly larger than the meteorite itself. Sunlight can then pass through this hole and impinge on a series of photocells, situated quite close to the foil, and which act as transducers converting the narrow beams of light into electrical signals proportional to the area of the pin-hole. An indication is thus given of the size of the meteorite falling on a given segment of the foil. Arrangements are made to mechanically move the foil across the detection area at a fairly slow speed, so that continuous information on the quantity and size of the meteorite stream can be telemetered from the satellite. Calibration is carried out by including a number of fixed position holes of a known diameter in certain detection areas. The signals produced by the sunlight passing through these calibration holes can be regarded as a reference for micro-meteorite size, which is independent of photocell aging or changes in detection and transmission sensitivity, since these are common to both calibration and micro-meteorite-formed holes.

When the transducer output is directly converted into a digital quantity (e.g. for telemetry transmission), the form of the calibration system must include a similar conversion system for the calibration reference standard. A hybrid system has been described by Enochson [6], which avoids many of the sampling errors involved in calibrating digital data. This is shown in fig. 3-15 for the calibration of potentiometric pressure transducers. A continuous excitation voltage, S is applied to the electrical input of the transducers so that when the measured force, $x(t)$ is applied it causes an output voltage, V_x, to be produced, which is proportional to the position of the wiper arm and hence of the applied

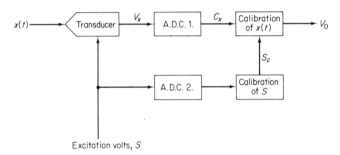

Fig. 3-15. Calibration of a potentiometric pressure transducer.

force. This output voltage is applied to an analog-to-digital converter 1, shown in the diagram.

Calibration is arranged by converting the excitation voltage into a standard-ising digital voltage, S_c via a similar analog-to-digital converter 2, using a measured counts-volts parameter P. If this conversion process is linear, the calibrated data signal, in terms of the binary count C_x is given as:

$$V_0 = (P.C_x.S/S_c)K \qquad (3\text{-}4)$$

where K is the transducer scale factor.

This represents a true calibration output in terms of the actual physical units involved and will be independent of changes due to the excitation voltage.

Scaling need not form part of the calibration procedure, e.g. where a substitution method is used. It is often carried out later in the main processing program prior to presentation of the final output, in order to simplify the intermediate calculations.

Timing for event identification and frequency calibration can be obtained from local generation of timing information and recording on a separate data channel. Standardised pulse code generators are available to recognised I.R.I.G. and N.A.S.A. formats an example of which is shown in fig. 3-16.

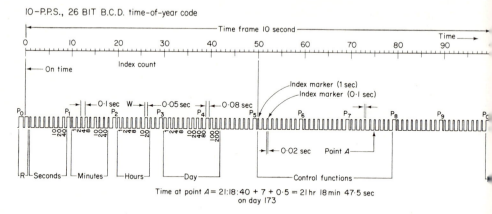

Fig. 3-16. I.R.I.G. time code format.

Data identification is important for the reasons stated earlier and, in its simplest form, will consist of a document listing the relevant data which will be retained with recorded information. Details required to be known should include:

Identification of experiment,
Record run number, date and time,
Environmental conditions,
Calibrated information for transducer system,
Filter setting, bandwidth, gain, etc.,
Special conditions distinguishing the record,
Duration of record,
Recording characteristics, e.g. channel number, carrier frequency, deviation ratio etc.

In the case of digital information, then additionally,

Sampling rate,
Sampling aperture,
Digital code,

will also require to be known.

Documents are likely to be separated from the relevant data, however, and it is preferable always to include this information recorded on the same media as

114

the data. Digital coding and recording can be used to store the identification information, set up on hand keys, or fed in from punched paper tape prior to the recording of the signal information. Similar coding information is recorded on the analog tape by using two discrete levels of the F.M. modulating voltage to represent binary digits so that the recorded data may be demodulated in the same way as the information signal. This method has an advantage that the recording system can remain the same for both identification and signal permitting the same F.M. channel to be used. Digital decoding will, of course, need to follow the demodulated F.M. information. Digital encoding systems such as this have advantages over voice track recording since they are unaffected by time scaling which can occur during playback of the recorded signal.

Where the data is recorded or converted into digital form then a number of blocks of data preceeding the signal data blocks are allocated to contain the identification information, which, in this situation, is referred to as the header information or header label (see fig. 3-17). The data forming this header label

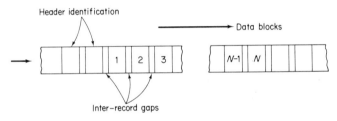

Fig. 3-17. Format of digitised data on magnetic tape.

would be input from punched cards or punched paper tape. In the case of converted data from analog magnetic tape containing its own form of digitised identification, then the digital decoder could be suitably interfaced into the digital computer to enable direct digital transfer to take place.

3-4 TREND REMOVAL

Acquired data is seldom representative solely of the varying parameter or parameters of interest. Among the many imperfections may be found a linear or slowly varying trend accompanying the particular time history of interest. A common trend introduced by the method of data acquisition or recording is the presence of a constant d.c. shift of the base-line for the recorded variable. This represents a pedestal and may be removed fairly easily, providing that its value is known. If unknown, then the mean may be estimated and subtracted from the signal during the pre-processing stage. A pedestal may be deliberately introduced prior to analog-to-digital conversion, where the process of conversion does not include a sign value. The added pedestal is made half the maximum expected

115

input range of the converter, and a similar value subtracted from the signal following the digital-to-analog conversion process. It is important to remove this trend before the signal undergoes any integration operation, such as a Fourier transform, since the error term will be integrated to produce an additive trend which can produce large errors in power spectral density or similar calculations.

Other trends may be introduced by the transducer and are related to its behavioural characteristics. These are in the nature of calibration errors, and providing they are known, can be accounted for in the calibration routine. Large trends of this type should be removed early in the processing of the data since

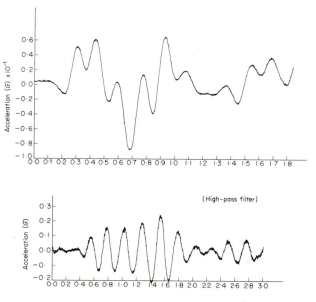

Fig. 3-18. Trend removal by high-pass filtering.

they can lead to other errors (e.g. amplifier non-linearity due to overloading). A particular induced type of trend generally results when the data is derived from vibration analysis equipment. This has a quasi-periodical nature and may be reduced by the use of tracking filters in the control loop discussed in chapter 10.

An example of removal of a low-frequency trend by means of digital high-pass filtering is shown in fig. 3-18. The periodicity of the signal and the nature of its build up and decay can be seen much more clearly following the removal of the base-line trend.

Some consideration should be given to the type of error which can arise due to the integration procedures used in power spectral density and related calculations. This error results in a magnification of the power corresponding to low frequency noise. Such integrated noise takes the form of a random trend which varies at a fairly slow rate (slow, that is, relative to the signal). It can be removed by high pass filtering.

It is particularly important in the case of digital data to recognise the presence of frequencies at or close to zero frequency. Analog equipment often does not respond all the way down to zero frequency, so that an apparent filter is present to these very low frequencies. This is not the case with discrete data where frequencies down to zero are inherent in the data. Power at or near zero frequency would be included in the lowest sampling band (one degree of freedom) of width; $\delta f = 1/2T$ where T is the period of the record. Other sampling bands $2\delta f$, $3\delta f$, up to the Nyquist frequency $n\delta f = f_N$, will be assumed to contain power samples relative to the recorded signal of interest. Zero frequency (d.c.) shift of the data, or a very low frequency drift can augment the lowest bands and considerably distort the power spectral density derived from the combined signal. High-pass filtering is used to remove these lowest frequencies including a slowly varying base-line trend. A cut-off frequency equal to $1/T$ can be chosen for this.

A significant trend associated with high sensitivity line amplifiers is that of additive sinusoidal oscillations at mains frequency and its harmonics. This can occur through inductive pick-up or poorly stabilised power supplies. If it cannot be eliminated at source using screening or other methods, then it may be possible to filter this out from the desired signal using a notch filter.

Polynomial trends may be removed by the method of least squares. An example of this has been described by Raitt [7] and used to remove the constant spin-rate modulation of experimental data recovered from a scientific satellite. This particular signal data concerned the measurement of electron temperature in the path of the satellite. Due to the rotation of the satellite within the earth's magnetic field, a sinusoidal trend voltage is added to the slow change of voltage resultant from variations in electron temperature. Since the angular spin frequency (ω) is known, the trend may be removed by fitting the data to a function which corrects for the induced sinusoidal voltage by adding a cosinusoidal term of the required amplitude, e.g.

$$T = A + B v_p + C \sin \omega t + D \cos \omega t. \qquad (3-5)$$

This fitting is carried out by a method of least squares where

$$\sum_1^N (T_{\text{theoret.}} - T_{\text{expt.}})^2$$

is a minimum for the number of data points N taken. A second curve fitting is necessary to derive the constants A and B to enable a trend-free estimate of the electron temperature to be made. The effectiveness of this correction for spin rate can be seen from fig. 3-19, reproduced from Dr Raitt's paper.

117

The existence of a definite trend implies some knowledge about its function. A general method which is usually applicable is the least squares method fitting to a low-order polynomial in t and described above. Often the trend is too slow to be represented by a polynomial in this way and a moving-average of trend is used. Here the procedure is to take the deviations of the observed values from a

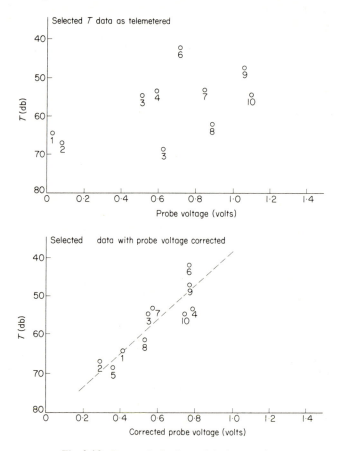

Fig. 3-19. Removal of spin-modulation trend.

moving-average trend. The means of groups of these deviations are taken, adjusted to sum to zero, as the estimate of trend. This method is easy to implement on an analog computer using an E.M.P. estimate for the moving average, (see chapter 5), and is shown schematically in fig. 3-20. Removal of the trend can then be obtained by subtracting from the original signal $x(t)$. An alternative method has been described by Wallis and Roberts [8] in which the ratios of the observed values to trend are taken instead of the differences. An experimental method of removing a background trend is to select the minimum values of the signal on the assumption that these represent the trend alone and to smooth these samples to form a continuous wave-form. The inverse of this may then be taken and added to the signal to cancel the estimated trend.

In some connotations trend has a slightly different meaning to the one assumed above. This is where trend determination refers to the testing technique used to obtain the equivalence of a measured distribution for sample data to some theoretical distribution. Since this is a statistical technique rather than a pre-processing operation, its consideration is deferred to chapter 5.

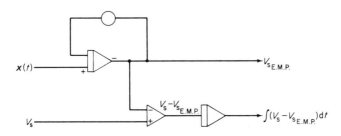

Fig. 3-20. E.M.P. trend removal.

3-5 PRE-WHITENING

As has been mentioned earlier, some distortion of the derived signal can occur at any stage of processing. This can take many forms. The chief amongst them are non-linearity, inter-modulation and noise. If we consider noise first, then we note that if this is at a fixed and uniform level, its average must add to *all* the desired signals, and therefore, does not prevent signal identification. Thus, a flat frequency spectrum is desirable, so that the effects of noise will be added uniformly in this way. Inter-modulation by two strong signals, which may fall outside the region of interest for the signal, can produce a third signal which may fall within the wanted region. This type of distortion can also be reduced if the frequency spectrum is flat. Another form of inter-modulation distortion is deliberately introduced with certain forms of heterodyne analysers. Here, the frequency band under consideration is heterodyned with an adjustable signal frequency to produce a fixed frequency band which is passed through a fixed band-pass filter. This avoids the necessity for tuning a filter over the entire band of interest. The difficulty here is that the shape of the filter characteristic is not perfect and, if an interfering frequency is powerful enough, it will produce a substantial effect, even though falling on the slope of the filter characteristic.

For all these reasons, there are advantages in introducing pre-emphasis, or smoothing, of the frequency spectrum before any analysis is attempted. The ideal is to bring the spectrum of the signal close to that of white noise (defined

119

as having constant spectral density from zero to the folding frequency f_N), and hence the process is known as 'pre-whitening'. Its main object is to make the rate of change of power spectral density with frequency relatively small, and is hence particularly valuable where intermodulation distortion is encountered.

Even if the processing of the acquired data is completely distortionless then some pre-whitening of the signal may be required. This is because in many engineering and physical situations the low frequency end of the spectrum will pre-dominate and result in amplitude differences of up to 100 db. Unless this is reduced then some fraction of the power contained within these low-frequency signals will be added to that of other frequencies well separated from them during the process of power spectral density estimation. To achieve this separation a high-pass filter is required, having a cut-off frequency close to that of zero frequency. A particularly useful design is given by Hillquist [20] which gives pass-band characteristics modelled on the subjective response of the human ear, and is therefore valuable for audio spectrum evaluation.

Pre-whitening is carried out as early in the measurement analysis sequence as possible. The scaling effects that are introduced can be compensated in the final analysis by an inverse modification based on the known pre-whitening characteristic.

3-6 PRE-PROCESSING FOR DIGITAL DATA

Raw data obtained from a process of digitisation or from a digital transducer will need to be processed before being entered into a digital computer. Apart from format and coding considerations, which are machine dependent, the common features of this pre-processing operation are:

1. Truncation of record length.
2. Reduction to zero mean.
3. Trend removal.
4. Tapering.
5. Calibration.
6. Decimation.

(2) and (4) form requirements for certain digital analysis procedures, e.g. spectral density evaluation, and (1), (3) and (5) are required for reasons similar to those given above for analog procedures.

Decimation (6), is a process of data reduction involving the selection of r samples of the digitised data at uniformly spaced intervals throughout the data sequence. Such data reduction may be necessary because too high a rate of digitisation was originally used or simply because of the limited resolution requirements of the analysis. It is important with digital processing to reduce the quantity of data to be analysed to a realistic minimum, since computing time is expensive and necessarily linearly related to data length (for example spectral

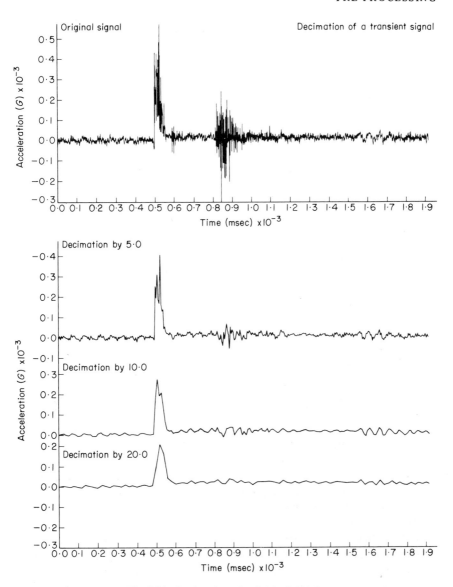

Fig. 3-21. Decimation of a digitised signal.

analysis time proportional to the square of the number of points). An example of decimation is given in fig 3-21 which shows clearly the reduction in resolution obtained by reduction of the total number of points representing the signal.

The program necessary to abstract every rth point from the digital signal is trivial but the decimation program will be made more complex since it will

121

be necessary to digitally filter the data before decimation. It will be shown later that for a sequence of data points, x_i, spaced at uniform intervals, h, then this can only represent constituent frequencies up to $\dfrac{1}{2h}$ Hz. If every rth point is retained then the new sampling rate will be $h' = rh$ and only frequencies up to, $\dfrac{1}{2rh}$ Hz can be represented. Unless frequencies higher than this are filtered out from the data then they will be effectively translated (aliased) into the band 0 to $\dfrac{1}{2rh}$ Hz and thus distort the decimated signal. A recursive low-pass filter to achieve the filtering required is described in chapter 7.

3-7 ANALOG FILTERING

Filters are used in signal processing for a number of reasons, some of which are:

(a) Smoothing of data.
(b) Event detection.
(c) Bandwidth limitation.
(d) Bandwidth selection.
(e) Signal–noise ratio improvement.

These operations can take place in any part of the analysis process and find a particularly important place in the pre-processing stage. The treatment of filter design given here will be derived solely from consideration of the required (and known) relationship between the input and output of the filter, i.e. its transfer function. This form is considered by the writer to be particularly suitable for continuous filter design using analog computer elements. Design from network analysis, using the notation of poles and zeros in the complex plane, will be deferred to chapter 7 when digital filters are discussed. Also, since the subject of design from lumped parameters (inductance, capacitance and resistance) is adequately treated elsewhere [9], only a passing reference to this will be given.

The simulation of wave filter characteristics on an analog computer is carried out using the basic properties of the operational amplifier. Lumped element passive filters are rarely used in signal processing operations due to the difficulty in re-design to meet a variety of processing requirements and also to the need to preserve a high signal-noise ratio and a very high rate of attenuation for the filter outside the pass-band. We are concerned therefore in this section with active filters, i.e. those using an active amplifier element as part of the filter design (which may also contain individual non-active elements, such as resistors and capacitors). In the analysis of active filters we shall find that the concept of a transfer function provides a valuable mathematical tool.

3-7-1   THE TRANSFER FUNCTION

A transfer function describes the operation of a physical system in an easily manipulative form. Using this definition we can treat a complicated system as a four-terminal 'black box' having a two-port entry and exit route. Referred to fig. 3-22 this system can be represented in transfer function notation as

$$H(p) = \frac{V_o(p)}{V_i(p)} \qquad (3\text{-}6)$$

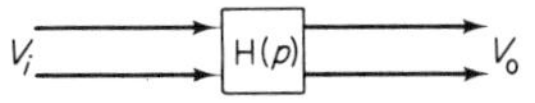

$$V_i \longrightarrow \boxed{H(p)} \longrightarrow V_o$$

Fig. 3-22. Representation of a physical system by its transfer function.

where $H(p)$ is the transfer function expressing the behaviour of the system in the frequency domain, i.e. its response characteristic, and where the function is expressed in terms of the operators:

$$p = d/dt \quad \text{or} \quad p^{-1} = \int dt.$$

This is a permitted form of operation in the context of analog computers where we are necessarily operating in the time domain and time is the independent variable of integration. For this reason we are able to replace expressions derived in terms of the Laplace transform:

$$H(s) = \frac{V_o(s)}{V_i(s)} \qquad (3\text{-}7)$$

where s is a complex variable, by the equivalent expression given in equation (3-6) in terms of the operator p. (The use of the Laplace transformation in the solution of linear constant-coefficient ordinary and partial differential equations will be referred to later when the use of an impulse function to define a system is considered.)

When calculating the frequency response from a transfer function the operator p is replaced by $(j\omega)$. This is applicable when considering sinusoidal inputs to the filter.

Thus if, $\quad x \quad = X \sin \omega t$

$\qquad\qquad px = dx/dt = X\omega \cdot \cos \omega t$

$\qquad\qquad p^2 x = -X\omega^2 \cdot \sin \omega t = -\omega^2 x$

therefore: $\qquad p^2 = -\omega^2$

$\qquad\qquad p \quad = j\omega \qquad (j = \pm\sqrt{-1}).$

123

Transfer function theory is applicable to linear systems only. It may be applied to a serial linear system consisting of a number of individual systems (fig. 3-23) so that the overall transfer function $H(p)$ becomes the product of the individual transfer functions:

$$H_1(p), H_2(p), H_3(p)$$

i.e.,
$$\frac{V_{0_1}(p)}{V_i(p)} = H_1(p)$$

$$\frac{V_{0_2}(p)}{V_{0_1}(p)} = H_2(p)$$

$$\frac{V_{0_3}(p)}{V_{0_2}(p)} = H_3(p)$$

and
$$\frac{V_{0_3}(p)}{V_i(p)} = \frac{V_{0_1}(p).V_{0_2}(p).V_{0_3}(p)}{V_i(p).V_{0_1}(p).V_{0_2}(p)} = H(p). \tag{3-8}$$

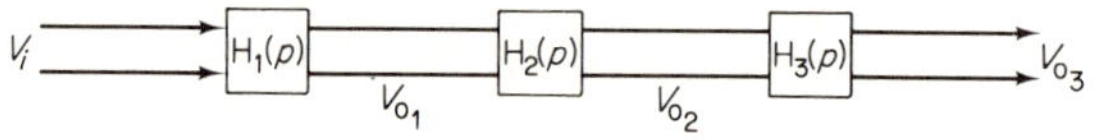

Fig. 3-23. Multiple transfer functions.

This is an important feature of transfer function manipulation and finds wide use in the determination of complete systems characteristic in terms of its sub-systems and elements.

Simulation of transfer functions expressing filter operation using active elements can be carried out either by direct simulation of the complex polynomial functions in (p), using a series of integrators, or by using complex impedances forming the feedback elements of a single operational amplifier. This second alternative will be considered first.

3-7-2   COMPLEX IMPEDANCE FORM

Figure 3-24 shows a high-gain amplifier associated with two passive networks described by the single impedances Z_i and Z_f. The amplifier is assumed to have a level response down to zero frequency and to provide an open loop gain of the order 10^6 to 10^8. If can be shown that the transfer function of the operational amplifier arrangement of fig. 3-24 may be represented with very little loss of accuracy as:

$$H(p) = V_0/V_i = -\frac{Z_f}{Z_i} \tag{3-9}$$

where the negative sign implies signal inversion by the amplifier.

124

As a consequence of the very high internal gain of the amplifier the difference current $i_s = i_i - i_f$ is extremely small so that terminal B (the true input to the amplifier) is practically at ground potential.

Thus Z_i and Z_f become the short-circuit transfer impedances:

$$Z_i = \frac{V_f}{i_i}; \quad \text{and} \quad Z_f = \frac{V_o}{i_f}. \tag{3-10}$$

This permits the numerator and denominator of the transfer function to be treated separately and considerably simplifies the determination of the complex impedance characteristic required in a given transfer function.

There are some practical difficulties associated with amplifier loading and phase shift if Z_i is made complex and, in the first instance, it will be assumed

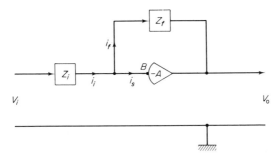

Fig. 3-24. An operational amplifier.

that Z_i is resistive so that the required transfer function will be specified precisely by Z_f. Practical considerations also preclude the use of L.C. or L.R. networks since the filtering requirements for signal processing will generally be designed for low frequency and hence require large L values. These in turn will result in the practical inductance values having considerable internal resistance and stray capacitance which can lead to instability when included in a feedback loop.

Table 3-1 lists a number of networks consisting of R.C. elements only, and suitable for inclusion as the impedance element in the operational amplifier circuit of fig. 3-24. Where $Z_i = R_i$ and Z_f is taken from this table then the transfer function is given as

$$H(p) = -\frac{Z_f}{R_i}. \tag{3-11}$$

As an example let us consider a low-pass filter having the transfer function:

$$H(p) = \frac{A}{1 + Tp}. \tag{3-12}$$

125

TABLE 3-1   Two-terminal transfer function networks

Network	Transfer Function
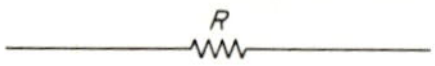 R (series resistor)	R
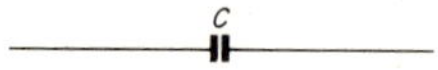 C (series capacitor)	$1/pC$
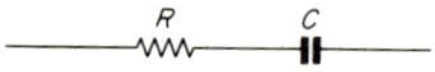 R, C (series resistor and capacitor)	$\dfrac{1 + RCp}{pC}$
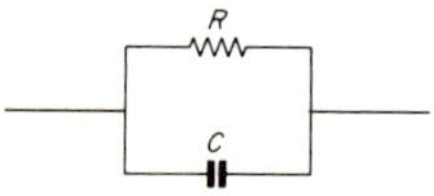 R parallel with C	$\dfrac{R}{1 + pRC}$
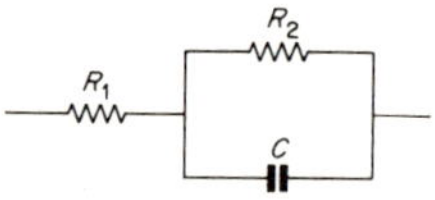 R_1 series, R_2 parallel with C	$(R_1 + R_2)\left\{\dfrac{1 + p\left(\dfrac{R_1 R_2 C}{R_1 + R_2}\right)}{1 + pR_2C}\right\}$
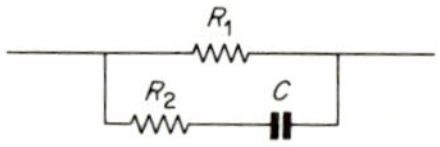 R_1 parallel (R_2 series C)	$R_1\left\{\dfrac{1 + pR_2C}{1 + p(R + R_2)C}\right\}$
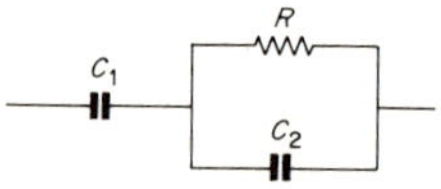 C_1 series, R and C_2	$\dfrac{1}{PC_1}\left\{\dfrac{1 + pR(C_1 + C_2)}{1 + pRC_2}\right\}$
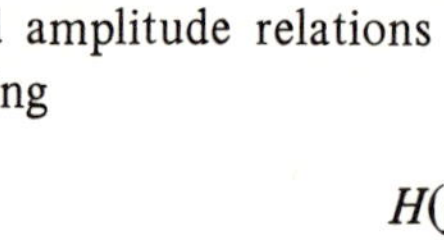 C_1, R, C_2 network	$\dfrac{1}{p(C_1 + C_2)}\left\{\dfrac{1 + pRC_2}{1 + p\left(\dfrac{RC_1C_2}{C_1 + C_2}\right)}\right\}$

This represents a first-order lag circuit where $A = R$ and $T = CR$. The phase and amplitude relations may be derived by setting $p = j\omega$ in equation (3-12) giving

$$H(j\omega) = \frac{R}{1 + CRj\omega} \quad \text{(where } \omega = 2\pi f\text{)}.$$

The modulus of $H(j\omega)$ gives the frequency response and the ratio of the real and imaginary parts the phase angle of lag:

$$|H(j\omega)| = \frac{V_o}{V_i} = \frac{1}{\sqrt{[1 + (CR\omega)^2]}} \tag{3-13}$$

and

$$\phi = \arctan(-CR\omega). \tag{3-14}$$

The phase angle is 0 for $\omega = 0$ and -90 as $\omega \to \infty$. At the corner frequency (defined later as $\omega = 1/CR$) the phase angle is $-45°$.

126

The amplitude ratio may be re-written in decibel notation as

$$20 \log_{10} | G(j\omega) | = 20 \log_{10} \sqrt{[1 + (CR\omega)^2]} \qquad (3\text{-}15)$$

When $CR\omega \ll 1$, then the gain characteristic $\simeq 20 \log_{10} 1 = 0$ which represents a level or zero db slope, and when $CR\omega \gg 1$ then $-20 \log_{10} \sqrt{[1 + (CR\omega)^2]} \simeq -20 \log_{10} (CR\omega)$, which represents a slope of -6db per octave frequency change.

The junction of the two slopes is known as the Corner Frequency when we have $\omega = 1/CR = 1/T$

or
$$f = \frac{1}{2\pi CR}. \qquad (3\text{-}16)$$

A phase/amplitude plot for this filter is given in fig. 3-25.

Filters derived from other networks given in table 3-1 can be similarly analysed in terms of slope, corner frequency and amplitude and phase characteristics.

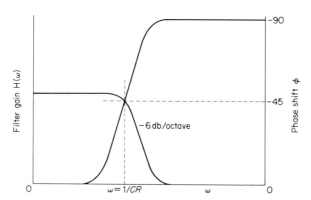

Fig. 3-25. Phase-amplitude response for a simple first-order filter.

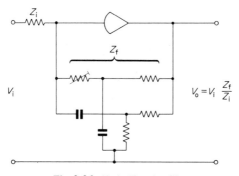

Fig. 3-26. Twin-T active filter.

127

The transfer function approach to the design of complex impedance filters is by no means the only one and in some cases it is simpler to select a frequency characteristic for Z_f having the inverse of that required for the filter (assuming Z_i is resistive). This is the case with the narrow band twin-T filter design shown in fig. 3-26. Analysis of this three-port circuit for Z_f in terms of its transfer impedance is complex and a practical alternative is to convert each T network into its equivalent π as shown in fig. 3-27. It will be seen that at one frequency, f_0 the series arm will behave as a parallel tuned circuit and exhibit a very high

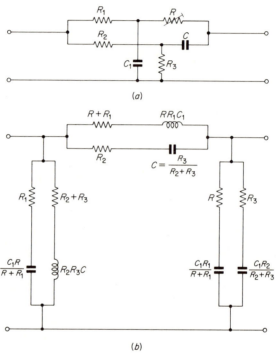

Fig. 3-27. Equivalent circuit.

impedance. Since the shunt reactance arms of the circuit will have very little effect on the value of this frequency then it is permissible to equate the imaginary components of the series arm to derive the resonant frequency,

$$\omega_0^2 = \frac{(R_2 + R_3)}{CRC_1 R_1 R_3} \cdot T_1 \qquad (3\text{-}17)$$

Expressing this in terms of time-constants $T = CR$, $T_1 = C_1 R_1$ and a reduction factor, $a = R_3 \,|\, (R_2 + R_3)$ gives

$$f_0 = \frac{1}{2\pi\sqrt{(T \cdot T_1 \cdot a)}} \qquad (3\text{-}18)$$

128

This is the centre frequency of a narrow band band-pass filter and in order to determine its bandwith a measure of its Q factor ($= \omega L/R$) is required.

This is given from fig. 3-27 as

$$Q = \frac{\omega_0 R.R_1 C_1}{(R_1 + R_2 + R)}. \qquad (3\text{-}19)$$

Substitution of equation (3-18) in (3-19) gives:

$$Q = \frac{R}{(R + R_1 + R_2)} \sqrt{[(T_1/Ta)]} \qquad (3\text{-}20)$$

if $R \gg R_1 + R_2$ then $Q_{max.} \simeq \sqrt{(T_1/Ta)}$, which is a useful figure for design purposes.

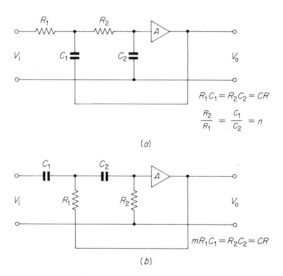

$$R_1 C_1 = R_2 C_2 = CR$$

$$\frac{R_2}{R_1} = \frac{C_1}{C_2} = n$$

(a)

$$m R_1 C_1 = R_2 C_2 = CR$$

(b)

Fig. 3-28. Three-port active filters.

It can be shown [10] that variation of the value of a single element R can alter the centre frequency by about 3:1 with very little effect on the Q factor and hence bandwidth. This represents an improvement in design flexibility over the symmetrical twin-T circuit where three filter elements need to be varied simultaneously to cover a range of frequencies.

This example is taken from a class of filters having three-port representation for one of the transfer impedances (in this case, Z_f). Three-port filter circuits have some advantages in terms of simplicity of design and reduction in number of passive elements required. Two simple versions having low-pass and high-pass characteristics respectively are shown in fig. 3-28. Assuming a high input/low

129

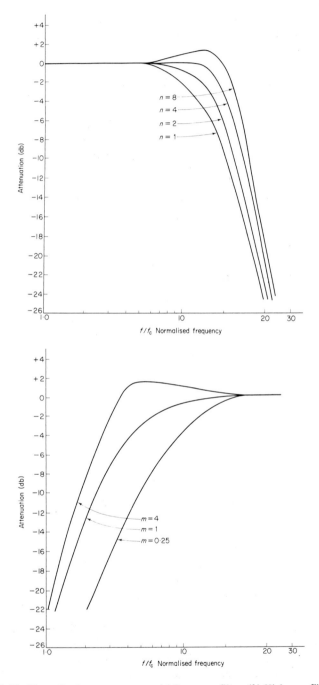

Fig. 3-29. Normalised response curves. (*a*) Low-pass filter. (*b*) High-pass filter.

130

output impedance amplifier of unity gain then analysis of these circuits will yield corner-frequency expressions:

$$f_c = \frac{1}{2\pi CR}\,[-F + \sqrt{(F^2 + 1)}] \qquad (3\text{-}21)$$

for the low-pass filter, and

$$f_c = \frac{1}{2\pi CR}\,[-G + \sqrt{(G^2 + m)}] \qquad (3\text{-}22)$$

for the high-pass filter,

where $F = (n + 1)/2n$ and $G = (m + 1)/2$.

Constants n and m determine filter shape in the region of f_c and their effect can be seen from the normalised response curves given in fig. 3-29. A more general treatment of three-port filters has been given by Hansen in terms of conformal transformation from the low-pass case [11].

TABLE 3-2 Complex impedance filter networks

Network		Transfer Function	Filter Type
Z_i	Z_f		
		$\dfrac{1}{1 + p2CR + p^2R^2C^2}$	Low-pass
		$\dfrac{p^2R^2C^2}{1 + p2CR + p^2R^2C^2}$	High-pass
		$\dfrac{pRC}{(1 + pRC)^2}$	Wide band-pass
		$\dfrac{pRC}{1 + p2RC + p^2R^2C^2}$	Narrow band-pass
		$\dfrac{p^2R^2C^2}{1 + p2RC + p^2R^2C^2}$	Band-stop

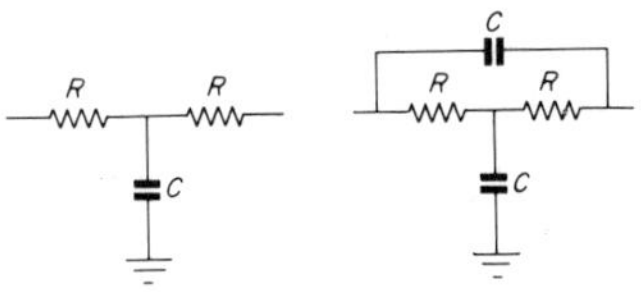

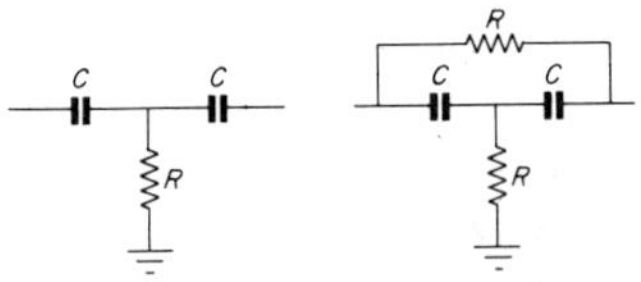

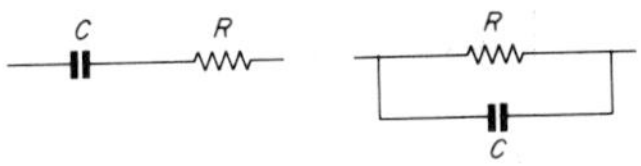

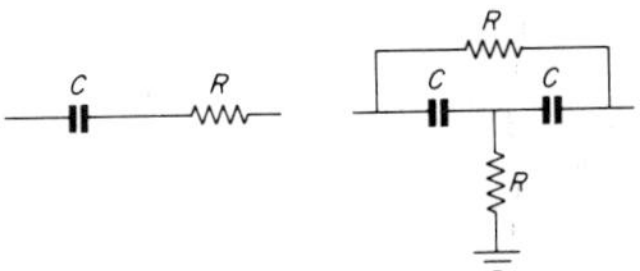

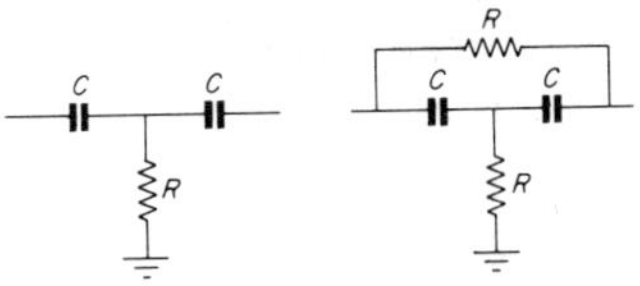

131

Where both Z_i and Z_f are made complex a wider range of filter characteristics, in terms of their transfer functions, become possible. In particular the simulation, using a single operational amplifier, can now extend to multiple-order filters, giving rise to a more favourable response characteristic, which can include band-pass transfer characteristics. Freedom of choice for the selection of the Z_f complex network is, however, limited to those networks whose impedance at any frequency is greater than that of the previous output stage (i.e. does not give rise to loading effects), and where the phase shift obtained does not result in an unstable amplifier configuration. Table 3-2 gives a series of pairs of values of Z_i and Z_f which may be selected to achieve a given filter characteristic and which avoid the difficulties mentioned above. Here the overall transfer functions can be obtained simply from division of the transfer function for Z_f by that for Z_i (equation (3-9)).

3-7-3   SIMULATION USING CASCADED INTEGRATORS

The complex impedance filters described in the preceding section have a number of disadvantages when used in signal processing work. The major difficulty is the complete lack of flexibility in respect of change of filter characteristics or frequency. These are determined by the value of the passive elements and require re-evaluation for each signal processing operation attempted. This could lead, for example, to a change in frequency and hence different passive elements required for each record to be analysed. In addition there are the practical limitations in accuracy of available passive elements, and the possibility of instability or loading difficulties with some configurations.

Direct simulation of the transfer function, considered as one or more differential equations, by the use of cascaded integrators enables a completely flexible approach to be made, having few operational limitations. The only disadvantage of this method lies in the large amount of operational amplifiers and potentiometers necessary to realize a given transfer function. This is less important in analog computing operations since the machine will probably be dedicated to signal processing work and the entire configuration available for use. As an example of the use of cascaded integrators let us consider the second order (quadratic) transfer function:

$$H(p) = \frac{1}{1 + \dfrac{2D}{\omega_n} \cdot p + \left(\dfrac{p}{\omega_n}\right)^2} \qquad (3\text{-}23)$$

where D is the damping factor and ω_n is the natural frequency. We can replace p by $j\omega$ (see section 3-7-1) thus:

$$H(j\omega) = \frac{1}{1 + j2D(\omega/\omega_n) - (\omega/\omega_n)^2}. \qquad (3\text{-}24)$$

The phase shift is given by

$$\phi = \arctan \left[\frac{2D(\omega/\omega_n)}{1 - (\omega/\omega_n)^2} \right] \qquad (3\text{-}25)$$

which approximates to

$$\phi = 2D(\omega/\omega_n) \qquad (3\text{-}26)$$

for $\omega \ll \omega_n$ (i.e. within the pass band). It is important to note that the phase shift of this filter increases proportional to frequency throughout the pass band until $\omega = \omega_n$ when the phase shift becomes

$$\phi = \pi/2.$$

At frequencies greater than this the phase shift is given by:

$$\phi = \pi - 2D(\omega_n/\omega). \qquad (3\text{-}27)$$

The frequency response is expressed by the modulus of equation (3-24) as

$$|H(j\omega)| = [[1 - (\omega/\omega_n)^2]^2 + 4D^2(\omega/\omega_n)^2]^{-\frac{1}{2}} \qquad (3\text{-}28)$$

If we make the damping factor, $D = \sqrt{2}/2 = 0\text{·}707$, then the expression (3-28) reduces to the simpler form,

$$|H(j\omega)| = [(1 + (\omega/\omega_n)^4]^{-\frac{1}{2}} \qquad (3\text{-}29)$$

which will be seen later to represent a second order Butterworth filter. A double integrator circuit for implementing this transfer function is shown in fig. 3-30. By a suitable selection of exit point it is also possible to simulate the performance of a high-pass or band-pass filter.

Thus from fig. 3-30 we can state the output values as

$$X_3 = \frac{Q_3 X_2}{p}$$

$$X_2 = -\frac{Q_1 X_1}{p}$$

and

$$X_1 = X_0 + Q_2 X_2 + Q_4 X_3,$$

therefore

$$X_1 = X_0 - \frac{Q_2 Q_1 X_1}{p} - \frac{Q_4 Q_3 Q_1 X_1}{p}$$

so that

$$X_0 = X_1 \left[\frac{p^2 + Q_2 Q_1 p + Q_4 Q_3 Q_1}{p^2} \right]$$

133

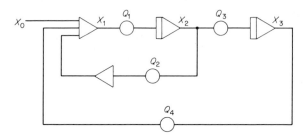

Fig. 3-30. Cascaded integrator filter.

giving transfer functions at the three possible exit points of

$$\frac{X_1}{X_0} = \frac{p^2}{(p^2 + Q_2Q_1p + Q_4Q_3Q_1)} \tag{3-30}$$

which corresponds to a high-pass filter,

$$\frac{X_2}{X_0} = \frac{-Q_1p}{(p^2 + Q_2Q_1p + Q_4Q_3Q_1)} \tag{3-31}$$

which corresponds to a band-pass filter, and

$$\frac{X_3}{X_0} = \frac{-Q_1Q_3}{(p^2 + Q_2Q_1p + Q_4Q_3Q_1)} \tag{3-32}$$

which is equivalent to the low-pass filter transfer function given by equation (3-23).

Thus if $Q_4 = 1$ then we can write

$$\frac{1}{Q_1Q_3} = \frac{1}{\omega_n^2} \quad \text{and} \quad \frac{Q_2}{Q_3} = \frac{2D}{\omega_n}$$

and if we put $Q_1 = Q_3 = \omega_n$, then the value of $Q_2 = 2D$.

It will be seen from this example that the characteristics and cut-off frequency of the filter are determined solely by the setting of the coefficient potentiometers. Thus any desired change can be carried out manually or under computer control (using servo-potentiometers or digital coefficient units), to achieve a high degree of flexibility.

A completely general method of determining the value of these potentiometer settings will now be developed and related to specific filter characteristics. Tables of potentiometer settings will be given to cover a wide range of requirements for filter design.

134

3-7-4 A GENERAL APPROACH TO FILTER SIMULATION

The simulation of filter characteristics using the method of cascaded integrators involves the mechanisation of a general form of transfer function given by

$$H(p) = \frac{b_0 + b_1(p/\omega_c) + b_2(p/\omega_c)^2 +, \ldots, + b_n(p/\omega_c)^n}{a_0 + a_1(p/\omega_c) + a_2(p/\omega_c)^2 +, \ldots, + a_n(p/\omega_c)^n} \qquad (3\text{-}33)$$

where a_n and b_n are constants determining the filter characteristics and $\omega_c = 2\pi$ x filter cut-off frequency. Stability requirements dictate that the order of the numerator must not exceed that of the denominator, and a simplification which permits most theoretical filters to be simulated is given below:

$$H(p) = V_0/V_i(p) = \frac{1}{1 + a_1(p/\omega_c) + a_2(p/\omega_c)^2 +, \ldots, + a_n(p/\omega_c)^n}. \qquad (3\text{-}34)$$

This expression is still completely general and will enable the characteristics of low-pass, high-pass, band-stop and band-pass filters to be obtained.

Two forms of equation (3-34) may be considered. They are

$$V_0 a_n(p/\omega_c)^n = V_i - V_0 - a_1 V_0(p/\omega_c) - a_2(p/\omega_c)^2, \ldots, a_{(n-1)}(p/\omega_c)^{n-1} \qquad (3\text{-}35)$$

and a 'nested form':

$$V_0 = V_i - (p/\omega_c)\{a_1 V_0 + (p/\omega_c)[a_2 V_0 + (p/\omega_c)(a_3 V_0 +, \ldots, (p/\omega_c)[a_n V_0])]\}. \qquad (3\text{-}36)$$

To illustrate the practical difference a fourth-order filter is shown in fig. 3-31a using equation (3-35), and fig. 3-31b using equation (3-36). The latter is of value in element economy where n is large. Referring to fig. 3-31a a further dichotomy in circuit arrangement may be seen. It will be apparent from the mechanisation of equation (3-35) that potentiometers Q_{01} to Q_{03} set the frequency term, ω_0 directly, whereas Q_{00} is set to the value (ω_c/a_4). Also $P_{00} = a_3$, $P_{01} = a_2$, $P_{02} = a_1$ and $P_{03} = 1\cdot0$. From this we can see that P_{03} is not required for direct mechanisation of equation (3-35). However, as will be seen later, for some filter conditions involving many stages and high Q-values then this could lead to very large loop gains. With the phase-shift margin of practical computer amplifiers instability would be precipitated. To avoid this when the coefficient/gain value becomes large, all terms on the right-hand side of equation (3-35) can be divided by a_n, thus normalising the coefficient/gain values. This allows the values of potentiometers, Q_n to set to the same value, ω_c for frequency determination only, whilst potentiometers P_n will be set to values:

$$a_i/a_n \text{ where } a_i = a_1, a_2, a_3, \ldots, a_{(n-1)}, \qquad (3\text{-}37)$$

135

and determine the characteristics of the filter. This latter method is carried out in the design of narrow-band Chebychev filters to be described later.

Since all the derivatives are available in fig. 3-31 then the selection of the type of filter depends on the exit point chosen, i.e.

V_o output gives low-pass type;

$V_o^4 p$ output gives high-pass type;

$V_o^2 p$ output gives band-pass type;

$V_o p$ and $V_o^3 p$ output gives asymmetrical band-pass type.

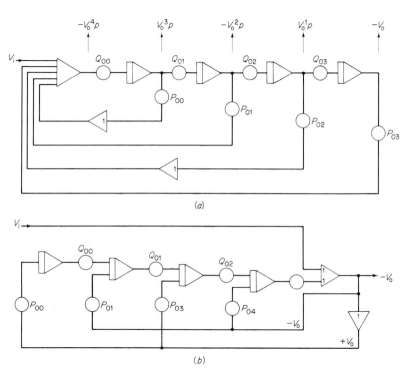

Fig. 3-31. Simulation of a fourth-order Butterworth filter.

Figure 3-32 illustrates this for a fourth-order filter. The slope of the L.P. and H.P. filters will be 6 db/octave per integrator used (24db/octave in this example). The slope of the symmetrical B.P. filter will be 6 db/octave per pair of integrators used (assuming n is an even number). The characteristics of the filter depend on the setting of potentiometers, P_n. Groups of settings can be calculated for a number of different types of filter characteristics. A number of these will be considered below and tables presented for the required potentiometer settings.

136

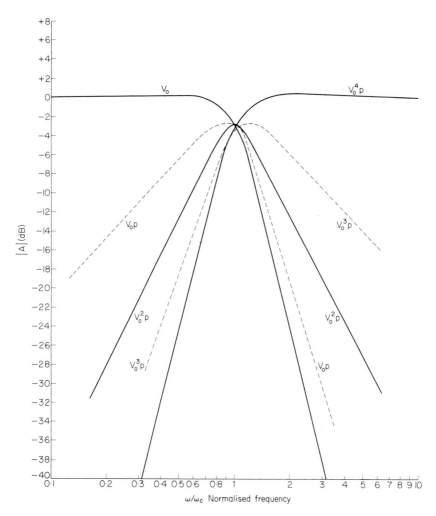

Fig. 3-32. Response at different exit points.

Butterworth filter. The transfer function for order n takes the following form:

$$|H(j\omega)|^2 = \frac{1}{1 + (\omega/\omega_c)^{2n}} \qquad (3\text{-}38)$$

where ω_c is the filter cut-off frequency.

This can be approximated by expanding the function, and taking n stages (ideally $n = \infty$ for a perfectly rectangular response). Over the pass-band, $(\omega/\omega_c)^{2n}$ should approximate to zero in the range $0 < \omega/\omega_c < 1$, and infinity beyond this. The Butterworth filter attempts to meet this requirement by

137

arranging that its first $(n - 1)$ derivates are zero at zero frequency. It concentrates its approximating ability near $\omega = 0$. The result is a filter of maximally flat low-frequency response with good response characteristics, approaching the ideal for large values of n. As n increases, however, the transient response becomes poor.

Potentiometer coefficients are shown in table 3-3a to correspond to those given in equations (3-35) and (3-36). Since $a = 1$ in all cases, they would be used directly as fractional coefficient settings followed by the appropriate number of gain decades.

TABLE 3-3   Potentiometer/gain coefficients for a Butterworth filter

(a)

n	a_1	a_2	a_3	a_4	a_5	a_6	a_7	a_8
1	1·00							
2	1·414	1·000						
3	2·000	2·000	1·000					
4	2·613	3·414	2·613	1·000				
5	3·236	5·236	5·236	3·236	1·000			
6	3·864	7·464	9·141	7·464	3·864	1·000		
7	4·494	10·103	14·606	14·606	10·103	4·494	1·000	
8	5·126	13·138	21·848	25·691	21·848	13·138	5·126	1·000

(b)

n	Denominator Polynomial
1	$(1 + p)$
2	$(1 + 1·414p + p^2)$
3	$(1 + p)(1 + p + p^2)$
4	$(1 + 0·7653p + p^2)(1 + 1·8477p + p^2)$
5	$(1 + p)(1 + 0·6180p + p^2)(1 + 1·6180p + p^2)$
6	$(1 + 0·5176p + p^2)(1 + 1·4142p + p^2)(1 + 1·9318p + p^2)$
7	$(1 + p)(1 + 0·4449p + p^2)(1 + 1·2465p + p^2)(1 + 1·8022p + p^2)$
8	$(1 + 0·3896p + p^2)(1 + 1·1110p + p^2)(1 + 1·6630p + p^2)(1 + 1·9677p + p^2)$

Table 3-3b gives the factorised form of the polynomial and is used in the derivation of Chebychev constants.

Chebychev filter. Here a function $F^2(\omega)$ is put in place of ω^{2n} in equation (3-38) to satisfy more closely the criterion given above:

$$| H(j\omega) |^2 = \frac{1}{1 + E^2 V_n^2(\omega/\omega_c)} \tag{3-39}$$

where V_n is a Chebychev polynomial of order n. This implies two parameters, E and n, both of which can be adjusted to approximate to the ideal response. The

138

gain over the pass band approaches unity, not at zero frequency, but at discrete frequencies distributed over the band. The resulting response gives a sharper roll-off near the cut-off frequency but the transient response is more oscillatory than for the Butterworth filter. The transfer function for a second-order or quadratic Chebychev filter can be shown to be

$$H(p) = \frac{\sigma_k^2 + \omega_k^2}{(p/\omega_c)^2 + 2\sigma_k(p/\omega_c) + \sigma_k^2 + \omega_k^2} \qquad (3\text{-}40)$$

where σ_k and ω_k are the roots of the Chebychev polynomial.

Hansen [11] gives the relationship between the roots of the Butterworth and Chebychev polynomials. If we take roots for the same angle, θ, we can write:

$$\theta = \arccos\left(\frac{k_{nm}}{2}\right)$$

where k_{nm} is the Butterworth coefficient with n expressing the order of the polynomial and m the number of the factorised quadratic given in Table 3-3b. Given the Chebychev constant E we can write:

$$\beta_{nE} = \frac{1}{n} - \sinh^{-1}\left(\frac{1}{E}\right).$$

The Chebychev roots are then defined as:

$$\left.\begin{array}{l} \sigma_k = \sinh \beta \cos \theta \\ \omega_k = \cosh \beta \sin \theta \end{array}\right\}. \qquad (3\text{-}41)$$

If those values are substituted in equation (3-40) for each quadratic factor and the products of the quadratics taken, then after normalising, an expression such as equation (3-34) can be obtained. This will give the value of the coefficients a_n for even-order filters and for different values of E. These are shown in tables 3-4, 3-5, 3-6 and 3-7. Two sets of values are given. The potentiometer/gain coefficients K_1 to K_n, corresponding to the coefficients a_1 to a_n in equation (3-34), and the normalised coefficient values N_1 to N_n corresponding to the values of a_i/a_n discussed earlier. The two sets of tables are separated by a set of $R(\text{db})$ values which are described later. Where the normalised N values are used then the setting of the first frequency-determining potentiometer, (Q_{00} in fig. 3-31) may be calculated from:

$$\omega_c/k_n \quad \text{or} \quad \omega_c N_n.$$

TABLE 3-4 Potentiometer/gain coefficients for a second-order Chebychev filter

| N = 2 | | | | | |
E	K_1	K_2	R(db)	N_1	N_2
0·100	0·599	0·199	0·043	3·008	5·025
0·200	0·794	0·392	0·170	2·024	2·550
0·300	0·905	0·575	0·374	1·575	1·740
0·400	0·966	0·743	0·645	1·301	1·346
0·500	0·994	0·894	0·969	1·112	1·118
0·600	0·999	1·029	1·335	0·971	0·972
0·700	0·989	1·147	1·732	0·862	0·872
0·800	0·968	1·249	2·148	0·775	0·801
0·820	0·963	1·268	2·233	0·760	0·789
0·840	0·958	1·286	2·319	0·745	0·778
0·860	0·952	1·304	2·404	0·730	0·767
0·880	0·947	1·321	2·491	0·717	0·757
0·900	0·941	1·338	2·577	0·703	0·748
0·920	0·935	1·354	2·663	0·691	0·739
0·940	0·929	1·370	2·750	0·678	0·730
0·960	0·923	1·385	2·837	0·666	0·722
0·980	0·916	1·400	2·923	0·655	0·715
0·990	0·913	1·407	2·967	0·649	0·711

TABLE 3-5 Potentiometer/gain coefficients for a fourth-order Chebychev filter

| N = 4 | | | | | | | | | |
E	K_1	K_2	K_3	K_4	R(db)	N_1	N_2	N_3	N_4
0·100	2·259	2·631	1·709	0·796	0·043	2·838	3·305	2·147	1·256
0·200	2·553	3·567	2·504	1·569	0·170	1·627	2·274	1·596	0·637
0·300	2·678	4·247	2·993	2·298	0·374	1·165	1·848	1·302	0·435
0·400	2·715	4·790	3·288	2·971	0·645	0·914	1·613	1·107	0·337
0·500	2·697	5·238	3·447	3·577	0·969	0·754	1·464	0·964	0·280
0·600	2·644	5·613	3·511	4·115	1·335	0·642	1·364	0·853	0·243
0·700	2·568	5·928	3·507	4·586	1·732	0·560	1·292	0·765	0·218
0·800	2·478	6·194	3·458	4·996	2·148	0·496	1·240	0·692	0·200
0·820	2·460	6·241	3·445	5·071	2·233	0·485	1·231	0·679	0·197
0·840	2·441	6·288	3·430	5·144	2·319	0·474	1·222	0·667	0·194
0·860	2·421	6·333	3·414	5·215	2·404	0·464	1·214	0·655	0·192
0·880	2·402	6·376	3·397	5·283	2·491	0·455	1·207	0·643	0·189
0·900	2·382	6·418	3·380	5·350	2·577	0·445	1·200	0·632	0·187
0·920	2·363	6·459	3·361	5·415	2·663	0·436	1·193	0·621	0·185
0·940	2·343	6·498	3·342	5·478	2·750	0·428	1·186	0·610	0·183
0·960	2·323	6·536	3·323	5·539	2·837	0·419	1·180	0·600	0·181
0·980	2·304	6·573	3·303	5·598	2·923	0·412	1·174	0·590	0·179
0·990	2·294	6·591	3·293	5·627	2·967	0·408	1·171	0·585	0·178

TABLE 3-6 Potentiometer/gain coefficients for a sixth-order Chebychev filter

$N = 6$

E	K_1	K_2	K_3	K_4	K_5	K_6	$R(db)$	N_1	N_2	N_3	N_4	N_5	N_6
0·100	4·153	8·804	11·531	11·220	6·405	3·184	0·043	1·305	2·765	3·622	3·524	2·012	0·314
0·200	4·460	10·640	14·629	16·719	9·576	6·274	0·170	0·711	1·696	2·332	2·665	1·526	0·159
0·300	4·557	11·871	16·277	21·049	11·553	9·192	0·374	0·496	1·291	1·771	2·290	1·257	0·109
0·400	4·545	12·812	17·116	24·674	12·760	11·880	0·645	0·383	1·078	1·441	2·077	1·074	0·084
0·500	4·464	13·568	17·430	27·759	13·426	14·305	0·969	0·312	0·948	1·218	1·941	0·939	0·070
0·600	4·341	14·188	17·389	30·395	13·708	16·456	1·335	0·264	0·862	1·057	1·847	0·833	0·061
0·700	4·191	14·703	17·111	32·645	13·719	18·342	1·732	0·228	0·802	0·933	1·780	0·748	0·055
0·800	4·027	15·134	16·681	34·563	13·546	19·979	2·148	0·202	0·757	0·835	1·730	0·678	0·050
0·820	3·993	15·211	16·582	34·911	13·495	20·280	2·233	0·197	0·750	0·818	1·721	0·665	0·049
0·840	3·959	15·286	16·480	35·248	13·440	20·571	2·319	0·192	0·743	0·801	1·714	0·653	0·049
0·860	3·925	15·358	16·375	35·575	13·381	20·854	2·404	0·188	0·736	0·785	1·706	0·642	0·048
0·880	3·891	15·428	16·268	35·891	13·318	21·128	2·491	0·184	0·730	0·770	1·699	0·630	0·047
0·900	3·857	15·495	16·158	36·198	13·252	21·395	2·577	0·180	0·724	0·755	1·692	0·619	0·047
0·920	3·823	15·560	16·046	36·494	13·182	21·653	2·663	0·177	0·719	0·741	1·685	0·609	0·046
0·940	3·789	15·623	15·933	36·782	13·110	21·904	2·750	0·173	0·713	0·727	1·679	0·599	0·046
0·960	3·755	15·684	15·818	37·060	13·036	22·148	2·837	0·170	0·708	0·714	1·673	0·589	0·045
0·980	3·721	15·742	15·702	37·330	12·959	22·385	2·923	0·166	0·703	0·701	1·668	0·579	0·045
0·990	3·705	15·771	15·644	37·461	12·920	22·500	2·967	0·165	0·701	0·695	1·665	0·574	0·044

TABLE 3-7 Potentiometer/gain coefficients for an eighth-order Chebychev filter

$N = 8$

E	K_1	K_2	K_3	K_4	K_5	K_6	K_7	K_8	$R(db)$	N_1	N_2	N_3	N_4	N_5	N_6	N_7	N_8
0·100	6·120	19·003	37·556	55·780	57·517	50·251	25·126	12·778	0·043	0·479	1·487	2·939	4·365	4·501	3·933	1·966	0·078
0·200	6·426	21·802	44·518	73·989	76·866	78·908	37·899	25·226	0·170	0·255	0·864	1·765	2·933	3·047	3·128	1·502	0·040
0·300	6·493	23·625	47·897	87·262	87·848	102·525	45·946	37·014	0·374	0·175	0·638	1·294	2·358	2·373	2·770	1·241	0·027
0·400	6·433	24·998	49·358	97·929	93·949	122·831	50·912	47·899	0·645	0·134	0·522	1·030	2·044	1·961	2·564	1·063	0·021
0·500	6·293	26·092	49·594	106·790	96·786	140·424	53·700	57·741	0·969	0·109	0·452	0·859	1·849	1·676	2·432	0·930	0·017
0·600	6·101	26·986	49·019	114·241	97·375	155·638	54·929	66·492	1·335	0·092	0·406	0·737	1·718	1·464	2·341	0·826	0·015
0·700	5·878	27·727	47·912	120·536	96·429	168·744	55·057	74·174	1·732	0·079	0·374	0·646	1·625	1·300	2·275	0·742	0·013
0·800	5·640	28·344	46·474	125·863	94·465	179·995	54·429	80·858	2·148	0·070	0·351	0·575	1·557	1·168	2·226	0·673	0·012
0·820	5·592	28·455	46·160	126·827	93·985	182·044	54·236	82·083	2·233	0·068	0·347	0·562	1·545	1·145	2·218	0·661	0·012
0·840	5·543	28·562	45·840	127·759	93·482	184·031	54·025	83·274	2·319	0·067	0·343	0·550	1·534	1·123	2·210	0·649	0·012
0·860	5·494	28·666	45·513	128·661	92·958	185·957	53·797	84·429	2·404	0·065	0·340	0·539	1·524	1·101	2·203	0·637	0·012
0·880	5·445	28·766	45·182	129·534	92·415	187·823	53·554	85·552	2·491	0·064	0·336	0·528	1·514	1·080	2·195	0·626	0·012
0·900	5·397	28·862	44·847	130·379	91·856	189·633	53·298	86·641	2·577	0·062	0·333	0·518	1·505	1·060	2·189	0·615	0·012
0·920	5·348	28·955	44·508	131·197	91·282	191·387	53·028	87·699	2·663	0·061	0·330	0·508	1·496	1·041	2·182	0·605	0·011
0·940	5·300	29·045	44·166	131·989	90·695	193·088	52·748	88·726	2·750	0·060	0·327	0·498	1·488	1·022	2·176	0·595	0·011
0·960	5·251	29·133	43·822	132·755	90·097	194·736	52·457	89·723	2·837	0·059	0·325	0·488	1·480	1·004	2·170	0·585	0·011
0·980	5·203	29·217	43·476	133·497	89·488	196·334	52·157	90·690	2·923	0·057	0·322	0·479	1·472	0·987	2·165	0·575	0·011
0·990	5·179	29·258	43·303	133·859	89·180	197·115	52·004	91·163	2·967	0·057	0·321	0·475	1·468	0·978	2·162	0·570	0·011

Bessel filter. This approximates the ideal phase characteristic in a similar manner to that attempted by the Butterworth in its amplitude-response. In the Bessel filter the first $(2n - 1)$ derivatives are, with the exception of the first, zero at zero frequency. Where accurate phase response is required, at the expense of poor attenuation characteristics, the Bessel filter would be chosen.

The denominator polynomial function for equation (3-34), as applicable to a Bessel filter, can be obtained from the expression:

$$F_n(p/\omega_c) = (p/\omega_c)^2 F_{n-2} + (2n - 1)F_{n-1} \qquad (3\text{-}42)$$

given that $F_c(p/\omega_c) = 1$, and $F_1(p/\omega_c) = (p/\omega_c) + 1$. Table 3-8a gives the coefficients of this expansion up to the sixth order.

TABLE 3-8 Potentiometer/gain coefficients for a Bessel filter
(a) Unnormalised

n	a_0	a_1	a_2	a_3	a_4	a_5	a_6
1	1	1					
2	3	3	1				
3	15	15	6	1			
4	105	105	45	10	1		
5	945	945	420	105	15	1	
6	10395	10395	4725	1260	210	21	1

(b) Normalised

n	a_0	a_1	a_2	a_3	a_4	a_5	a_6
1	1	1					
2	1	1	0·3333				
3	1	1	0·4000	0·0667			
4	1	1	0·4285	0·0952	0·0952		
5	1	1	0·4434	0·1011	0·0159	0·00106	
6	1	1	0·4546	0·1212	0·0202	0·00202	0·000096

The large loop gains involved for orders $>n = 4$ render this method of doubtful value for stable simulation on the computer. An improvement is possible if the coefficient values are normalised with respect to the a_0 coefficient, as shown in table 3-8b. The first frequency-potentiometer will now have to be set to a potentiometer gain coefficient of $\omega_0 a_0$ (a_0 is taken from table 3-8a). This implies a large gain localised in one or a few amplifiers preceding the integrators, and the filter will remain stable for a higher order of n.

Paynter filter. The Paynter filter [13] approximates to the ideal phase vs. frequency characteristic in a similar manner to the Chebychev by matching the phase angle at specific frequencies spaced throughout the pass band. Its transient

143

TABLE 3-9 Potentiometer/gain coefficients for a Paynter filter

n	a_1	a_2	a_3	a_4	a_5	a_6	a_7	a_8
1	1·000							
2	1·571	1·000						
3	2·145	1·865	1·000					
4	2·721	3·333	2·041	1·000				
5	3·297	4·895	4·539	2·157	1·000			
6	3·874	7·001	7·363	5·755	2·239	1·000		
7	4·451	9·248	12·161	10·028	6·975	2·301	1·000	
8	5·028	12·005	17·533	18·800	12·833	8·198	2·348	1·000

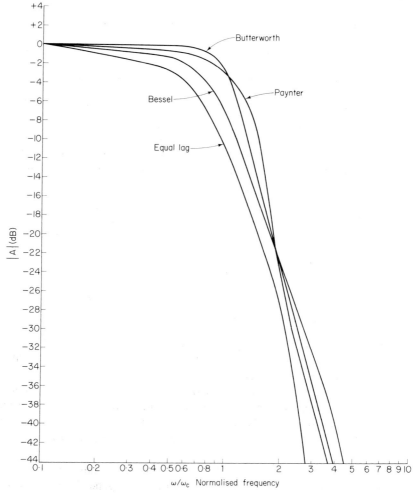

Fig. 3-33. Low-pass filter amplitude response.

response is superior to that of the Bessel filter. The form of the transfer function is as given in equation (3-34). The potentiometer/gain coefficients are given in table 3-9.

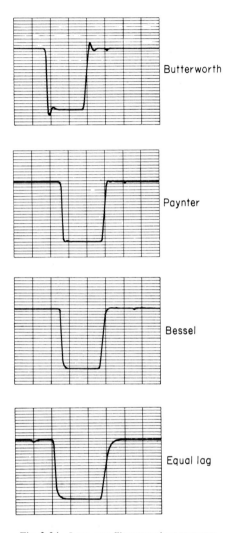

Fig. 3-34. Low-pass filter transient response.

A comparison between several four-stage low-pass filters is shown in fig. 3-33. An equal lag filter is defined by the integer coefficients having equal spacing e.g.

$$H(p) = (1 + 4p + 6p^2 + 4p^3 + p^4)^{-1}. \qquad (3-43)$$

145

The superiority of the Butterworth in respect of its gain *vs.* frequency characteristic may be seen in this diagram. The transient response is shown in fig. 3-34, also for a fourth-order filter. The Bessel filter is seen to give the closest approach to the ideal Gaussian impulse response, although very little overshoot is apparent with the Paynter filter.

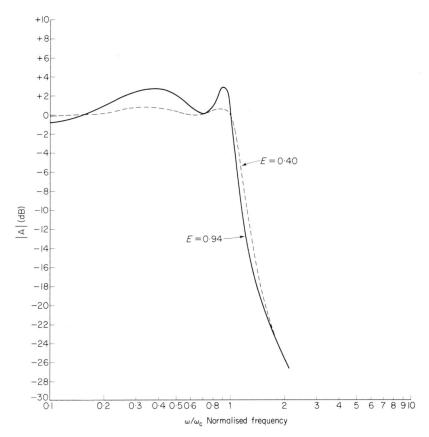

Fig. 3-35. Fourth-order Chebychev filter.

The Chebychev low-pass filter approaches the ideal rectangular response (fig. 3-35) but is subject to a ripple in the pass band of maximum amplitude:

$$R = 20 \log_{10} \sqrt{(1 - E^2)} \text{ db.} \tag{3-44}$$

This had been calculated and included as the R value shown in the computed tables 3-4 to 3-7.

146

3-7-5 CASCADED FILTERS

The simulation of high-order filters using cascaded integrators can lead to instability due to undesired phase shifts at the higher frequencies. This is particularly the case for even-order filters, where a signal is present at the summing amplifier output which will include phase-shifted higher frequency components.

One way of overcoming this is to include a first-order low-pass filter within the feedback loops. This will attenuate the higher frequency components before they enter the main filter and thus extend the permissible dynamic range of amplitude excursion within it.

A similar technique is in wide use for recursive digital filtering where a high-order filter can comprise a series of filters, each limited to the first or second order. In this case, however, the filters are strictly cascaded having no overall feedback. This can be carried out with analog filters but becomes uneconomic in hardware utilisation.

3-7-6 DIGITALLY-CONTROLLED ANALOG FILTERS

The cut-off frequency of cascaded integrator filters is dependent on the setting of one or more potentiometers. In certain types of repetitive analysis (e.g. Fourier or spectral analysis), where a variable frequency filter is used, it will be necessary to alter these settings during the course of computation and unless an automated procedure is adopted then the calculations are likely to be long and tedious. An analog method of control using multipliers in place of potentiometers has been suggested earlier. Digital control is attractive since in many signal processing operations a small process-control digital computer will be linked to the analog machine so that digital control by software can be included.

Direct control of servo-potentiometers may be carried out with some linked systems. The procedure is slow, however, and not well-suited for automatic control. A more practical alternative is to use digital coefficient units (D.C.U.) in place of the servo-potentiometers. The principle of the D.C.U. is shown in fig. 3-36. It consists of a single buffered digital-to-analog converter capable of producing an output current proportional to a digital input word K multiplied by an analog input voltage X. The binary weighted resistors R_0 to R_{14} shown in the diagram are connected to the amplifier summing junction via electronic switches, each controlled by a single bit of the digital K register. Thus by loading the appropriate digital word into the register the effective setting of the potentiometer can be varied over its full range at a speed determined by control and register logic. Some versions of the D.C.U. incorporate a second buffer register in order to allow updating during execution giving a further decrease in sequential setting time.

Control of a sequence of D.C.U.s incorporated in a cascaded integrator or other type of filter is thus feasible from a digital computer at a rate determined

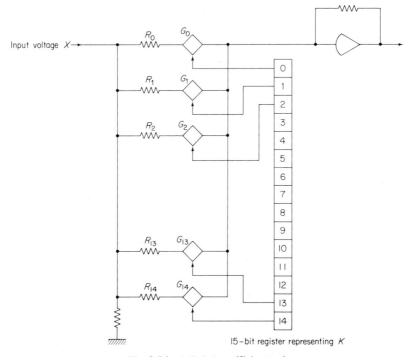

Fig. 3-36. A digital coefficient unit.

by the analysis requirements. The major disadvantage lies in the amount of equipment required for control purposes, including the D.C.U.s themselves.

A sampling method is described in the next section which uses fixed value potentiometers and obtains frequency control by adjustment to the relative integrator operate and hold periods.

3-7-7 INTEGRATOR MODE CONTROL

A sampling method that is applicable to an iterative analog computer is described below. This is known as Integrator Mode Control and using this method frequency control can be obtained, either by suitable logic programming on the iterative machine, or from the linked digital computer. Control under this method is obtained by adjusting the relative periods that the integrators dwell in the operate and hold modes during a repetitive high-speed sampling cycle.

We have seen from the previous treatment of cascaded integrator filters that the cut-off frequency is determined by the integrator time-constant, amplifier gain and setting of the series potentiometers (Q_0 to Q_3 in fig. 3-31), whilst the filter characteristics (Chebychev, Bessel, Butterworth, etc.) are determined by the feedback potentiometers (P_0 to P_3 in fig. 3-31). Now if the integrators shown are capable of being switched sufficiently rapidly between their operate

148

and hold modes an attenuation factor is introduced in the integrator operation which is equal to the duty cycle:

$$\theta = \frac{T_o}{T_o + T_h} \tag{3-45}$$

where T_o = operate period and T_h = hold period.

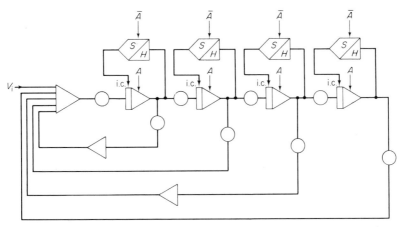

Fig. 3-37. Addition of sample/hold amplifiers to provide operate-hold control.

Thus the frequency to which the filter is tuned is now dependent on a further factor, θ, which is capable of electrical control and we can write for equation (3-34):

$$H(p) = [1 + a_1(p/\theta\,\omega_c) + a_2(p/\theta\,\omega_c)^2 +, \dots, + a_n(p/\theta\,\omega_c)^n]^{-1} \tag{3-46}$$

and

$$f_c = \frac{bG\theta}{2\pi\,T_i} \text{ Hz,} \tag{3-47}$$

where f_c = filter natural frequency (Hz), b = attenuator setting ($b_1 = b_2 =, \dots b_n = b$), G = integrator input gain ($G_1 = G_2 =, \dots, = G_n = G$), and T_i = integrator time-constant (seconds).

Where the analog computer only permits repetitive operation between operate and initial condition, it will be necessary to include a sample/hold amplifier at each integrator stage to retain the previously reached output value for reference as the initial condition voltage during the subsequent operate cycle (fig. 3-37).

Adjusting the filter frequency in this way permits of several applications in signal processing some of which are discussed in this and subsequent chapters.

149

The extent to which the frequency can be adjusted and the accuracy obtained are dependent upon sampling theory and consideration is given to this below.

The action of a mode-controlled filter can be considered as a form of zero-order hold circuit [14], in which two sample states alternate during the process of sampling the filtered signal. These states correspond to the periods of operate and hold for the integrators, and are shown in fig. 3-38 for a sinusoidal output sampled wave-form. During the operate period, T_o, the output of the filter follows that of the input signal, modified by the frequency characteristics of the filter. During the hold period, T_h, the output of the filter is maintained constant at the final value obtained during the previous operate period. This latter state corresponds quite closely to the sampling operation of a zero-order hold circuit (sample/hold amplifier), and becomes identical to it for small values of θ.

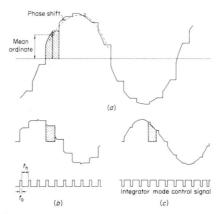

Fig. 3-38. Sampled output wave-form of the mode-controlled filter.

Let us consider therefore, the operation of such a circuit reproducing a stepped function of a sinusoidal wave-form, as shown in fig. 3-39. A signal sample of this wave-form is shaded in the diagram and for this we can write:

$$y_i(t) = H(t - nh) - H[t - (n + 1)h],$$
(3-48)

where $y_i(t)$ denotes a sampled function of (t).

If we put $nh = 0$ then

$$y_i(t) = H(t) - H(t - h).$$
(3-49)

A Laplace transform allows the transfer function of the zero-order hold circuit to be expressed as

$$H(s) = h\left(\frac{1 - \exp(-sh)}{sh}\right).$$
(3-50)

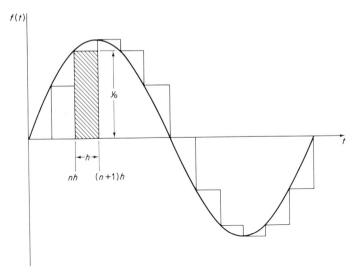

Fig. 3-39. Sampling of a sinusoidal wave-form.

The magnitude and phase angle of this are derived by letting $s \to j\omega$ and also substituting $\omega_s = 2\pi/h = 2\pi \times$ sampling frequency.
Thus:

$$|H_0(j\omega)| = h \left| \frac{\sin \pi\, (\omega/\omega_s)}{\pi\, (\omega/\omega_s)} \right| \qquad (3\text{-}51)$$

and

$$\phi_o = -\pi(\omega/\omega_s),$$

where, $\omega/2\pi =$ input signal frequency.

It will be proved later that if certain forms of frequency distortion are to be avoided then the sampling rate must be equal to or greater than twice the highest frequency of interest, i.e. $\omega_s \geqslant 2\omega$.

At low values of the filter tuned frequency,

$$T_0 \to 0 \quad \text{and} \quad \theta \to 0$$

The filter then behaves almost exactly as a zero-order hold circuit, (fig. 3-38b). At high values of ω_c, $T_h \to 0$ and $\theta \to 1$, i.e. little distortion of the output wave-form takes place (fig. 3-38c). If we assume $\theta = 1$ when $\omega_s/\omega_c = 2$, the mean ordinate for the summated area, shown shaded in fig. 3.38a, can be stated as

$$|H| = (1 - \theta)|H_0| + \theta, \qquad (3\text{-}52)$$

151

and the mean phase angle as

$$| \phi | = (1 - \theta)\phi_0. \tag{3-53}$$

These are the effective magnitude and phase angle for the mode-controlled filter and are plotted in fig. 3-40 for comparison with those for the zero-order hold circuit. It is interesting to note that with this form of sampling, operations on continuous wave-forms may be carried out with little distortion much closer to the Shannon limit than is possible with other systems.

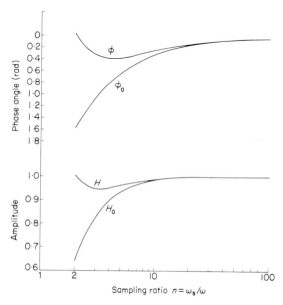

Fig. 3-40. Comparison between zero-order hold and mode-controlled filter response.

The control waveform required for the mode-controlled integrators is clearly that of a rectangular wave-form with the mark/space ratio modified in accordance with the frequency required. Iterative logic for achieving this during a complete spectral analysis operation is described in chapter 8.

BIBLIOGRAPHY

1. TRUSCOTT, J. R. The Eskdalemuir seismological station. *Geo. J. Roy. Astr. Soc.* **9**, 1, 59, 1964.
2. KEEN, C. G. *et al.* The British seismometer array recording system. *I.E.R.E. J.* **30**, 5, 297, 1965.
3. Ariel 1—The first international satellite. N.A.S.A. pub. SP.-43, 1963.
4. CREVELING, C. J. Automatic data processing. *I.R.E. Trans. (Space Elec. Telem.)* Set-8, 124-34, June 1962.

5. POUNDS, K. and WILLMORE, A. P. X-ray measurements on the Ariel satellite. *Proc. Int. Conf. on ionosphere,* London, July 1962.
6. ENOCHSON, L. D. Modern aspects of data processing applied to random vibrations and acoustics. Conf. on applications and methods of random data analysis. Southampton University, July 1969.
7. RAITT, W. J. Analysis of electron probe data from Ariel, 1. Symposium on Data Processing for space research, *Nat. Inst. Res. Nuclear Sci.,* Chilton, England, October 1963.
8. WALLIS, N. A. and ROBERTS, H. V. *Statistics—A New Approach.* Glencoe, Ill. Free Press, 1956.
9. GUILLEMIN, E. A. *Synthesis of passive networks.* John Wiley, New York, 1957.
10. BEAUCHAMP, K. G. A twin-T filter design having an adjustable centre frequency. *Elec. Eng.* **39,** 384, 1967.
11. HANSEN, P. D. New approaches to the design of active filters. *Simulation* **32,** 223-336, May, 1966 also **38,** 388-98, June, 1966.
12. NORONHA, L. G. Analog simulation of transfer functions. *Proc. A.I.C.A.* **7,** 232, 1964.
13. PAYNTER, R. The Paynter delay line. *The Lightning Empiricist* **11,** 3, 10, 1963.
14. BEAUCHAMP, K. G. Mode control filters and their use in automative spectral analysis. Proc. Fifth A.I.C.A. Congress, Lausanne, Switzerland, 90-6, September 1967.

Additional references

15. BURROW, L. R. Tracking filters standardise sinusoidal vibration tests. *Test Eng.* 2-6, April 1964.
16. BEAUCHAMP, K. G. The simulation of wave filters having polynomial transfer functions on an analog computer. *Computer J.* **10,** 4, 352, 1968.
17. VICHNEVETSKY, R. Experimental calibration of bandpass filters by the impulse method. *Proc. A.I.C.A.* **5,** 88-89, April 1963.
18. HOWE, R. M. The digital coefficient unit. *R and H Appl. Dynamics Rep.* 2/25/69, 1969.
19. BENDAT, J. S. A general theory of linear prediction and filtering. *J. Soc. Ind. Appl. Maths.* **4,** 131-51, Sept. 1956.
20. HILLQUIST, R. K. Objective and subjective measurement of truck noise. *J. Sound Vib.* **2,** 8, April 1967.

Chapter 4

ANALOG-TO-DIGITAL CONVERSION PROCESSES

4-1 INTRODUCTION

Conversion of continuous signal information into a sampled digitally-coded form can be carried out with a number of variations in which accuracy, economy and format provide the essential features.

The process implies certain requirements which need to be met if statistically equivalent quantities are to be realised for the continuous analog signal before conversion, and the digital coded data after conversion. These will be considered fully in this chapter. Other requirements for digital analysis such as record length, sampling frequency and bandwidth are closely associated with analysis methods and are more properly discussed in the relevant chapter.

In order to determine the characteristics of analog-to-digital conversion for specific data we need to consider the following:

1. Data acquisition.
2. Pre-processing.
3. Multiplexing.
4. Quantising.
5. Analog–digital conversion.
6. Formatting.
7. Storage or immediate processing.
8. Digital–analog conversion.
9. De-multiplexing.
10. Errors in all these processes.
11. Economy of conversion.

Not all of these are relevant to a particular process.

In considering the effects of these operations, and in particular (4) and (5), it will be necessary to anticipate some of the analysis terms defined in later chapters. However the brief introduction, already given in chapter 1 should enable the text to be followed without any difficulty.

If the acquired signals are to be subjected to digital conversion later then awareness of this can influence the methods of data capture. Such requirements as length of background signal prior to a recorded transient, the length of the record, and such indications as channel indentification and event markers become important in these circumstances.

Digital conversion is a non-informative task and represents an overhead on the process of analysis. As a consequence the efficiency of the process is paramount and this usually means performing the conversion in the least possible time–preferably faster than real-time for the case of conversion from stored analog information. Pre-processing plays an important part in this process, not only because of the necessity to filter the signal to avoid aliasing, but also due to the economy in conversion time resulting from editing of the signal prior to conversion.

Much data is acquired in multi-channel form and for reasons of hardware economy or simplicity will need to be multiplexed before being applied to a single fast analog-to-digital converter. The process of multiplexing will give rise to errors which can be serious if the data is later subject to analysis in the frequency domain. Multiplexing can also generate an identification problem for which a solution must be found.

Fig. 4-1. Sampling and quantisation of an analog signal.

Conversion of a continuous analog signal into a equivalent numerical form requires that the processes of sampling in the time domain and quantisation in the amplitude domain be carried out (fig. 4-1). The limitations and applicability of these processes will first be considered and particularly the consequences arising from the selection of a given sampling rate and number of quantising levels. These will often be chosen to be higher and finer than is apparently demanded by the known characteristics of the analog signal in order to allow for unforeseen contingencies. As has been mentioned earlier, digital conversion may be a non-reversible process, e.g. from a real-time signal, and once digitised the limiting characteristic of the signal in both frequency and amplitude domain will be determined precisely at a level below that possible with the original continuous signal.

4-2 SAMPLING OF TIME HISTORIES

Conversion of a continuous analog signal to a discrete form involves the selection of a series of narrow impulses or 'slices' of the signal, spaced at equal time

155

intervals. A unique number is ascribed to each impulse and represents the mean amplitude of the sample taken over the area of the impulse. Sampling in the time domain is assumed here to take place at a uniform rate. This need not be so and there may be certain advantages in sampling in accordance with a linear function of time (e.g. a sinusoidal function known as Cyclic Rate Sampling [1]), or a rate related to the signal being sampled, or even in a random manner. These methods are not of direct value for signal processing purposes although they have led to the study of adaptive sampling systems where the sampling rate is adjustable [2], and are thus of value in studying the effect of perturbation on a uniform sampling rate [3]. Only the uniform method of sampling will be considered in this chapter.

Ideally, we would like each sample to be taken over an infinitely short period of time but, in a practical case, it is necessary to estimate an averaged quantity

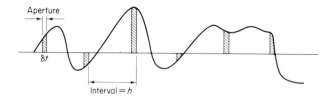

Fig. 4-2. Sampling a continuous signal.

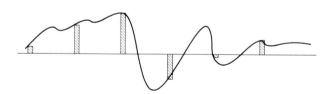

Fig. 4-3. Effect of Sample and Hold.

over the sampling period. The length of time over which the data are averaged is known as the aperture. The width of this is important and to obtain a minimum conversion error the aperture width should be small compared with the sampling period (see fig. 4-2). Aperture errors can be minimised by the use of very fast multiplexors and analog-to-digital converter devices so that the time shift in the input–output sequences is negligible compared with the smallest signal period. The use of parallel sample/hold circuits having very small aperture is easier to achieve and will generally precede the conversion process. The purpose of the sample/hold circuit is to produce a rectangular section from the signal having a constant amplitude over the aperture period, at a value defined by the instantaneous amplitude of the signal at the time of sampling. This is shown in

156

fig. 4-3. A ratio of sampling interval/aperture of the order of 400:1 is realisable with present-day equipment.

We have assumed that sampling of the continuous signal is to be carried out at regular intervals. Unless this is done spurious frequencies related to the uncertainty in time location of the sampling interval will be introduced into the sampled data. This uncertainty is referred to as jitter of the sampling pulse. It can occur due to an unstable or inaccurate pulse generator controlling the sampling rate. When jitter is present it can introduce progressive errors in the phase information associated with high frequency signals.

The most important of the difficulties that arise when a continuous signal is sampled is undoubtedly that of aliasing. The nature of this problem is illustrated in fig. 4-4 which shows that the same set of sampled data points can describe a number of time series histories which are indistinguishable to the digital

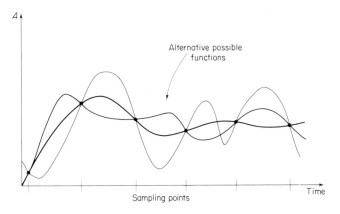

Fig. 4-4. Aliasing.

computer. It will be shown later that any time history can be expressed as the summation of a series of cosinusoids. This enables a simplified model of a real signal to be assumed, and provide a means of quantitatively analysing this aliasing effect.

Consider the sampled time history shown in fig. 4-5. Here, the signal wave-form, $x(t)$, is sampled at regular intervals, h, and the amplitude of the points plotted as before. We will assume the function $x(t)$ to represent a cosinusoid of frequency f_0. The same points could equally well be taken to represent a cosinusoid of frequency f_1 or f_2 which are multiples of f_0. These 'aliased' frequencies are obviously related to the sampling period h and the aliased sequence of frequencies can be written as

$$f_0, \left(\frac{1}{h} - f_0\right), \left(\frac{1}{h} + f_0\right), \left(\frac{2}{h} - f_0\right), \left(\frac{2}{h} + f_0\right), \cdots, \left(\frac{n}{h} \pm f_0\right). \quad (4\text{-}1)$$

157

The fundamental frequency, f_0 is known as the Principal Alias. The range of frequencies below which this effect is not present extends from $f_0 = 0$ to $f_0 = f_N$. This maximum frequency, f_N, is known variously as the folding or Nyquist frequency and is also referred to as defining a frequency limit–the Shannon limit [4] to the sampled data, above which an unambiguous reconstruction of the signal is not possible. It can be described as the lowest frequency coinciding with one of its own aliases [5]. This can be seen from fig. 4-6 and is clearly ($1/h = f_0$). Thus, given a signal containing no frequency components at and beyond a frequency f_N, then the lowest sampling frequency necessary to preserve the

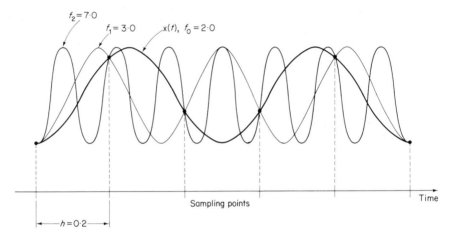

Fig. 4-5. Analysis of the aliasing effect.

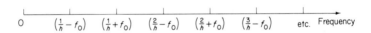

Fig. 4-6. Aliased frequency spectrum.

information contained in a sampled version of this signal is given as $f_S \geqslant 2f_N$ or since, $f_S = 1/h$, then $f_N = \frac{1}{2}h$. This is known as the Sampling Theorem. It follows that for a given frequency spectrum, the individual frequency components lying between $f = 0$ and $f = f_N$ can be separately examined, but if the signal contains components having frequencies $f > f_N$ they will not be distinguishable. Since in a practical case, these higher order components can contribute some power to the frequency spectrum, their contribution will be indistinguishable from those lying between $f = 0$ and $f = f_N$.

This is further illustrated in fig. 4-7 which compares the true frequency spectrum for a signal (fig. 4-7a) with its sampled frequency spectrum (fig. 4-7b). We have no way of deciding which of the frequency axis shown is correct, since

158

all are equally valid. The phenomenon is often known as Nyquist folding or frequency folding for reasons which are apparent from the diagram.

A further consequence of sampling a continuous signal is that its spectrum is now represented by a set of identical coefficients, repeated for successive

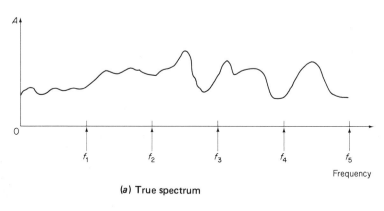

(a) True spectrum

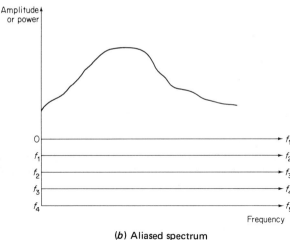

(b) Aliased spectrum

Fig. 4-7. Comparison between the aliased and true spectrum.

frequency bands equal in width to the value of the Nyquist frequency. This follows directly from the aliased frequency spectrum shown in fig. 4-6. The repetition must be considered when manipulating the generated spectral series as we shall see later when digital filters are discussed.

The sampling theorem also applies to the time domain so that if the signal of interest lies within a frequency band extending from 0 to B Hz then the

159

minimum length of record necessary in order that we may recover this signal from the sampled data is

$$T \geqslant \frac{1}{2B} \text{ sec,} \qquad (4\text{-}2)$$

which is known as Rayleigh's criterion.

From the preceding discussion we see that the sampling theorem consists of two parts which may be restated as:

1. signals having a *finite bandwidth* up to and including B rads/sec can be completely described by specifying the values of the time history series at particular instants of time separated by π/B sec;

2. if the signal is band-limited and contains no frequency greater than B rads/sec, it is theoretically possible to recover completely the original signal from a sampled version when the sampling interval is equal to or smaller than π/B sec.

This concept of finite bandwidth is important. Consider signal $x(t)$ to contain no frequencies higher than B rad/sec. We can represent this in the frequency domain by its Fourier series, $F_x(\omega)$ (defined in chapter 6)

$$F_x(\omega) = \frac{1}{2\pi} \sum_{k=-\infty}^{+\infty} C_k \exp\left(-j\pi \frac{k\omega}{B}\right) \quad -B \leqslant \omega \leqslant B$$

where

$$C_k = \frac{1}{2B} \int_{-B}^{+B} F_x(\omega) . \exp\left(j\pi \frac{k\omega}{B}\right) . d\omega, \qquad (4\text{-}3)$$

and represents the complex Fourier amplitudes of the series $F_x(\omega)$. If C_k is known, $F_x(\omega)$ fully defines the spectrum and the equivalent time function $x(t)$ can be obtained from the inverse transform:

$$x(t) = \int_{-B}^{+B} F_x(\omega) . \exp(j\omega t) . d\omega. \qquad (4\text{-}4)$$

If t is defined as $n\pi/B$ then equation (4-4) becomes

$$x\left\{\frac{n\pi}{B}\right\}h = \int_{-B}^{+B} F_x(\omega) . \exp\left(jn\pi \frac{\omega}{B}\right) . d\omega. \qquad (4\text{-}5)$$

From equations (4-3) and (4-5) we can deduce that if a band-limited function, $F_x(\omega)$ is sampled at times $t = n\pi/B$ the original signal is completely recoverable from the sampled signal with no loss of information.

Hence the sampling period must be $h \leqslant \pi/B$.

In practice this recovery is not perfect and the reason lies in the inadequacy of filter design. This may be seen if we consider the Fourier representation of a

160

spectrum of limited bandwidth (fig. 4-8), shown here as a two-sided spectrum. This necessarily includes an infinite series of spectra on either side of the original spectra so that a filter having the characteristics, shown dotted in fig. 4-8, is required for full recovery of the original spectrum. As we have seen such perfect filter characteristics are unattainable and some distortion due to sampling is inevitable.

The necessity for finite bandwidth is therefore a direct consequence of the aliasing effect. Unless practical steps are taken to limit the bandwidth of the sampled signal to below f_N it will not be possible to decompose frequencies lower than this from the composite signal. Thus from equation (4-1) if we substitute the aliasing frequency term $(n/h \pm f)$ for the signal frequency, f, noting that t can be replaced by h and n/h is an integer, we have $\cos 2\pi f t = \cos 2\pi [(n/h) \pm f]h = \cos 2\pi f h$. In other words, all data at frequencies $(n/h \pm f)$ will have the same cosine function as data frequency f when the data is sampled at intervals of length h.

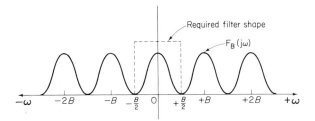

Fig. 4-8. Fourier representation of a limited bandwidth signal.

From the considerations given above it is seen to be essential that the signal be subjected to a low-pass filtering operation prior to digitisation to ensure that all frequencies greater than f_N are excluded. This is necessary, not only to avoid the aliasing effects of the actual signal content, but also to reduce the contribution of higher frequency order noise components to the digitised data, which will otherwise be accepted as noise components falling within the Nyquist bandwidth. The inadequacy of practical filters used for band-limiting modifies our choice of sampling rate such that a rather higher rate is required than is suggested by the sampling theorem. We may, for example, find that the necessary attenuation (equal to the required dynamic range of the measurement) will not be reached until a frequency three octaves higher than the cut-off frequency. In such a case the sampling rate will need to be six times the highest frequency of interest. In mitigation of this we may note that this addition or 'folding' into the accepted band is actually, vector addition. Where the phase characteristics of the higher frequency signal are precisely known then a lower choice of sampling frequency can be tolerated. Finally when considering the complete analog-to-digital-to-analog conversion process we shall see later in connection with

161

digital-to-analog conversion that we will need to consider sampling rates considerably higher than the Nyquist limit if high overall conversion accuracy is required.

4-3 QUANTISATION

The representation of a variable amplitude series of discrete sample values as an equivalent series of discrete numbers representing their amplitude values is termed Quantisation. The process can only be an approximation since, whilst the analog signal can assume an infinite number of states, the number of bits in a digital representation is limited. The numerical values of the quantised variable

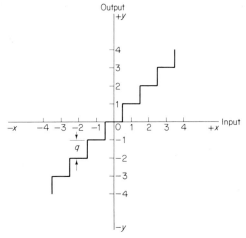

Fig. 4-9. Quantiser transfer characteristic.

may be represented by some form of binary code to permit entry into a digital computer or device.

The unit used to carry out the separation of the input signal into a limited number of level states is termed a Quantiser. In fig. 4-1 the input to the quantiser is shown in sampled form and this is convenient for the purpose of analysis. Many of the conclusions drawn are applicable to the quantisation of continuous signals when the quantiser takes the form of an analog-to-digital converter which combines both the functions shown in fig. 4-1.

In broad terms, quantisation is a non-linear operation that is carried out whenever a physical quantity is represented numerically. The resultant numerical value is given as an integer corresponding to the nearest whole number of units. For this reason it is convenient to think of the process in statistical terms [6], and to regard the quantised series as representing the probability density distribution of the input series. The transfer characteristics of a quantiser can be represented as in fig. 4-9. An input value lying between the midpoint values of

162

two consecutive unit values will produce an output value at the level corresponding to the higher of the two.

Quantisation is thus seen as the process of sampling the statistics of the input signal. Referring to fig. 4-10, this shows the probability distribution function $p(x)$ sampled at amplitude intervals h to give a discrete series of probabilities $P_d(x)$. The process is entirely analogous to sampling in the time domain, where uniform intervals of time are replaced by intervals in amplitude. Similar properties exist for sampling in the amplitude domain and a sampling theorem for quantisation exists which states that if the dynamic range of the input signal extends over several intervals then the statistical properties of the signal are recoverable.

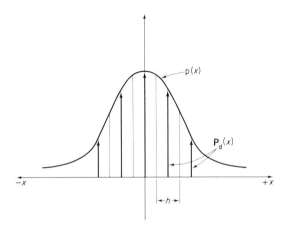

Fig. 4-10. Quantisation of a probability distribution for a signal $x(t)$.

The frequency characteristics of this quantised signal can be studied by regarding it as the result of modulating the continuous input signal, $x(t)$ by a series of impulses spaced uniformly along a time-axis, $D(n) = \Sigma\ (\delta t - nh)$ [7]. This gives a convolution summation;

$$x_q(t) = \sum_{n=-\infty}^{\infty} x(t)\ \delta(t - nh) \qquad (4\text{-}6)$$

where $\delta(t - nh)$ represents an impulse of unit area (Delta function) delayed by nh seconds and h is the sampling interval. As we shall derive later, the spectrum for a rectangular pulse series can be obtained as

$$D(n) = AB/h \left\{ \frac{\sin n\pi(B/h)}{n\pi(B/h)} \right\}$$

where A represents the pulse height and B its width. For a Delta function $B \to 0$ when $AB = 1$ so that

$$\lim_{B \to 0} D(n) = \frac{1}{h}$$

and the inverse Fourier transform of $D(n)$ can be substituted in equation (4-6) to represent the Delta series in the time domain, i.e.

$$x_q(t) = x(t)\frac{1}{h} \sum_{n=-\infty}^{\infty} \exp{(j2\pi nt/h)}. \tag{4-7}$$

Thus the quantised signal can be expressed as the product of the Fourier series for the impulse train and the continuous signal, $x(t)$. It will consist of an infinite

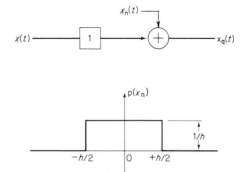

Fig. 4-11. Equivalent representation of the quantisation process.

number of harmonically related sinusoidal carriers having a uniform frequency spacing $1/h$ which are modulated by the input signal to produce a pattern of identical sidebands about each carrier frequency. It is only necessary to filter the complex modulated signal with a low-pass filter of cut-off frequency close to that of the modulating signal to recover the original signal. This assumes, of course, that the signal is band-limited so as to exclude frequencies higher than the Nyquist limit, $1/2h$ as discussed previously.

Under these conditions it has been shown [8] that the effect of a quantiser on the input signal can be considered as that of a non-phase-shifting device having unit gain plus added white noise which has rectangular probability distribution (fig. 4-11). Thus we can represent the quantised output as the sum of the input signal, $x(t)$ and noise component $x_n(t)$, i.e.

$$x_q(t) = x(t) + x_n(t) \tag{4-8}$$

where $x_n(t)$ can be precisely defined if the dynamic range of the input and number of quantising levels are known.

164

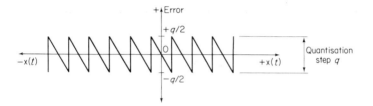

Fig. 4-12. Quantisation noise characteristic.

This noise represents a quantisation error, shown in fig. 4-12 as a function of the quantiser input. The quantisation error is clearly related to the least value of quantisation, q and can be given as the ratio of the magnitude of the exact sample number A, and the approximate number expressed as a power of 2,

$$\epsilon = \pm \frac{A}{2} \cdot \frac{1}{2^N} = \pm \frac{q}{2} \tag{4-9}$$

To assess the value of this error let $p(x)$ be the quantisation error probability density function defined by

$$p(x) = 1 \text{ for } -0.50 \geqslant x \geqslant +0.50$$
$$= 0 \text{ otherwise.}$$

The variance for this error is

$$\sigma_x^2 = q^2 \int_{-\infty}^{\infty} (x - \bar{x})^2 p(x) \mathrm{d}x$$

and since the mean value must be zero,

$$\sigma^2 = q^2 \int_{-0.5}^{0.5} x^2 \mathrm{d}x = q^2 \left(\frac{x^3}{3} \right)_{-0.5}^{0.5} = \frac{q^2}{12}, \tag{4-10}$$

giving a standard deviation for a single unit of quantisation of $\sigma_x = 1/\sqrt{12} = 0{\cdot}29$ level units. This gives a value for the quantisation noise which is added to the desired signal. As an example if we quantise a signal to 256 level units (i.e. 2^8 or 8 binary bits), then the signal/noise ratio at the output of the quantiser will be

$$\frac{256}{0{\cdot}29} = 1000 \text{ or } 60 \text{ decibels.}$$

For many purposes this dynamic range will be adequate to represent the input signal so that providing at least 256 possible levels are taken this particular source of error will have a negligible effect.

To express this error in more general form, we have, from equations (4-9) and (4-10):

$$\sigma_x^2 = \left(\frac{A^2}{12}\right) 2^{-N} \tag{4-11}$$

or

$$A^2 = 12\sigma_x^2 2^N, \tag{4-12}$$

but the dynamic range of the input signal expressed in decibels is $R = 20 \log_{10} A$ db so that from equation (4-12)

$$R = 10 \log_{10} 12 + 20 \log_{10} \sigma_x + 10N \log_{10} 2$$

and

$$N \geqslant \frac{1}{3} [R - 10(1 + 2 \log_{10} \sigma_x)]. \tag{4-13}$$

This allows the number of binary bits, N, or word length of the input signal to be defined for a given input dynamic range (in decibels). The quantisation error must be expressed as a standard deviation, σ_x.

4-4   MULTIPLEXING AND DEMULTIPLEXING

The process of sampling a number of parallel signal inputs periodically in order to provide an interleaved train of sampled values introduces a number of new factors and errors into the analog-to-digital conversion operation. Where many signal channels are to be converted time-division multiplexing is necessary to ensure efficient use of the computer and conversion hardware. A digital computer is a serial device and whilst selection or demultiplexing from a mixed single input stream presents no software problems to the digital machine, the provision of multiple streams of parallel input data implies multi-access operation which is complex and expensive and rarely justified by the nature of the input data. Additionally, each data stream would require its own analog-to-digital converter, which is also expensive in hardware terms. A generally accepted solution to the problem of multiple analog inputs (e.g. from multi-track analog magnetic tape), is to sample each input sequentially using a time-division multiplexing technique and to utilise a single fast analog-to-digital converter capable of converting the sampled output of the multiplexor. The arrangement is shown in fig. 4-13. This can impose stringent demands on the design of both multiplexor and converter, particularly with regard to speed of operation, since for N inputs the sampling rate must be $2N$ times as fast as the highest frequency of interest present in any given channel. This will be especially important where the data is derived from a dynamic environment, since the frequency content will be considerably higher than that obtained in a static data-logging situation.

Where the multiplexed data is digitised and stored for subsequent analysis it will be found useful to rearrange the order of retention so that data from a given channel is stored in sequential order. This is contrary to the order in which the samples are acquired but will reduce computer access time considerably for iterative operations e.g. Fourier transformation or digital filtering.

The operation of a demultiplexor which accepts a single channel and re-distributes this into a number of parallel time sharing analog channels is the exact inverse of the multiplexor and the same considerations apply. The problems of designing multiplexors and converters to achieve the performance

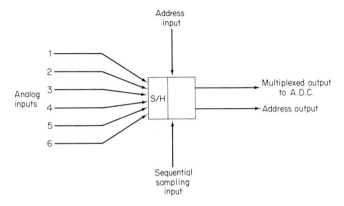

Fig. 4-13. Time-division multiplexing.

requirements for signal processing will not be considered here in any detail except where needed to indicate attainable limits.

A fundamental source of error in multiplexing operations is due to the time delay between samples taken from successive channels. Subsequent analysis using the converted digital data will be made on the assumption that all the samples were taken simultaneously and therefore refer to the same instant in time. This will not be the case with the multiplexor as described above. The time delay error is known as Skewing, and will introduce unwanted phase shifts into the sampled data. One technique for avoiding this error is the use of sample/hold techniques (see chapter 3). One sample/hold unit is required for each channel. All channels can then be sampled simultaneously and their values held at the time of sampling so that the converted data will refer to the same sampling instant, despite the sequential conversion carried out at the outputs. Other errors are related to the imperfection of the circuits used. These include the effects of switching transients, cross-talk between channels and common-mode pick-up, all of which contribute to the random noise added to the multiplexor output. Mean

167

level drift and a fixed shift can also occur and will modify the low frequency performance of the device.

A multiplexor can be considered as carrying out two functions; that of switching between channels, and the addressing or identification of the switched channels. The switching function in its simplest form simply consists of a series of electrical switches closed in some defined order (usually sequential). High reliability of an electro-mechanical multiplexor is obtained by the use of reed relays. These are often controlled by means of a diode matrix when the arrangement is known as a 'cross-bar' multiplexor. It is used as an economical method of switching 1000 or more lines at a fairly slow rate. Speed of operation for relay devices is limited and such methods do not find wide use for signal processing purposes. Instead a solid-state switch matrix is used which is considerably faster in operation. Scanning rates of 1 MHz or more can be obtained with signal/noise ratios of better than 60 db.

An addressing function for the multiplexor is necessary to permit channel connection in a given order (which may not be sequential). A simple 'rotary switch' type of multiplexor requires only a sequential series of pulses to permit closure of each switch in turn. Solid-state and cross-bar types include switching logic such that each channel can be given an identifying address in binary form and the addressing logic arranged to carry out the decoding of this into a control signal operating on the corresponding switch. A detailed discussion of addressing functions is given by Susskind [9] together with a treatment of scanning errors.

4-5 ANALOG-TO-DIGITAL CONVERSION

The conversion of an analog quantity into a digital number consists of obtaining a time limited sample of the continuous signal and converting this into a binary-coded word equivalent to the amplitude of the analog sample. The two processes involved, namely sampling and quantising are found combined in the analog-to-digital converter shown diagramatically in fig. 4-14. Methods of design for analog-to-digital converters fall outside the scope of this book. It is necessary however to describe certain features of design which have an important bearing on their use and accuracy.

The converter must be capable of carrying out quantising of a continuous analog signal at a uniform rate, and to replace a succession of digital numbers sequentially in its output register. In order to ensure that these numbers can be read at the correct times it is necessary to know when the process of conversion for a given sample is completed. Conversion will take a finite length of time and knowing this conversion delay t_d, it is, of course, possible to arrange for the output register to be read at a time not less than t_d seconds after the sampling time t_s. This calculation is generally carried out within the analog-to-digital converter itself which provides a 'digitisation complete' pulse to be used as an interrupt signal for the digital computer.

Additional facilities currently found in many convertors and indicated in fig. 4-14 are an input buffer in the form of a sample/hold circuit, introduced to reduce aperture error, and a means of detecting sign and indicating this by an additional binary bit forming part of the output word. It is fairly common practise to link a multiplexing device with the analog-to-digital converter so that both are available in the same hardware unit. Since this will require its own addressing and triggering inputs the combined multiplexor-convertor can be a fairly complex unit.

Analog-to-digital converters introduce a number of errors into the converted signal which need to be considered. Some are appropriate to any analog device; e.g. those due to component tolerances, zero shift, and drift errors. Others refer specifically to the process of conversion such as aperture error, quantising error and round-off errors. These errors will be considered in detail in a later section when methods of conversion for minimum error are discussed. Since these errors are closely associated with the method of coding used, the coding requirements for conversion will first be considered.

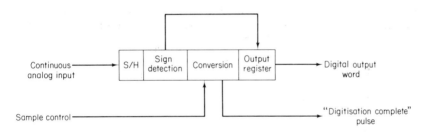

Fig. 4-14. Analog-to-digital conversion.

4-5-1 CODING REQUIREMENTS

The form of coding used to convert the data is important since this must match the code forms of associated equipment if code conversion operations (hardware or software) are to be avoided. This applies not only to the code selected, but also the form of output (parallel or serial) and the significance of the digit order. Other considerations, dependent on the subsequent processing or storage of the data, can also influence the choice of conversion code.

We can consider conversion codes as consisting of two groups; pure binary and binary-coded decimal (B.C.D.). A binary analog-to-digital converter samples a continuous signal into any one of 2^N different levels, where N is the number of binary bits. A binary-coded decimal converter produces a digitised output in the form of a set of 10^D numbers, where D is the number of decimal digits. The pure binary form is obviously more economical in the number of binary bits used and would be selected in cases where this is the most important factor. If large volumes of data are to be handled and stored then the advantages can be clearly seen. If we assume for example a dynamic range of 60 decibels, i.e. a scale of 1000:1 in the input information, then 10 binary digits (bits) would be adequate to express numbers in the range 0 to $(2^{10} - 1)$. By comparison 4 binary digits will be needed to represent a decimal number in the range 0 to 9 so that to cover the decimal scale 0 to 999 a 12-bit B.C.D. number would be required.

If the purpose of conversion is to retain the data on digital magnetic tape, then since the data are written in blocks there are advantages in the binary format, as programs written to handle this will permit the forming of longer blocks and thus speed up the reading/writing process. This advantage is derived from the considerations of a repeated tape transport movement which stops a few milliseconds at the end of each block. Longer blocks mean a shorter aggregate stopping time and hence faster reading by the digital computer.

B.C.D. format is, however, easier to interpret in the interface hardware. It also has the advantage on some machines that if it is properly formatted on to the digital mangnetic tape, it can be read directly by a computer program written in the Fortran language. This is a very real advantage since the programming for the digitised data can represent a high proportion of the total

TABLE 4-1 B.C.D. Fixed weight codes

Digit	Decimal weighting					
D_1	8	7	6	5	4	3
D_2	4	4	3	3	2	2
D_3	2	2	2	2	1	2
D_4	1	1	1	1	1	1

cost of acquiring and processing of experimental data. Numerous Fortran programs exist for signal processing and can easily be incorporated as subroutines in the users main processing programs.

B.C.D. is a fixed weight code for which a variety of weighting possibilities exist. Some of these are shown in table 4-1. Each decimal digit is represented by a group of four binary bits. The weighting of 8 4 2 1 shown has the advantage that it is not possible to represent the same decimal number by more than one coding arrangement. This is not the case, for example, with the code 4 2 1 1 where the decimal numbers 1 to 7 can be expressed by several different binary combinations. For this reason the 8 4 2 1 weighting is chosen to represent the numerical character set in most digital computer codes.

A number of variable weight codes are in use for specific data acquisition purposes. One example of these is the Lippel Code, which has advantages when used to express mechanical displacement since it simplifies the practical coding arrangements needed [10]. This is shown in table 4-2 and illustrates the main feature of this code which is that groups of five consecutive ones are found in the three most significant columns.

TABLE 4-2 Lippel variable weight code

Decimal	Code
0	0000
1	0001
2	0010
3	0011
4	0110
5	1111
6	1110
7	1101
8	1100
9	1001

A third type of code is the single-digit change code, of which the Gray code [11] is the most widely used (table 4-3). Here the decimal value of a binary 1 in the digit column k is given by

$$(\text{sign}) \sum_{i=0}^{i=k} 2^i$$

where (sign) is considered positive for the most significant 1 and negative for the next lower significant 1, positive for the next and so on. The weighting of the four binary columns shown in table 4-3 is thus 15, 7, 3, 1 with sign according to the presence and situation of the binary ones. E.g. the decimal value of the Gray code, $1\ 1\ 0\ 1_G$ is

$$+ \sum_{i=0}^{i=3} 2^i - \sum_{i=0}^{i=2} 2^i + \sum_{i=0}^{i=0} 2^i = 15 - 7 + 1 = 9,$$

and for $1\ 1\ 1\ 1_G$ we have $15 - 7 + 3 - 1 = 10$.

TABLE 4-3 Gray single-digit change code

Decimal	Code	Decimal	Code
0	0000	8	1100
1	0001	9	1101
2	0011	10	1111
3	0010	11	1110
4	0110	12	1010
5	0111	13	1011
6	0101	14	1001
7	0100	15	1000

The Gray code is used for digital transducers such as shaft encoders. Its advantage is that only one bit is changed between any two successive decimal values. Thus if an analog-to-digital converter is used to provide a sequence of digital values for a continuous signal a mis-reading during the processing of the data is likely to result in a digital value close to the correct value, since the original analog signal will be changing slowly. This is in contrast to a pure binary code which can alter several of its bit values in changing from one consecutive digital number to the next. An error in any one of its digits is likely to produce a wildly different equivalent decimal value to that expected.

4-5-2 ERROR CHECKING

A validity check is carried out on the converted data to check system functioning. This is required for example where digital magnetic tape is involved since the inhomogenities of the oxide coating on the tape can result in

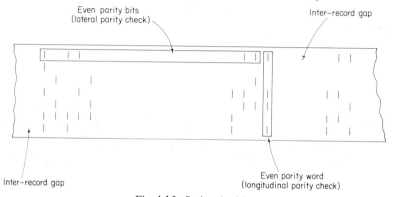

Fig. 4-15. Parity checking.

'drop-outs' perhaps affecting only a single binary bit of the digital sample word. Methods of error checking are in use to detect this and in some cases even correct the error found. The most important of these are parity checking and check sum addition.

Parity checking involves adding to the end of a word a binary bit the value of which depends on the number of the binary 1's present in the word. In the case of even parity this additional binary bit has the value of 1 if the number of 1's in the word is even and 0 if the number of 1's is odd. The converse holds for the case of odd parity checking. When referred to data recorded on magnetic tape this is known as lateral parity. A form of longitudinal parity may also be used in which the 1's along each digit track of the length of tape forming a complete data block can be added and a further digit included in the next stripe position to indicate if the sum is even or odd. This will be carried out for each bit track position across the tape to give an additional parity check word. Both of these methods are illustrated in fig. 4-15.

172

The advantage of the parity checking method is that it is simple to apply and the check will be infallible for the detection of errors occurring in any single bit position. If two or more errors occur simultaneously in a given word then the method breaks down and if this eventuality is to be covered a checksum of the block is carried out.

In one realisation of the checksum method the addition is carried out in a register equal to the word length. Overflow is dealt with by detecting this and transferring the overflow digit to the least significant digit position. Upon reading the tape the same procedure is carried out so that the checksum obtained may be compared with that previously recorded in order to detect any error present.

4-6 DIGITAL-TO-ANALOG CONVERSION

Digital-to-analog conversion will occasionally be required for signal processing purposes, either to permit analog processing of digitally acquired data or as one conversion operation in a hybrid processing system. The digital-to-analog

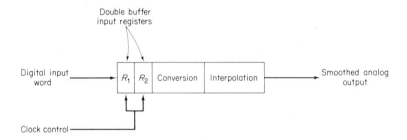

Fig. 4-16. Digital-to-analog conversion.

converter accepts a number in digital form and generates at its output a voltage level proportional to this number. If the digital word contains a sign bit then the electrical polarity of the analog signal is adjusted accordingly. Since the output from the converter represents a continuous function sampled at a rate of $1/h$ times per second it is necessary to include smoothing of the resultant series of voltage levels obtained. The effectiveness of the smoothing operation is considerably enhanced if arrangements are made so that the digital number on the output channel does not return to zero between successive samples. For this reason a second buffer register is often included in the input section of the converter (see fig. 4-16). This also applies to the digital-to-analog multiplier which can also be used as a converter by maintaining the analog input at a constant reference value.

173

Digital-to-analog converters are constructed from a ladder network of precision weighted resistors switched into or out of the circuit in accordance with the value of binary bits comprising the digital input word. This is followed by some form of extrapolation or prediction circuit to obtain a smoothed value of the output approximating to the equivalent continuous value of the signal. Resolution is determined by the digital word length and code employed. A resolution equivalent to fifteen bits is about the limit that can be obtained with present day equipment.

Static accuracy is determined mainly by the quality of the network-forming resistors and associated switches. Dynamic errors include those caused by jitter of the timing impulses, slow drift of the d.c. level at the output of the operational amplifier used and skew error. Where a number of separate digital-to-analog converters are used, one for each output of a demultiplexor, the possibility of skew error arises due to the time delay between individual outputs to the converters. This form of error can be completely eliminated by the provision of a double buffer register at the input to each converter.

4-6-1 INTERPOLATION ERRORS

A major source of error is due to imperfections in the interpolation between samples. As described earlier, the sampled signal contains not only the Fourier components of the original signal but also the repetitions of the signal on either side of the desired band (see fig. 4-8). It is necessary to separate the original signal from these repetitions by means of a low-pass filter, which in the ideal case, should have rectangular characteristics. Since this is unrealisable its place is taken by various forms of hold filters, which function by generating a signal between samples to approximate as closely as possible the continuous function present in the original signal.

A hold filter is a smoothing filter which is automatically matched to change in the sampling rate. It is often referred to as a sample/hold unit and acts as a low-pass filter, passing those constituent frequencies of the original signal which are below the Nyquist limit. Two practical hold circuits are termed Zero-order Hold and First-order Hold. They carry out a smoothing operation by extrapolation from n preceding points to produce a polynomial approximation to the input signal. The transfer function for the zero-order hold circuit shown in fig. 4-17 is given as

$$H_0(p) = \frac{1}{p}\,(1 - \exp(-ph)).\tag{4-14}$$

The output is a stepped version of the original signal with the output value held constant between samples at the level previously reached. The filter is a low-pass one having zero phase distortion and characteristics similar to a Butterworth analog filter. The approximating function for a first-order hold circuit consists of

174

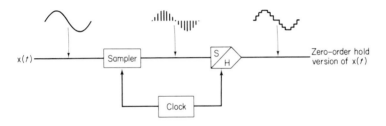

Fig. 4-17. Zero-order hold.

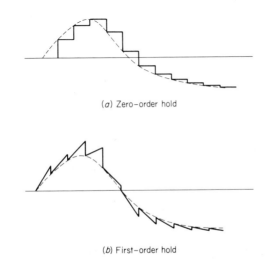

(*a*) Zero–order hold

(*b*) First–order hold

Fig. 4-18. Comparison between zero-order and first-order hold.

a series of ramps introduced in an attempt to extrapolate between data points (fig. 4-18). The transfer function is:

$$H_1(p) = h(1 + ph)\left[\frac{1 - \exp(-ph)}{ph}\right]^2 \qquad (4\text{-}15)$$

which represents a low-pass filter having a considerable amount of over-shoot, corresponding to a Chebychev filter. The amplitude characteristics of the two filters are shown in fig. 4-19, which compares their performance against the ideal rectangular characteristic. Various combinations of these two basic circuits have been suggested [12] to obtain optimum results. A mode-controlled version having some advantages over a zero-order hold has been referred to earlier [13].

175

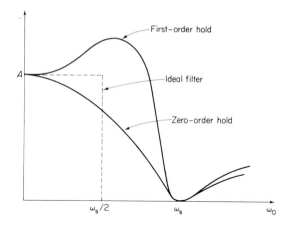

Fig. 4-19. Amplitude characteristics of hold circuits.

The transfer function given in equation (4-14) can be expressed as

$$H_0(j\omega) = h\left[\frac{\sin \omega h/2}{\omega h/2}\right]\exp\left(-j\omega\frac{h}{2}\right) \qquad (4\text{-}16)$$

where h = sampling period = $1/f_s$.

If we now define the number of samples taken per cycle of the signal frequency as s then $\omega h/2 = \pi f/f_s = \pi/s$ and (4-16) can be written

$$H_0(j\omega) = \frac{h.\sin \pi/s}{\pi/s}.\exp\left(-j\,\pi/s\right). \qquad (4\text{-}17)$$

A first-order hold circuit replaces the rectangular step form by a ramp whose slope is dependent on a simple backward difference obtained from the previous pair of sample values (fig. 4-18b). This leads to a better overall approximation to the desired function by emphasising the higher frequency content. Equation (4-15) gives the transfer function for the first-order hold circuit in terms of the operator, p. Rearranging this in terms of $H_0(p)$ we can write.

$$H_1(p) = \frac{1+ph}{h}(H_0(p))^2. \qquad (4\text{-}18)$$

Replacing p by $j\omega$ and recognising that $h = 2\,\pi/s$ then

$$H_1(j\omega) = \frac{H_0^2(j\omega)}{h}\left[1+\left(4\frac{\pi}{s}\right)^2\right]^{\frac{1}{2}}. \qquad (4\text{-}19)$$

From equations (4-17) and (4-19) it is apparent that considerable errors in signal reconstruction will occur unless s is large so that the data is sampled at a rate considerably in excess of the Shannon limit. This can clearly be seen if we consider the nature of the instantaneous errors obtained.

Let the original function be

$$y = A \sin \omega t \qquad (4\text{-}20)$$

which is sampled at periods h seconds apart corresponding to a rate of s samples per cycle of the signal frequency. Therefore

$$sh = \frac{1}{f} = 2\frac{\pi}{\omega} \quad \text{so} \quad h = \frac{2\pi}{\omega s}. \qquad (4\text{-}21)$$

For the zero-order hold case, we see from fig. 4-20 that the error will be a maximum for the sample close to $\omega h = 0$ degrees, so that for $t = h$, $\epsilon = A \sin \omega h = A \sin 2\pi/s$.

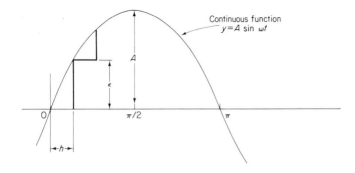

Fig. 4-20. Instantaneous error of a zero-order hold circuit.

If the sampling rate is large,

$$\epsilon \simeq \frac{2\pi A}{s}. \qquad (4\text{-}22)$$

Substituting values in equation (4-22) we see that approximately 628 samples per cycle of the signal frequency will be required to limit the instantaneous error to less than 1% of the full scale value.

In the case of first-order hold (fig. 4-21) the error will be a maximum at the region where the second derivative of equation (4-18) is a maximum, i.e. $\omega t = \pi/2$. If we consider two samples symmetrically disposed about this point the error ϵ, at point $h(n + 1)$ will result from the formation of a ramp, commencing from the centre sample point, h_n and having a slope proportional to the difference in level of this and the preceding point $h(n - 1)$. Thus the error will be seen to have a value of twice the difference between the ordinate values obtained at points $A \sin \omega h n$ and $A \sin [\omega h(n - 1)]$.

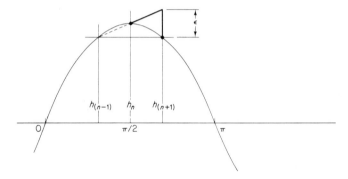

Fig. 4-21. Instantaneous error of a first-order hold circuit.

But since $\omega hn = \pi/2$, $n = \pi/2\omega h$, and we can write

$$\epsilon = 2A\left[\sin \pi/2 - \sin \omega h\left(\frac{\pi}{2\omega h} - 1\right)\right]$$

$$= 2A\,(1\text{-}\cos 2\pi/s) = 4A\,\sin^2\,(\pi/s).$$

If the sampling rate is large

$$\epsilon \simeq A\left[\frac{2\pi}{s}\right]^2. \tag{4-23}$$

The reduction in error compared with the zero-hold case is $s/2\pi$ and, for the example given above, only 63 samples per cycle will be required for the same accuracy.

4-7 DIGITISATION

The transformation of analog signals, recorded in multi-track form on F.M. analog magnetic tape, into blocked digital records on digital magnetic tape is commonly referred to as Digitisation. The essential features of the transformation can be summarised as:

1. Parallel demodulation of the multi-track F.M. signals.
2. Analog filtering to minimise aliasing.
3. Time-division multiplexing into a single channel.
4. Analog to digital conversion.
5. Addition of parity checking bits.
6. Blocking data into suitable record lengths.
7. Writing blocked data onto digital magnetic tape.

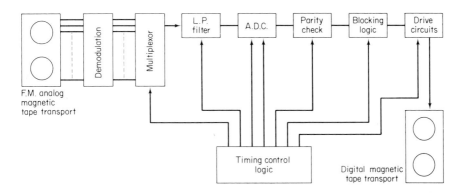

Fig. 4-22. An outline digitisation system.

The main features of a digitisation system are shown in fig. 4-22. Additional features that will be required in many circumstances include:

1. Pre-processing of the analog signals.
2. Automatic calibration signal generation.
3. Signal overload indication.
4. Analog channel identification.
5. Checksum calculations.
6. Ensemble identification.
7. End of ensemble indication.

A calibration signal may already have been recorded in analog form on one track of the F.M. tape. This will be digitised, together with the signal channels, and accessed by the pre-processing digital program operating on the resultant formatted data. This is a convenient method of overall calibration since any variation in stage gains for the F.M. demodulator, multiplexor, converter, etc., will be common to all signal channels including the calibration channel.

The choice of an acceptable recording level from a given transducer is not always an easy one. Whilst it is desirable to record at a fairly high level to obtain a good signal/noise ratio the dynamic range of the signal may be unknown and the possibility of overload cannot be ruled out. How the overload condition is to be treated depends on the signal analysis required and a common method is to detect the excess digital value and replace this by a special character (such as an asterisk) which can be recognised by the digital program and acted upon.

In some cases the overload may be caused by a rapid transient or drop-out and have a duration equal to one sample only. Since low-pass filtering to avoid aliasing will have been carried out previously, it is unlikely that the sampling frequency will have been set too close to the Nyquist limit, so that a single sample transient is inadmissable as a component of the original data. A common treatment for single transients of this type is to replace the overload sample by the value reached by the previous or following sample. A better method to avoid such discontinuities in the data is to arrange linear interpolation replacement of the faulty sample.

Signal identification will be required both in terms of individual channel identification and the complete multi-channel ensemble. For sequential channel sampling this may be obtained by allocating one multiplexor channel to some easily identifiable signal, e.g. zero or maximum voltage excursion. This can result in ambiguity since an adjacent channel value can also assume this value. Several channels can be allocated for this purpose and suitable checking software devised. A more reliable method, however, is to use the addressing feature of the multiplexor. Where the multiplexor hardware includes a digital channel address word this can be included in the data stored on digital magnetic tape. For sequential sampling this address information need only precede each sampling sequence. For non-sequential sampling an address will be required for each converted sample.

Ensemble identification represents the heading information discussed in chapter 3. This can be made available directly in digital form e.g. as data from punched paper tape and read into the formatting routine as the first block or blocks preceeding the digitised signal.

The characteristics of analog-to-digital conversion can be considered in terms of coding and methods of handling the sign of the analog sample. The sign may be included as an additional bit to the sample word or, in some cases, the signal in analog form is raised to a given positive pedestal level, equivalent to at least half the expected dynamic range. This latter method is equivalent to a reduction in accuracy of one binary bit, but simplifies subsequent handling of the converted sample in the digital part of the digitisation hardware.

Some of the important considerations for a digitisation process are listed below:

1. Speed of replay for the analog tape.
2. Number of multiplex channels.
3. Sampling rate.
4. Filter cut-off frequency.
5. Choice of digital code.
6 Method of sign inclusion.
7. Length of digital sample word.
8. Error checking method.
9. Length of digital block.
10. Packing density of digital magnetic tape.

A number of these are discussed below. Others such as sample size and block length are a function of analysis requirements and will be considered in later chapters.

180

Digitisation of analog signals must be regarded as a real-time operation from the point of view of the digital computer or digitisation hardware. The analog magnetic tape replay system provides a continuous signal having no repeating capability other than reverting to the beginning of the entire analog recording. Consequently if a digital computer forms part of the digitisation process it must be capable of accepting real-time information, which means that its interrupt capabilities should permit immediate attention to the demands of the digitisation process. This will be required even if the digital computer is a general purpose computer carrying out parallel time-shared batch processing work. Not all machines will permit this type of working and the controlling or executive program may be designed to refuse interruption whilst it is carrying out peripheral transfer instructions (e.g. data file manipulation) and render digitisation impossible without adequate buffer storage capability, independent of the digital computer. Many digital tape transport systems operate on a continuous basis for each block of data. Consequently the digitisation operation is such that the conversion hardware must control two real-time systems, the continuous analog tape transport and the digital transport operating continuously within the block length, each of which operates independently of the other. This gives rise to complex control problems which are not easy to overcome at high digitisation rates. These are eased considerably if an incremental digital transport is used but at a prohibitively large increase in conversion time and loss of real-time operation for most requirements.

4-7-1 DIGITISATION RATE

The overall sampling rate at which digitisation of analog signals can proceed will be limited by the characteristics of the signal, design of the hardware and (where applicable) by the controlling software. The theoretical limit will be given by

$$\frac{FA}{2N} \text{ samples per second,} \tag{4-24}$$

where F is the highest constituent frequency of the signal that it is required to digitise, N represents the total number of multiplexed channels, and A is the speed-up factor between playback and recording of the analog magnetic tape.

Hardware limitations will invariably reduce this figure, which can only be regarded as an upper bound. Actual analog-to-digital conversion rates of 1 MHz and a multiplexing rate of some ten times this figure are now practicable, so that the limitation will not be found in his area. Where the digitising equipment is specifically designed for this purpose then the limiting factor is likely to be the transfer rate of the digital tape unit. This in turn will depend on the length of the digitised block and will be found to approach the continuous maximum transfer rate as the physical block length becomes long compared with the inter-record-gap.

181

If a small digital computer of the process-control type is used for format control it will generally have the desired interrupt facilities and can be programmed directly in machine code. Its instruction execution time may, however, be rather slow, setting a low limit to the digitisation rate. This can be seen if the role of the digital computer is considered as an intermediatory between the converter output register and the digital magnetic tape. We will consider the least efficient form of conversion, namely B.C.D. coding, and assume a B.C.D. word length of 12 bits for the transducer data. This gives a three

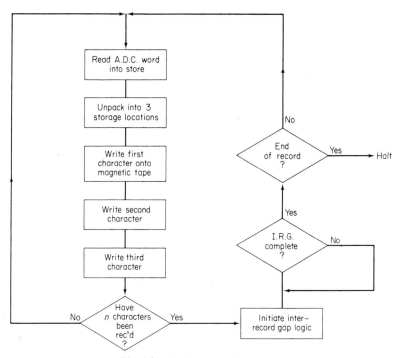

Fig. 4-23. Digitisation software.

decimal digit representation of an analog sample. However, since the data will need to be written on to the digital magnetic tape in character format a parallel serial unpacking routine will be necessary. This, together with a routine for writing the data in blocks separated by an inter-block gap, can be considered as the minimum required for successful digitisation (see fig. 4-23).

If this is programmed in machine code for the digital machine at least 100 instructions will have to be obeyed for each three-digit character sample transferred to magnetic tape. Typical instruction times of 5 μsec for this type of machine lead to a maximum digitisation rate of $10^6/500$ = 2kHz, which is well below the limitations set for other hardware contained in the digitisation loop. It

TRACK NUMBER 6

NUMBER OF BLOCKS/TRACK = 20
NUMBER OF WORDS/BLOCK = 2001

```
  512 - 357 - 212 - 347 - 896 - 335 - 213 - 332 - 902 - 355 - 233 - 326 - 892 - 365 - 212 - 318 - 902 - 355 - 201 - 328 - 902 - 353 - 214 - 330 - 886 - 351 - 217 - 320 - 893 - 354
 -892 - 376 - 215 - 318 - 902 - 370 - 209 - 324 - 898 - 359 - 215 - 329 - 896 - 370 - 218 - 321 - 898 - 382 - 207 - 323 - 902 - 380 - 212 - 329 - 896 - 371 - 215 - 330 - 896 - 370
 -896 - 406 + 056 + 134 - 900 - 405 + 078 + 148 - 898 - 410 + 068 + 120 - 904 - 439 + 073 + 125 - 885 - 433 + 054 + 139 - 896 - 438 + 062 + 135 - 900 - 390 + 058 + 132 - 900 - 385
 -899 - 344 + 054 + 079 - 896 - 349 + 070 + 090 - 900 - 344 + 071 + 104 - 904 - 346 + 082 + 094 - 901 - 347 + 056 + 079 - 888 - 356 + 060 + 084 - 892 - 350 + 076 + 095 - 893 - 347
 -900 - 369 + 029 + 131 - 893 - 391 + 048 + 116 - 902 - 361 + 013 + 129 - 899 - 387 + 031 + 132 - 899 - 376 + 043 + 112 - 905 - 376 + 034 **** - 896 - 385 + 064 + 123 - 898 - 403
 -900 - 454 - 002 + 089 - 898 - 447 + 010 + 085 - 904 - 444 + 010 + 071 - 902 - 415 - 011 + 084 - 896 - 460 + 039 + 087 - 907 - 426 + 006 + 090 - 911 - 433 + 036 + 079 - 889 - 426
 -880 - 412 + 068 + 148 - 892 - 415 + 067 + 146 - 900 - 387 + 064 + 149 - 899 - 385 + 088 + 132 - 900 - 385 + 072 + 155 - 897 - 385 + 080 + 151 - 899 - 400 + 097 + 151 - 908 - 399
 -901 - 340 + 010 + 084 - 896 - 350 + 022 + 082 - 875 - 357 - 031 + 082 - 899 - 347 - 009 + 088 - 902 - 330 + 002 + 079 - 896 - 335 - 021 + 086 - 890 - 351 + 004 + 094 - 890 - 327
 -900 - 343 + 105 + 148 - 906 - 345 + 093 + 150 - 896 - 323 + 070 + 154 - 898 - 318 + 064 + 156 - 897 - 357 + 094 + 157 - 897 - 343 + 136 + 138 - 893 - 337 + 122 + 148 - 897 - 350
 -896 - 399 + 037 + 142 - 912 - 382 + 048 + 127 - 898 - 363 + 023 + 132 - 890 - 357 + 034 + 140 - 900 - 360 + 051 + 136 - 901 - 370 + 059 + 132 - 912 - 374 + 044 + 124 - 898 - 396
 -883 - 325 + 082 + 174 - 896 - 325 + 083 + 174 - 896 - 295 + 064 + 169 - 898 - 297 + 086 + 170 - 896 - 302 + 081 + 186 - 896 - 317 + 071 + 189 - 909 - 285 + 065 + 162 - 900 - 277
 +079 + 151 - 887 - 365 + 086 + 156 - 900 - 356 + 060 + 145 - 913 - 349 + 042 + 143 - 886 - 330 + 056 + 144 - 890 - 341 + 075 + 145 - 897 - 339 + 038 + 149 - 905 - 328 + 069 + 151
 -897 - 333 + 030 + 144 - 912 - 337 + 032 + 142 - 896 - 327 + 040 + 143 - 886 - 331 + 036 + 152 - 900 - 311 + 051 + 143 - 910 - 319 + 086 + 135 - 881 - 304 + 074 + 142 - 890 - 349
 +070 - 900 - 378 + 022 + 066 - 905 - 379 + 000 + 056 - 896 - 377 - 019 + 071 - 897 - 367 - 013 + 071 - 905 - 364 - 001 + 059 - 900 - 377 + 001 + 064 - 916 - 339 - 022 + 064 - 891
 +106 - 896 - 496 + 059 + 120 - 904 - 418 + 043 + 098 - 898 - 387 + 033 + 117 - 890 - 394 + 051 + 108 - 897 - 421 + 040 + 128 - 905 - 434 + 035 + 112 - 904 - 405 + 057 + 104 - 896
 +053 - 877 - 402 + 052 + 074 - 896 - 413 + 064 + 070 - 909 - 420 + 042 + 056 - 910 - 399 + 036 + 066 - 901 - 407 + 075 + 059 - 892 - 444 + 062 + 073 - 898 - 447 + 070 + 061 - 912
 +102 - 899 - 383 + 091 + 099 - 901 - 404 + 071 + 113 - 897 - 407 + 085 + 106 - 903 - 415 + 070 + 117 - 898 - 391 + 090 + 094 - 896 - 387 + 073 + 105 - 897 - 397 + 085 + 120 - 892
 -310 - 909 - 333 - 188 - 314 - 902 - 332 - 201 - 319 - 890 - 333 - 209 - 308 - 896 - 345 - 195 - 309 - 908 - 337 - 196 - 319 - 904 - 319 - 204 - 314 - 898 - 319 - 203 - 304 - 896
```

Fig. 4-24. Summary details of a digitisation process.

183

must be emphasised that this represents the simplest B.C.D. digitisation program. Error checking alone will effectively double the processing time required. An improvement of almost an order of magnitude is obtainable if the sample is treated as a whole binary word within the machine. Digitisation can be carried out using a suitable general purpose digital computer having real-time capabilities. Regarded as a time-sharing operation for this machine the instruction rate can be considerably increased so that the computer software need not be the limiting factor. As stated previously the limitation under these conditions is likely to be the speed of the digital tape transport. This difficulty can be overcome and the digitisation process made to conform to the operating system in use of the output is first directed to magnetic disc file storage. Transfer of the data on to magnetic tape can proceed at a much slower rate as a further time-sharing operation after digitisation has been completed.

There are some advantages in linking a small dedicated process-control digital computer to a large general purpose digital computer to form a digitisation facility, particularly if the general purpose computer is not specifically designed as a real-time machine. The simpler operations of data unpacking can be carried out on a small machine so that a continuous character stream may be conveyed across the interface to the larger machine. Here the editing, formating and error checking can be carried out at high speed and the blocked data written directly on to magnetic tape. The peripheral availability of the larger machine gives a number of advantages here. In particular, the heading information can be entered from punched paper tape or cards and the line printer used to list summary details of the process (e.g. heading information, number of blocks digitised, listing of the first 100 characters of each digitised data block). To some extent this digitisation complex is less susceptible to peripheral transfer delays caused by the time-sharing activities of the larger machine, since a buffer storage area is available in the smaller machine. The value of this will depend, of course, on the rate of digitisation permitted, the size of the buffer, and the transfer delay characteristics of the larger machine. A digitisation facility based on a linkage system of this type is described in the next section.

4-8 A LINKED DIGITISATION PROCESS

A schematic diagram for a system which has been used at Cranfield is shown in fig. 4-25. The three linked computers shown form part of a hybrid system for which digitisation plays a comparatively small time-sharing role. The continuous signals for conversion originate at a multi-channel F.M. replay tape unit. Initial pre-processing is carried out if required using the analog computer. The signals are then multiplexed and converted into digital form by an analog-to-digital converter. In order to simplify the ordering of the data a small process-control computer is programmed to deliver the data in serial character form via a recognised standard computer interface to the ICL 1905 buffer storage. The data

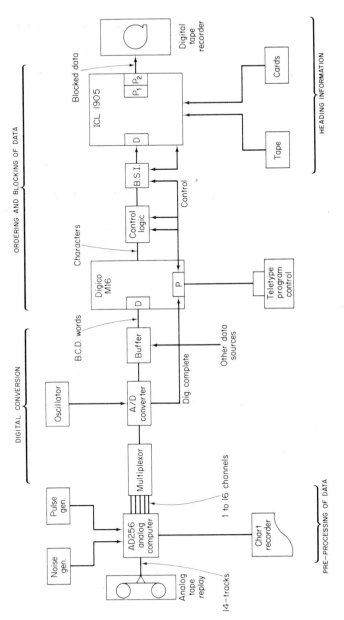

Fig. 4-25. The Cranfield digitisation facility.

185

are later retrieved from buffer storage and blocked into a standard form suitable for writing on to digital magnetic tape. Advantage is taken of the other peripherals of the ICL 1905 to permit entry of calibration and identification information into the final digitised record, and to obtain intermediate output listing. This system is described in some detail in the following pages, commencing with a description of the operation of the interface linking the two digital computers.

4-8-1 THE BRITISH STANDARD INTERFACE (B.S.I.)
There are obvious difficulties in transferring data between two digital computers of essentially different design. The computers will have dissimilar operating periods, control methods and hardware logic levels. These difficulties have led to the formation of standard techniques of data interchange and the embodiment of these in an interface design form known as the British Standard Interface. Only a brief description of its mode of operation can be given here. The reader is referred to the complete specification [14] and a paper by Blake and Leighton [15] for more detailed information.

Referring to fig. 4-26 the method of control used is to provide two

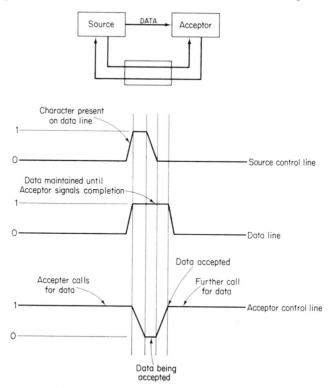

Fig. 4-26. The 'hand-shake' mode of data transfer.

TABLE 4-4 B.S.I. Interface lines

Signal		Abbreviation	Condition	Function
Control line	Acceptor Operable	AO	= 1	Acceptor is operable and able to transfer data under control of SC and AC
	Source Operable	SO	= 1	Source is operable and able to transfer data under control of SC and AC
	Acceptor Control	AC	= 1	Acceptor is ready to accept a data character
	Source Control	SC	= 1	A valid character is present on the data lines
	Source Terminate	ST	= 1	The character being transferred is the last character of a block
	Acceptor Error	AE	= 1	Acceptor has detected an error in the data transferred
	Parity Valid	PV	= 1	The parity bit is being used in the data transfer
Data lines	Parity Bit	P		Odd parity hardware check on data
	Data b_1	2^0		
	b_2	2^1		
	b_3	2^2		
	b_4	2^3		Data available for transfer
	b_5	2^4		
	b_6	2^5		
	b_7	2^6		
	b_8	2^7		

187

sets of binary signals across the interface, one in each direction, so that one set of parallel bits can be transferred at a time determined by the readiness of the data source to provide the data, and the data acceptor to accept it. A logic level 1 on the source or acceptor control lines indicates that the data is available for transmission or acceptance respectively. Thus the transfer of a single data word on a set of parallel lines commences with a logic 1 being set on the acceptor line. As soon as a logic 1 appears on the source line the data is transmitted and the acceptor line drops to zero in recognition of this.

The level signals on the data lines are maintained until they have been accepted. In the meantime the level on the source line has dropped to zero and

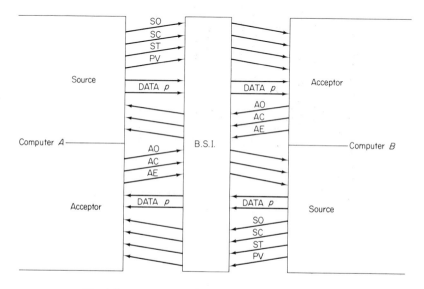

Fig. 4-27. British Standard Interface duplex data transfer.

after acceptance of the set of parallel bits the acceptor line returns to its awaiting data 1 level. This method of data transfer is known as the handshake mode of operation and is independent of the timing requirements of either of the linked computers. In the B.S.I. the method is developed to include a number of additional control lines whose functions are indicated in table 4-4. Two sets of identical lines are used so that data interchange can proceed as a full duplex mode of working between two computers, as shown in fig. 4-27.

The operation of the complete interface may be described with reference to the timing diagram of fig. 4-28. The data will be transferred across the interface in blocks and, in the case of continuous data availability, buffer storage will be required elsewhere in the system. Assuming that a single block of data is to be transferred from computer A to computer B the procedure is as follows.

188

At the commencement of the block data transfer the source operable line is placed at a logic 1 level by computer *A*. After a short delay, whilst computer *B* recognises and acts on this interrupt, the computer raises its acceptor operator line level to logic 1 level to indicate that the process of transfer can begin. A further delay ensues whilst computer *A* transfers its data to the data lines. Computer *B* then raises its acceptor control line and, after a short delay to allow for logic settling time, computer *A* raises its source control and the data is accepted by computer *B*. When it has accepted the data character it lowers its

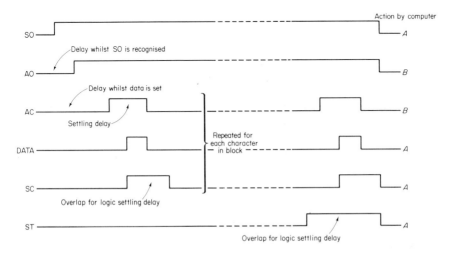

Fig. 4-28. Timing diagram.

acceptor control line level to zero in readiness for the next character. Shortly afterwards computer *A* lowers its source control line and all is in readiness for the transfer of the next character of the block. The process of alternately raising and lowering the source acceptor control lines and transferring data characters under a strict 'hand shaking' procedure continues until the last character is about to be transmitted. The source terminate line is then raised by computer *A* and, after the last character of the block has been transferred, the source acceptor operable lines and the source terminate line are simultaneously lowered and the link returns to its quiescent state.

If the data being transferred is to be accompanied by a parity bit then the parity valid line is raised by the computer *A* at the same time as the source operable line and remains in this condition until the end of the block. Should a parity error be detected then computer *B* will raise its acceptor error line which will be recognised by computer *A* as an instruction to terminate the transfer by raising its source terminate line or some other alternative action.

189

4-8-2   DIGITISATION – HARDWARE DESCRIPTION

The analog inputs for conversion are derived from multi-channel recordings replayed by an F.M. magnetic tape system. Arrangements for the control of this input information can be made using the logic capability of the analog computer, e.g. start/stop, shuttle, speed change etc. Pre-processing of the signals is carried out on the analog machine, which may be programmed in various ways to simulate filter operation. This may be simple low-pass filtering for each channel to minimise aliasing effects or may be more complex selective filtering or pre-whitening. Trend removal can also be included at this stage. The analog computer peripherals are available to the program and are valuable to provide a 'quick look' at the signals prior to conversion. Post recording calibration or the inclusion of locally generated timing signals can also be multiplexed with this input signal.

The signal channels are sampled by the time-division multiplexor at a rate determined by a manually set local oscillator or a clock track pre-recorded on the analog tape. Addressing is either sequential (the usual case) or random access, in which case the particular channel required may be uniquely addressed by an 8-bit code. Analog-to-digital conversion is made to a B.C.D. code permitting a 60 db dynamic range, adequate to match the accuracy of most transducers and recording equipment. The convertor includes a sample/hold input to reduce aperture error. Where manual control of sampling rate is in operation, software constraints are included to inhibit digitisation should transfer not be possible at the rate selected due to hardware limitations. Control of the rate of acceptance is provided by a conversion-complete pulse, originating from the converter each time an input sample has been digitised and assembled in its output register.

The function of the intervening digital computer is to control the transfer of data from the converter, to the B.S.I. The ICL 1905 character set consists of 64 6-bit words which are stored as a four character set within the 24-bit word length of the machine. It is therefore convenient to retain this format for the digitised data and to convert the B.C.D. word and sign into sequential 6-bit characters. This is carried out by means of an unpacking routine in the small computer. The four characters are then transferred serially via the B.S.I. to the ICL 1905 computer. Transfer rate is limited by the instruction speed of the small computer for reasons stated earlier. The transmission of this character data stream into the ICL 1905 is carried out under the control of a resident executive program as a time-shared operation. This is obtained under 'hesitation' program control such that once a transfer has been initiated by the executive program it proceeds autominuously character-by-character placing the data in sequential storage locations. From this storage area the data is retrieved for writing to magnetic tape in blocks, separated by an inter-block gap. The digitisation control program permits specification of block length and includes the writing of run identification and calibration information from punched paper tape or card input. The digitisation operation is carried out concurrently with the normal batch-processing load of the larger machine.

4-8-3   DIGITISATION – SOFTWARE DESCRIPTION

Format requirements for digitisation vary in accordance with specific data needs so that in order to allow a flexible digitising system suitable for a variety of user specifications, the programming is arranged in modular form as three separate programs.

The first of these programs is an input program written for the small computer which handles the output from the analog-to-digital converter and services interrupts from the B.S.I. Input parameters are entered from the teletypewriter keyboard, such as the number of characters per digitised block and the number of blocks per analog record. Once digitisation commences the analog-to-digital converter provides a string of data which begins to fill up sequential core storage locations. The process continues until the allocated core area is full. The B.S.I. is then initiated and the stored words are unpacked into four separate characters contained in four temporary storage locations. When a word has been unpacked, the unpacked numbers are sent to the B.S.I. serially under program control. The speed of operation will depend on the sampling rate which may be manually set or depend on a pre-recorded clock track signal. The maximum rates of converter and B.S.I. are comparable so that a check is necessary to ensure that the speed of conversion does not overtake that of the B.S.I. In the event that the B.S.I. requires unpacked information from the store area and finds this clear, i.e. before the converter has read the information into store the B.S.I. is held up until the converter has been serviced. This will not result in jitter error since the converter is always permitted to operate at a uniform rate. The converse event, namely when the converter attempts to operate at a faster rate than the maximum B.S.I. speed, results in an error message displayed at the teletypewriter advising the operator to manually reduce the sampling rate or reduce the replay speed of the analog tape unit. The process can be halted manually from the teletypewriter at the end of the analog record, as indicated from visual inspection of the replay tape unit, or in response to a warning message originating from the program and displayed on the teletypewriter. This message is determined from a block count and indicates that the last digitised block of the analog record is being written.

The second program is located in the ICL 1905 and is written to handle the data as a character stream through the B.S.I. The program accepts the characters and stores them in a buffer area within core storage. This area is made large enough to permit read-out of the data in blocks during the third stage. Interrupts generated by the B.S.I. are serviced by the program which also produces operator messages required for console display.

The third program is also arranged to be resident within the ICL 1905 and is

191

largely concerned with formatting requirements. Several versions are necessary to meet different data format specifications. One such version, designed for entry into ICL 1900 Fortran programs is described.

A magnetic tape is opened and a standard ICL magnetic tape header label written automatically on to the selected tape. Before any data is transferred a series of header blocks are written on to magnetic tape which consists of the number of blocks of characters specified by the user. These will contain such details as number of tracks digitised, number of blocks per track, number of tracks, sampling rate, calibration information etc. Digitised data may then be accepted from the output of the previous program. The characters are checked for the presence of a special character generated in the first program. This indicates the end of the analog record from a count of the specified number of blocks completed. Parity checking is automatically provided by inclusion of the seventh bit and an inter-block gap included between blocks written on magnetic tape. When an end of analog record special character is detected, the remainder of an unfilled block is filled with zeros, and a final information block is copied from paper tape. Further heading blocks and data blocks can be added to the magnetic tape if the process is recommenced from the control teletypewriter keyboard of the small machine. When the final data block is completed the magnetic tape is closed automatically and the digitisation program deleted.

BIBLIOGRAPHY

1. FRIEDLAND, B. Sampled data control systems containing periodically varying members. *Dept. Elec. Eng. Tech. Rep.* Columbia University, New York, T 39/B, Nov. 1959.
2. BEKEY, G. A. and TOMOVIC, R. Sensitivity of discrete systems to variation of sampling interval. *I.E.E.E., Trans.* (Auto. Control) AC-11, 284-7, April 1966.
3. BEKEY, G. A. and KARPLUS, W. J. *Hybrid Computation.* John Wiley, New York, 1968.
4. SHANNON, C. E. A mathematical theory of communication. *Bell. Syst. Tech. J.* **27**, 623–56, 1948.
5. CAMPBELL, G. A. and FOSTER, R. M. Fourier integrals for practical applications. *Bell. Syst. Tech. J.* **7**, 639, 1928.
6. WIDROW, B. Statistical analysis of amplitude quantised sampled data systems. *I.E.E.E. Trans.* (Appl. and Ind.) **52**, 555, Jan. 1961.
7. WIDROW, B. A study of rough amplitude quantization by means of Nyquist sampling theory. *I.R.E. Trans.* (Circuit Theory) CT-2, 266-76, 1956.
8. LINVILL, W. K. Sampled data control systems studies through comparison with amplitude modulation. *I.E.E.E. Trans.* **70**, 11, 1779-88, 1951.
9. SUSSKIND, A. K. *Notes on Analog–Digital Conversion Techniques.* John Wiley, New York, 1958.
10. LIPPEL, B. A decimal code for analog to digital conversion. *I.R.E. Trans.* (Elec. Comp.) EC-4, 158-9, Dec. 1955.
11. GRAY, F. Pulse code communication. U.S. Pat. 2.632.058, 1953.
12. GILOI, W. Error-corrected operation of hybrid computer systems. Proc. Fifth A.I.C.A. Congress, Lausanne, Switzerland, Sept. 1967.

13. BEAUCHAMP, K. G. Mode control filters and their use in automatic spectral analysis. Proc. Fifth A.I.C.A. Congress, Lausanne, Switzerland, 90-6, Sept. 1967.
14. British Standards Specification, BC68-5083, Feb. 1968.
15. BLAKE, C. C. and LEIGHTON, C. C. Linking two alien digital computers via a British Standard Interface. *B.C.S. Comp. Bull.* **14**, 106-11, April 1970.

Additional references

16. JAGERMAN, D. L. and FOGEL, L. J. Some general aspects of the sampling theorem. *I.R.E. Trans.* (Inf. Theory) IT-2, 4, 139-46, 1956.
17. RAGGAZINNI, J. R. and ZADEK, L. A. The analysis of sampled data systems, *I.E.E.E. Trans.* **71**, 11, 225-34, 1952.
18. BENNETT, W. R. Spectra of quantized signals. *Bell, Syst. Tech. J.* **27**, 446-72, July 1948.
19. MONROE, A. J. *Digital Processes for Sampled Data Systems,* John Wiley, New York, 1962.
20. LEES, A. B. Interpolation and extrapolation of sampled data. *I.R.E. Trans.* (Inf. Theory) IT-2, March 1956.
21. GILBERT, E. G. Dynamic error analysis of digital and combined analog-digital computer systems. *Simulation* **6**, April 1966.
22. GRANDINE, J. D. and HAGAN, L. G. A parallel/sequential stored program hybrid signal processor. *Simulation* **5**, Jan. 1965.
23. HAMMING, R. W. Error detecting and error correcting codes, *Bell. Syst. Tech. J.* **26**, 147-60, April 1950.
24. VIDAL, J. and KARPLUS, W. J. Characterization and compensation of quantisation errors in hybrid computer systems. *I.E.E.E. Int. Conv. Rec.* **3**, New York, 1965.
25. VIDAL, J., KARPLUS, W. J. and KELUDJIAN, G. Sensitivity coefficients for the correlation of quantisation errors in hybrid computer systems. I.F.A.C. symposium on sensitivity analysis, Dubrovnik, Yugoslavia, 1964.

Chapter 5

STATISTICAL TECHNIQUES

5-1 INTRODUCTION

Many problems in engineering and science require concepts of probability and statistics for their solution. Physical measurements are rarely exact and repetition is likely to produce a slightly different answer for each measurement taken. The probability of obtaining a particular answer cannot usually be derived analytically, since the mathematical form of the measurement will in general, not be known. Instead as many repetitions of the measurements are taken as is practical in the circumstances and statistical analysis made from these. An example would be the response of a mechanical system in a random vibration environment.

Considering first a single record taken from the system. This could be analysed to produce various statistical properties such as mean value, probability distribution, power spectral density etc. The accuracies of these estimations will depend on the fidelity of the acquisition methods which we have already discussed, the length of the record and the method of analysis. Even if we succeed in carrying out the acquisition and analysis processes to a high degree of accuracy we are still left with the uncertainty arising from the limited length of the record since the statistical properties of the process only apply to this particular period in time. From these results we may wish to determine the accuracy of this description and also to extrapolate the measurement into a future time for preduction of the expected behaviour of the system.

Where the signal is periodic in form then a limited duration of record is adequate to provide a highly descriptive accuracy of estimation. Random signals on the other hand, demand unlimited length of record for exact estimation of their properties and only a limited accuracy is possible in a practical case. It is for this reason that multiple sample records, taken at different periods, are often acquired so that joint analysis of these can enable a higher statistical accuracy to be obtained.

Another range of problems which require the application of statistical techniques are those concerned with signature or pattern recognition. This generally implies recognition or classification of the signal into recognisable types. The accuracy of this classification is required to be specified in terms of the signature or pattern characteristics. These may be specified in terms of probability values over a given range or require a correlation estimation with a given waveform.

The basic descriptions of random data are derived from physical measurements having a certain measure of uncertainty and will be therefore statistical in form. In this chapter we will consider the statistical properties of the signal in the amplitude domain. Subsequent chapters will consider the signal characteristics in the frequency domain (spectral calculations) and in the time domain (correlation calculations).

5-2   SOME STATISTICAL PROPERTIES

First we will consider a number of simple statistical properties of a random series which will be needed later in the text:

Mean value. The average or mean value of continuous data is

$$\bar{x} = \lim_{T \to \infty} \frac{1}{T} \int_0^T x(t) \, dt. \tag{5-1}$$

The corresponding value for a discrete set of N measurements, x_i, is

$$\bar{x} = \frac{1}{N} \sum_{i=1}^{N} x_i = \sum_{i=1}^{N} \frac{x_i}{N}. \tag{5-2}$$

Mean-square value. A single figure describing the intensity of random data is the mean-square value. This is the average of the squared values of the time-history record, which for continuous data is

$$\overline{x^2} = \lim_{T \to \infty} \frac{1}{T} \int_0^T x^2(t) \, dt, \tag{5-3}$$

and for discrete data:

$$\overline{x^2} = \frac{1}{N} \sum_{i=1}^{N} (x_i)^2. \tag{5-4}$$

Root mean-square (r.m.s.) value. The positive root of the mean-square value is called the r.m.s. value of the signal.

Variance. A signal can consist of a slowly variable trend or low-frequency component upon which are superimposed rapidly changing higher-frequency components. In such a case the mean value will represent the average base level

195

of the signal about which it is fluctuating at the higher rate. A measure of the mean value of a signal about its average mean value is given by the variance of the signal. It is described for continuous signals as the mean-square value about the mean, viz

$$\text{var} = \sigma_x^2 = \lim_{T \to \infty} \frac{1}{T} \int_0^T [x(t) - \bar{x}]^2 . \, dt \tag{5-5}$$

and for discrete data:

$$\sigma_x^2 = \frac{1}{N} \sum_{i=1}^N (x_i - \bar{x})^2 . \tag{5-6}$$

Standard deviation. In order to specify the instantaneous extent of the deviation of the signal from a mean value we define the deviation as

$$\delta = x_i - \bar{x}. \tag{5-7}$$

The sum of the deviations for all the data is likely to be zero so that it will be necessary to use another method if a single figure is required to identify the deviation value of the entire signal. This is obtained by taking the root of the sum of the squares of the instantaneous deviations, since this will always be a positive real number. This value is called the Standard Deviation and will be seen to be the positive square root of the variance, i.e.

$$S = \sigma_x \tag{5-8}$$

or

$$S = \sqrt{\left[\frac{1}{N} \sum_{i=1}^N (x_i - \bar{x})^2 \right]}. \tag{5-9}$$

Note that the standard deviation for a signal having a zero mean value is the r.m.s. value of that signal. An iterative form of equation (5-9) suitable for digital calculations is:

$$S = \frac{\sqrt{(S' - x_i^2/i)}}{i} \qquad (i = 1, 2, \ldots, N) \tag{5-10}$$

where S' is the previous calculated value of S.

5-3   PROBABILITY

A random data sample can be considered as a number representing the output of a stochastic process at a particular time. If the process or experiment is repeated a large number of times then a particular result will be obtained for a given fraction of the total number of repetitions of the process. For example if a card is selected at random from a standard pack of fifty-two playing cards then this can take one of fifty-two values. If the selection is carried out a very large number of times then we could expect a particular card to be selected on 1/52 of the total number of selections. We say that under these conditions each card has a probability of 1/52 of being selected.

The results of a random process will often be found to centre about certain constant values as the number of repetitions of the process becomes large. This characteristic of the process is called Statistical Regularity. In general terms we can express the actual probability of finding a specific random variable, ξ in a process in terms of the total number of variable values, or events, N, present, and the number of times the specific variable occurs, namely M. The probability is given as

$$P(\xi) = M/N. \tag{5-11}$$

The ratio of M/N is sometimes termed the frequency ratio for the variable. Thus the frequency ratio or probability of choosing an ace from a pack of fifty-two cards will be seen to be 4/52 from equation (5-11).

This is not, however, the situation found in many problems in science and engineering where the total number of variable values may be known but not the subdivision of these into equally likely events. In the general case we must carry out a large number of trials and determine the frequency ratios empirically. The results are likely to be obtained in the form of a probability function which defines the probability that a random variable will assume a specific value x, i.e.

$$P(\xi = x) = P_\xi(x). \tag{5-12}$$

Thus if the results of the process ξ is a function containing a number of mutually exclusive values $x_1, x_2, \ldots x_n$ then their respective probabilities are

$$P_\xi(x_1), P_\xi(x_2), \ldots, P_\xi(x_n)$$

which leads to a general expression

$$P(\xi = x_k) = P_\xi(x_k). \qquad (k = 1, 2, \ldots, n) \tag{5-13}$$

The function (5-13) gives the probability of the specific random variable, ξ, taking on a complete set of n values and so describing a distribution of probable values of the resultant process.

When considering a series of probability results of this kind we may note that if ξ consists of a number of mutually exclusive values $x_1, x_2, \ldots x_n$, then ξ can be regarded as a compound event and we can define $P_\xi(x_k)$ as the sum of a series of probable values for $x_1, x_2, \ldots x_n$ i.e.

$$P_\xi(x_k) = P_\xi(x_1) + P_\xi(x_2), \ldots, + P_\xi(x_n) \tag{5-14}$$

which is known as the theorem of total probability.

5-3-1 PROBABILITY DISTRIBUTION FUNCTION

Very often a large number of possible values of x will be defined and it is necessary to express probability either as a set of tables or in graphical terms. The functions used are Probability Distribution and Probability Density.

Probability distribution is defined from equation (5-13) and the theorem of total probability as the probability that ξ will assume a value less than or equal to x, i.e.

$$P(\xi \leq x) = \sum_{k=0}^{n} P_{\xi}(x_k). \qquad (x_k \leq x, \text{ and } k = 0, 1, 2, \ldots, n) \qquad (5\text{-}15)$$

This is known as the Cumulative Probability Distribution Function, or simply the Probability Distribution Function. Note the fact that we obtain a zero

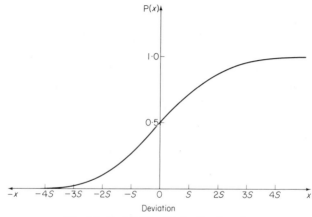

Fig. 5-1. Probability distribution function.

probability of obtaining a definite value x from a continuous random variable. Owing to the limits of our measurement techniques, we can never measure exactly, e.g. we measure 7·68 volts, but really all we know is that the measurement lies between 7·675 and 7·685 volts. However, we do not say that it is impossible to obtain exactly the value of 7·68 volts, only that the distinction is unrealistic since we are always concerned with probability connected with intervals and not with isolated points. The probability distribution function is shown as a continuous function in fig. 5-1. Since only a discrete number of possible values exists then a histogram form of the probability function is generally given. As indicated in this diagram, the probability distribution function is always positive and has a maximum value of unity so that we can write:

$$P_{\xi}(+\infty) = 1$$

$$P_{\xi}(-\infty) = 0. \qquad (5\text{-}16)$$

This is axiomatic. As long as x exists there must also exist a probability that it will take on a particular value. The sum of all the probabilities of x assuming values over the entire range of the variable must equal 1.

Random variables derived from physical experiments are not restricted to a limited range of discrete values but will be obtained in continuous form. Where the range of possible values for ξ extends over a continuous range giving an infinite number of possible values the probability function is meaningless, since the probability of a particular value for x is zero, as noted previously. A continuous probability distribution function is viable, however, and is defined as

$$P_\xi[x(t) \leqslant x] = \int_{-\infty}^{t} p_\xi(x)\, \mathrm{d}x \qquad (5\text{-}17)$$

where $p_\xi(x)$ is called the Probability Density Function.

5-3-2 PROBABILITY DENSITY FUNCTION

This describes the probability that a random variable will take on a value within some defined amplitude range at any instant in time. Thus with reference to fig. 5-2 the probability that $x(t)$ will take on a value in the range between x_1 and x_2 is given as the ratio of the time that x dwells within this range and the total time of the record, i.e.

$$P(x_1 < \xi < x_2) = \lim_{T \to \infty} \frac{t_n}{T} \qquad (5\text{-}18)$$

where

$$t_n = \sum_{1}^{k} t_k$$

n is the level number and $k = 1, 2, \ldots$

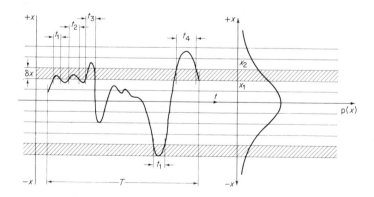

Fig. 5-2. Derivation of probability density function.

The probability density function, $p_\xi(x)$ may be defined as

$$p_\xi(x) = \lim_{\delta x \to 0} \left[\frac{P(x < \xi \leqslant x + \delta x)}{\delta x} \right]$$

$$= \lim_{\delta x \to 0} \left[\frac{P_\xi(x + \delta x) - P_\xi(x)}{\delta x} \right]$$

$$p_\xi(x) = \frac{dP_\xi(x)}{dx} \simeq \frac{t_n}{T \, \delta x} \tag{5-19}$$

and δx is the level interval (i.e. $|x_{(n+1)} - x_n|$.)

It has the same relationship with the single-level pair described by equation (5-18), as the probability distribution function of equation (5-15) has with the single point probability of equation (5-13), namely that it provides a continuous measure of the function with respect to x. The probability density function indicates the probability of x lying between level pairs each separated by an infinitely small value of δx, and extending over the entire range of x. The relationship between the probability density function and the distribution function for continuous functions was given in equation (5-17). As we have seen earlier, the distribution function is a positive increasing monotonic function since $p_\xi(x)$ is non-negative.

Using this idea of probability density we can now redefine the mean and mean-square values for a random variable in more precise terms. The random variable, $x(t)$ can assume any value in the range $-\infty$ to $+\infty$ so that its expected (or mean) value is obtained by integrating all the values of $x(t)$ multiplied by its probability of occurrence, thus

$$E[x(t)] = \int_{-\infty}^{\infty} x.p(x) \, dx = \bar{x}. \tag{5-20}$$

If $x(t)$ is replaced by a power of x then the following relationship holds:

$$E[x^n(t)] = \int_{-\infty}^{\infty} x^n.p(x) \, dx = \overline{x^n}. \tag{5-21}$$

If $n = 2$ then the mean-squared value of $x(t)$ is obtained. Statisticians sometimes refer to the results of equations (5-20) and (5-21) (when $n = 2$) as the First and Second Moment of $x(t)$. These form part of a larger set of moments ($n \geqslant 1$) termed the first-order statistics for $x(t)$.

The shape of the probability density function will enable the signal to be classified broadly as a stationary or a non-stationary process and also indicate the presence of periodic signals contained in random data. Classification of probability density functions into different categories can also assist in the identification of a particular process. For example, it is of particular value to know if a process has a normal density distribution since this will enable certain analytical procedures to be carried out. A number of these classifications will now be described.

5-3-3 EXAMPLES OF PROBABILITY DENSITY FUNCTIONS

Where a finite set of values is considered then the distribution of the values over the range of all possible values refers to the actual values themselves, i.e. a certain number of unit values, two unit values and so on. For an infinite set of values the distribution of values is expressed as a distribution of probability density values. Since most signals for analysis are only finite in the sense of constituting a finite section of a continuous time-history, then we nearly always refer to a distribution as a probability density distribution. For this reason the following examples will be discussed in terms of a probability density function and the term probability distribution used only as an abbreviated form for the cumulative probability distribution function discussed earlier.

The variation in types of density function for random data is fairly limited and identification is helpful in determining the optimum form of analysis to be applied. We shall consider three groups of density functions, namely:

(a) Boolean functions,
(b) continuous functions,
(c) sampled functions.

Boolean functions are those in which only two possible results can be obtained, which will be referred to as a 1 result and a 0 result.

Continuous functions form the most important class for signal processing applications, and include several 'natural' distributions which are characteristic of many physical processes.

Sampled functions are applicable where data is acquired in discrete samples, statistically independent of each other. The characteristics of some of these functions will be considered briefly in the following pages. A fuller discussion and derivation of these functions will be found in the many standard text books on statistics, some of which are listed at the end of this chapter.

(a) Boolean density functions

1. *Binomial.* The conditions under which a function is said to be binomial are:

(i) only two possible results of the measurement can be obtained, namely 1 or 0.

(ii) the probability of a 1 being obtained is constant for all records and is given by P. The probability of obtaining a 0 is therefore $(1 - P)$.

(iii) N records are available which are statistically independent. Since the records are independent then the probability of obtaining a given sequence of results, say all values of $x = 1$ followed by all values in which $x = 0$, equals the product of the probabilities of the two conditions,

$$P^n(1 - P)^{N-n} \tag{5-22}$$

where n indicates the number of 1 results obtained. For example, if the probability of $x = 1$ in a given record is $p(x) = 0.8$, then in N records the probability of obtaining a sequence of 1's (remembering from (ii) that the probabilities are equal for all records) will be $P(x)^n = (0.8)^n$, so that the probability of obtaining n ones followed by $(N - n)$ zeros will be the product of the probabilities given in equation (5-22).

To find the probability of obtaining n values of $x = 1$ in any order, during the N records, then we have to multiply equation (5-22) by the number of ways in which the n 1's can be distributed in the N records. This is given as

$$p(x, n) = \left\{ \frac{N!}{n!\,(N - n)!} \right\} P^n(1 - P)^{N-n} \tag{5-23}$$

which is the probability of obtaining n values of $x = 1$ in a total of N records. This is known as the Binomial Probability Density Function. The factor $N!/[n!(N - n)!]$ is the binomial coefficient. The use of the binomial function can be seen from the following example. Let us assume that we have a given mechanical structure containing ten identical sections, and that this structure is subject to a random vibration. We are interested in the peak displacement for a given section of the structure. The probability of peak displacement being reached over a given time period is given as $p = 0.8$ and this is assumed to be the same for all sections. Thus if we wish to know the probability of peak displacement occurring simultaneously on three sections of the structure during ten test runs we can write from equation (5-23).

$$p_{(x)} = \frac{10!}{3!\,7!} (0.8)^3 . (0.2)^7 = 0.0008.$$

To find a probability that *at least* three out of ten sections will reach peak displacement we must add the probabilities, since these are mutually exclusive events, viz:

$$p(x_3) + p(x_4) + p(x_5), \ldots, p(x_{10}) = \sum p(x_N). \tag{5-24}$$

If N is large then the calculation of binomial probability functions can be fairly lengthy. For such cases we use the Poisson density function.

2. *Poisson density function.* This is derived on a similar basis to the binomial function substituting $P = k\delta T$ so that the probability is given as a constant k multiplied by a very small increment, δT of the record length, T. Using the

202

Poisson distribution we wish to find the probability of $x = 1$ in a small period of the record – small enough to exclude the possiblity of x assuming a value of 1 more than once. If we make $\delta T = T/N$ where T is the record length and $N =$ the no. of records (which can be very large), then substituting kT/N for P in equation (5-23), and letting $x = n$ values of $x = 1$,

$$p(x) = \frac{N!}{x! \, (N - x)!} \left(\frac{kT}{N} \right)^x \left\{ 1 - \frac{kT}{N} \right\}^{N-x} \tag{5-25}$$

$$= \frac{\left(1 - \frac{1}{N} \right) \left(1 - \frac{2}{N} \right), \ldots , \left(1 - \frac{x - 1}{N} \right)}{x!} (kT)^x \left\{ 1 - \frac{kT}{N} \right\}^{N-x}. \tag{5-26}$$

If $N \to \infty$ then the numerator of the first term tends to unity and $[1 - (kT)/N]^{N-x}$ tends towards an exponential factor $\exp(-kT)$. Replacing kT by a single parameter, λ we can write for the Poisson probability function:

$$p(x) = \frac{(\lambda)^x}{x!} \exp(-\lambda). \qquad (x = 0, 1, 2, \ldots) \tag{5-27}$$

which describes the probability that the variable will assume a value of x during the record length of T seconds. The constant k describes the average rate of unity events ($x = 1$), as will be seen from the following:

Since δT is considered infinitely small (for an infinite value N) then there are an infinite number of possible unity events in time T.
Hence we can write

$$\text{total no. of events, } N_\varrho = \sum_{x=0}^{\infty} xp(x),$$

and from (5-27):

$$N_\varrho = \exp(-\lambda) \sum_{x=0}^{\infty} (\lambda)^x \cdot \frac{x}{x!}. \tag{5-28}$$

Re-arranging and recognising that the $x = 0$ term is zero we have:

$$N_\varrho = \lambda \exp(-\lambda) \sum_{x=1}^{\infty} \frac{(\lambda)^{x-1}}{(x - 1)!}.$$

But $\qquad \displaystyle\sum_{x=1}^{\infty} \frac{(\lambda)^{x-1}}{(x - 1)!}$ is the series for $\exp(\lambda)$

so that equation (5-28) reduces to

$$N_\varrho = \lambda, \exp(-\lambda) \cdot \exp(\lambda) = \lambda, \tag{5-29}$$

and the average rate of unity events is

$$\frac{\text{total number}}{\text{record time}} = \frac{N_\varrho}{T} = k. \tag{5-30}$$

The Poisson density function is an approximation to the binomial density function and would be used, for example to reduce the amount of calculation required where N is large and the single record probability, P, is small. Fig. 5-3

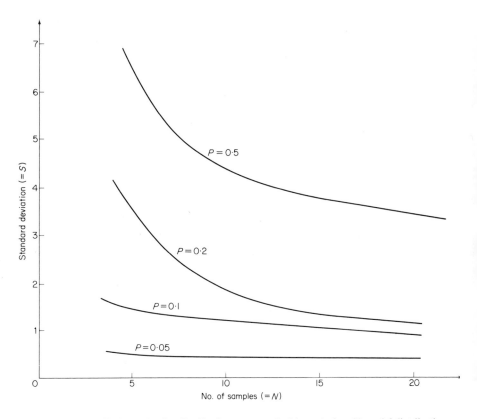

Fig. 5-3. Error of Poisson density distribution compared with equivalent binomial distribution.

indicates the error obtained as the value of N is reduced. From this we see that a choice of Poisson rather than binomial could usefully be made when $N \geqslant 20$ and $P \leqslant 0.2$.

Representative curves are shown in fig. 5-4 for both Boolean functions. In each case the number of records is equal to 10 and the probability equal to 0.2. A disparity between the two functions is apparent for small values of x resulting from the approximations taken in the Poisson case.

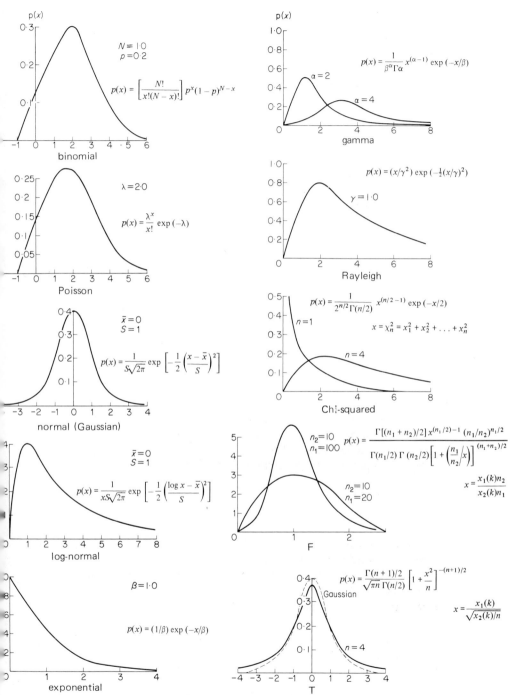

Fig. 5-4. Some probability density functions.

(b) Continuous value functions.

The first two examples are concerned with purely random variables represented in continuous and discrete form.

1. *Uniform.* The simplest case is that of the Uniform Probability Density Function, fig. 5-5 which is given by:

$$p(x) = \frac{1}{b-a} \quad \text{for } a \leqslant x \leqslant b \quad = 0 \text{ otherwise.} \quad (5\text{-}31)$$

This would give, for example, the probability of choosing any point randomly within two fixed limits a and b – any point is as likely to occur as any other.

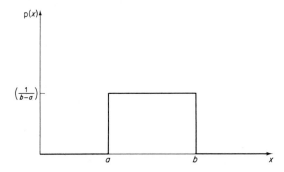

Fig. 5-5. Uniform probability density function.

2. *Discrete.* Where only a discrete number of possible values can be obtained with the equal probability of obtaining any one of the values, then the probability density at each of these discrete values will be infinite by virtue of the definition given in equation (5-19).

The discrete probability density function is given by:

$$p(x) = A_1\, \delta(x - a_1) + A_2\, \delta(x - a_2), \dots, + A_n\, \delta(x - a_n) \quad (5\text{-}32)$$

where $\qquad\qquad\qquad A_1 + A_2 +, \dots, A_n = 1.$

Here $\delta(x - a_1)$ etc., are Delta functions and $a_1, \dots, a_n$ represent the several discrete values which can be assumed by the random variable.

The next set of examples concern the probability density functions of particular random variables, which are themselves functions of another random variable having a known probability distribution.

3. *Sinusoidal waveform.* If we let $x = A \sin(\omega t + \phi)$, where ϕ is a phase angle and x has a uniform probability density function:

$$p(\phi) = 1/2\pi \quad \text{for } -\pi < \phi < +\pi,$$

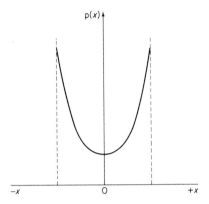

Fig. 5-6. Probability density function – sinusoidal wave-form.

we can consider this wave-form as a random variable, since we will not know its initial phase angle, ϕ, although the frequency and amplitude may be fixed. The probability density function may be shown to be [1] :

$$p(x) = \frac{1}{\pi\sqrt{(A^2 - x^2)}} \quad \text{for } |x| < A$$
$$= 0 \quad \text{for } |x| \geqslant A \tag{5-33}$$

which is illustrated in fig. 5-6.

It is important to note that the probability density function is independent of frequency or time and hence phase angle.

Sinusoidal wave-form plus random noise. It will be shown shortly that the probability density function of a random noise wave-form is Gaussian in shape (see fig. 5-4). When the probability of a sinusoidal wave-form having added noise

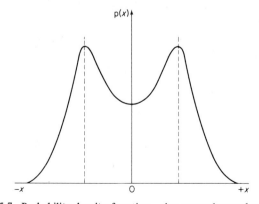

Fig. 5-7. Probability density function – sine-wave plus random noise.

is determined, then the dish-shaped wave-form of fig. 5-6 is modified by the Gaussian response of the noise wave-form to result in the shape shown in fig. 5-7. The shape of the probability density function can be used therefore to determine the sinusoidal content of a mixed sine-wave plus noise signal.

4. *Normal or Gaussian probability.* Gaussian processes are found in many physical situations so that the corresponding probability density function is an important one. This is particularly so since a Gaussian process also results from a linear transformation of a process, which is itself Gaussian. This means that data acquired directly or indirectly from a physical system will still retain its original probability density distribution, if this is Gaussian, despite the distortions obtained in the process of data capture. A further advantage is that the response of such a system can be determined analytically, which is not the case with many non-Gaussian processes.

The importance of the Gaussian probability density distribution arises as a consequence of the Central Limit Theorem, which states that under very general conditions if $x_1, x_2, \ldots, x_n$ are independent random variables whose probability density functions are different (and may be unknown), then the function of the sum of these random variables

$$x = \sum_{k=1}^{n} x_k \tag{5-34}$$

approaches a Gaussian density distribution as $n \to \infty$. This result applies on the assumption that no particular constituent random variable dominates the behaviour of the summed random variable x. In practice many random processes result from the summation of a large number of independent random inputs, e.g. atmospheric turbulence, pressure from a jet exhaust, vibration of a complex structure and many of the forms of random noise. We will consider first a standardised form of this density function.

Standardising a random variable, $x(t)$, means the process of obtaining a new series of values for $x(t)$ where the mean is assumed to be zero and the deviation-squared from the mean is unity (i.e. unit variance). We shall see later that this is a common requirement and simplifies subsequent analysis. If we denote this new series by $z(t)$ then we can write:

$$z(t) = \frac{x(t) - \bar{x}}{S}. \tag{5-35}$$

Using this normalised standard version of the series the Gaussian, or normal probability density function can be defined as

$$p(z) = \frac{1}{\sqrt{(2\pi)}} \exp\left(-\tfrac{1}{2}z^2\right). \qquad (-\infty < z < +\infty) \tag{5-36}$$

This is a symmetric bell-shaped curve and is shown in fig. 5-4. It is sufficient to know this function for positive value of z since

$$p(z) = p(-z).$$

The Gaussian probability distribution function is, by definition

$$P(z) = \int_{-\infty}^{z} p(z) \, dz \tag{5-37}$$

and defines the probability that $z(t) \leqslant z$.

The width of the Gaussian probability density curve can be expressed in terms of standard deviation, S using equation (5-35). For a zero mean value of $x(t)$ then the area under a Gaussian probability density curve of width $\pm 2S$ will be 95% of the total theoretical value thus providing a useful reference point for comparison with other results.

The probability that a random variable $z(t)$ lies between two limits A and B can also be defined as the definite integral of equation (5-36):

$$P(z)_{AB} = \frac{1}{\sqrt{(2\pi)}} \int_{B}^{A} \exp\left(-\tfrac{1}{2}z^2\right).dz. \tag{5-38}$$

Such integrals are difficult to evaluate using exact methods and a difference technique using standard tables is used [1]. These tables generally give area under the standardised normal distribution function (equation 5-37). To use these to obtain $P(z)_{AB}$ we make use of the relationship

$$P(A \leqslant z \leqslant B) = P(A) - P(B).\ dz. \tag{5-39}$$

Taking the difference between the two distribution functions obtained at the limiting values $z = A$ and $z = B$, a value for the integral is found (fig. 5-8).

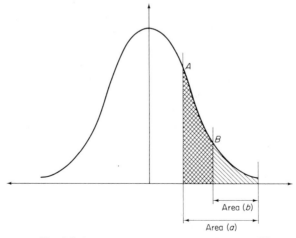

Fig. 5-8. Derivation of $P(z)_{ab}$ = area (a) − area (b).

209

If the random variable does not have zero mean and unit variance then $x - S.z + \bar{x}$ is the value of the random variable whose probability density function is obtained from equations (5-35) and (5-36), thus;

$$p(x) = \frac{1}{S\sqrt{(2\pi)}} \cdot \exp\left[-\tfrac{1}{2}\left(\frac{x - \bar{x}}{S}\right)^2\right] \qquad (5\text{-}40)$$

which is the usual form of the Gaussian probability density distribution function.

The Gaussian curve shown in fig. 5-4 is drawn for the normalised standard version where $\bar{x} = 0$ and $S^2 = 1$. Where the mean value and variance depart from

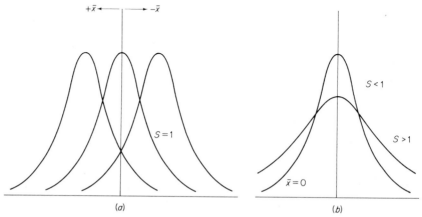

(a)   (b)

Fig. 5-9. (a) Gaussian curve – shift of mean value. (b) Gaussian curve – change of variance.

these values then we can expect some alteration in the position and shape of this curve. For a fixed value of S then a change of mean value, $\bar{x}$ will merely shift the curve laterally without any change of shape (fig. 5-9a). The value of S will, however, considerably effect the shape of the curve. As S becomes larger then the curve will tend to flatten and as S becomes smaller then the curve will become steeper (fig. 5-9b). In many practical situations we find that although a process is Gaussian its probability curve may be distorted in a number of ways. The two major forms of distortion are known as Skewing and Kurtosis. The former describes the lack of symmetry of the curve about its mean value (fig. 5-10a) and the latter the steepness of the curve about its peak value (fig. 5-10b). These effects can be included in an analytical expression by including a third-order term, S^3 to define the skewing effect and a fourth order term, S^4 to define the kurtosis effect. A general expression for a Gaussian curve having a zero mean value can be stated as:

$$p(x) = \exp\left(-a_1 S^{-1} - a_2 S^{-2} - a_2 S^{-3} - a_4 S^{-4}\right) \qquad (5\text{-}41)$$

where a_1, a_2, a_3, a_4, are constants defining the peak amplitude, Gaussian moment, skewing and kurtosis, respectively.

210

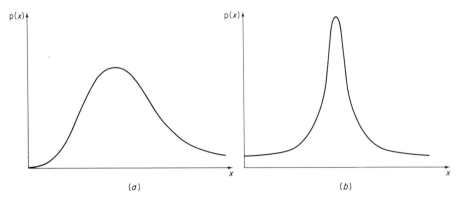

Fig. 5-10. (a) Effect of skewing. (b) Effect of kurtosis.

5. *Log-normal.* This type of density distribution occurs where the logarithm of the random variable has a Gaussian function. The probability density function of a log-normal distribution is given by

$$p(x) = \frac{1}{xS\sqrt{(2\pi)}} \exp\left[-\tfrac{1}{2}\left(\frac{\log(x) - \bar{x}}{S}\right)^2\right]$$ (5-42)

and is also shown in fig. 5-4.

6. *Gamma density functions.* A number of continuous functions form a particular class of probability density functions known as Gamma Functions. A general definition is given by

$$p(x) = \frac{1}{\beta^\alpha \Gamma \alpha} \cdot x^{(\alpha-1)} \exp(-x/\beta)$$ (5-43)

where $\Gamma(\alpha)$ is the value of a gamma function defined by

$$\Gamma(\alpha) = \int_0^\infty x^{(\alpha-\frac{1}{2})} \exp(-x).dx.$$ (5-44)

For integer values of α the gamma function can be obtained from

$$\Gamma(\alpha) = (\alpha - 1)!$$ (5-45)

Examples of this function are given in fig. 5-4, from which we see that the distributions are positively skewed with a skew value proportional to $1/\alpha$. This function, and those derived from it, are valuable to describe waiting times for random events, such as the time intervals between successive breakdowns in physical systems, or the time intervals between random 'hits' for a given process.

211

A special gamma density function for $\alpha = 1$ is the exponential function shown in fig. 5-4 for which

$$p(x) = \frac{1}{\beta}. \exp(-x/\beta). \tag{5-46}$$

The example shown is for a value of $\beta = 2$.

Another set of density functions closely related to the gamma functions include the chi-squared function, Rayleigh function and Maxwell function. These, together with two other functions, the F function and the Student-T function form a third set of continuous functions called the Sampled value functions.

(c) Sampled value functions

1. *Chi-squared*. It was remarked earlier that when only one random variable is considered then in many physical situations the density function is likely to be Gaussian. This is also the case where processes consist of the summation of the effects of a large number of statistically independent random variables, irrespective of the density function of the individual variables. Where the number of contributory independent random variables is finite and each random variable is Gaussian, having zero mean and unit variance then we can define a new random variable as

$$\chi_n^2 = z_1^2 + z_2^2 + z_3^2, \ldots, z_n^2 \tag{5-47}$$

here the new random variable χ_n^2 is known as a chi-squared variable having n degrees of freedom.

Degrees of freedom will be defined more specifically later in connection with power spectral density evaluation. For the moment we can consider n to be equal to the finite number of random variables involved in equation (5-47). This form of density distribution is particularly relevant to sampled data where it is permissible to consider each sample as being statistically independent and random. It approaches normal distribution as $n \to \infty$.

The probability density function is given by

$$p(x) = \frac{1}{\{2^{n/2}\Gamma n/2\}}(x)^{(n/2-1)}. \exp(-x/2) \tag{5-48}$$

where

$$x = \chi_n^2.$$

Substitution of $n/2 = \alpha$, $2 = \beta$, reduces equation (5-48) to the more general case of the gamma density function given by equation (5-43). The set of gamma curves given in fig. 5-4 are in fact χ^2 functions expressed in terms of a single

212

parameter (degrees of freedom). This parameter is indicative of the sharpness of the probability selection and is allied to the definition of system performance in the frequency domain (see chapter 8).

2. *Rayleigh.* This is another form of skewed gamma function. As with other functions of the gamma class it is extremely non-symmetrical, being zero for negative signal values. The probability density function for a Rayleigh density distribution is given by

$$p(x) = \frac{x}{\gamma^2} \cdot \exp\left(-\tfrac{1}{2}(x/\gamma)^2\right). \tag{5-49}$$

One use for this distribution is to describe analytically the noise output of an envelope detection system used in radio reception. In such a system the Gaussian noise characteriestic of the random noise is modified by the narrow bandwidth so that the probability density of the signal peaks follows a Rayleigh distribution law. In equation (5-49) the factor γ is related to the standard deviation, S of the variable x, by a constant $\gamma = 1 \cdot 526\, S$.

3. *F-function.* A sampling distribution which is used to determine the ratio of the variances of two independent random series is known as the F-density function. If $x_1(k)$ and $x_2(k)$ are statistically independent random variables, each having a chi-squared probability function with n_1 and n_2 degrees of freedom respectively, then a new random variable can be defined as

$$x = \frac{x_1(k).n_2}{x_2(k).n_1}. \qquad (k = 1, 2, \ldots) \tag{5-50}$$

The probability density function for x is given as

$$p(x) = \frac{\Gamma[(n_1 + n_2)/2]x^{(n_1/2-1)}(n_1/n_2)^{n_1/2}}{\Gamma(n_1/2)\Gamma(n_2/2)[1 + (n_1/n_2/x)]^{(n_1+n_2)/2}} \tag{5-51}$$

where the gamma functions are defined as

$$\left.\begin{array}{l}
\Gamma\left(\dfrac{n_1 + n_2}{2}\right) = \displaystyle\int_0^\infty x\left(\dfrac{n_1 + n_2}{2}\right)^{-1} \exp\left(-\dfrac{n_1 + n_2}{2}\right) dx \\[4mm]
\Gamma(n_1/2) = \displaystyle\int_0^\infty x^{(n_1/2-1)} \cdot \exp\left(-n_1/2\right) \cdot dx \\[4mm]
\Gamma(n_2/2) = \displaystyle\int_0^\infty x^{(n_2/2-1)} \cdot \exp\left(-n_2/2\right) \cdot dx.
\end{array}\right\} \tag{5-52}$$

The F-distribution is applicable to system gain factor (chapter 10):

$$\left|H(f)\right|^2 = \frac{G_y(f)}{G_x(f)} \tag{5-53}$$

213

since $G_x(f)$ and $G_y(f)$ are both expressed in terms of chi-squared distribution. This is also the case for the calculation of coherence function (chapter 9), which is again expressed as the ratio of chi-squared variables.

4. *Student-T.* This is also defined for two independent random variables, $x_1(k)$ and $x_2(k)$, where $x_1(k)$ has a normal distribution and $x_2(k)$ has a chi-squared density distribution. The new random variables is given as

$$x = \frac{x_1(k)}{\sqrt{\dfrac{x_2(k)}{n}}} \qquad (5\text{-}54)$$

which is described as a student-T variable having n degrees of freedom. The probability density function is given as

$$p(x) = \frac{\Gamma(n+1)/2}{\Gamma(n/2)\sqrt{\pi n}} \cdot \left[1 + \frac{x^2}{n}\right]^{-(n+1)/2}. \qquad (5\text{-}55)$$

This probability density function is compared in fig. 5-4 with that of the Gaussian function. Both are bell-shaped and symmetrical about the mean. It can be shown that the Student-T Function approaches that of a Gaussian function as $n \to \infty$, and differs from it by a negligible amount for n values of 50 or more.

The major advantage of the T distribution is that useful predictions can be made from it using a small number of sample values.

5-3-4 JOINT PROBABILITY FUNCTIONS

The probability functions described in the preceding sections refer to one random variable or a composite variable derived from two independent random variables. It is frequently desirable to describe the output of a process in terms of two or more processes. Examples are the collision occurences arising from two adjacent spring coil segments vibrating randomly, but with some measure of interdependence, or the probability of two transverse waves reaching a target area at the same instant. In these situations we need to establish a probabilistic description for an event which arises as a result of the interaction of two or more sets of correlated random information.

If x_1, x_2, $\ldots$, x_n are the values of n random variables then the probability that, $a_1 \leqslant x_1 \leqslant b_1$, $a_2 \leqslant x_2 \leqslant b_2$, $\ldots$, $a_n \leqslant x_n \leqslant b_n$, is given in terms of a function, $f(x_1, x_2, \ldots, x_n)$ as

$$\int_{a_1}^{b_1} \int_{a_2}^{b_2}, \ldots, \int_{a_n}^{b_n} p(x_1, x_2, \ldots, x_n) \, dx_1 \, dx_2, \ldots, \, dx_n \qquad (5\text{-}56)$$

where $p(x_1, x_2, \ldots, x_n)$ is the joint probability density of the random variables.

Note that this is equal to unity if the integration is carried out over an infinite range of possible values for the variables, i.e.

$$\int_{-\infty}^{\infty} \int_{-\infty}^{\infty} p(x_1, x_2, \ldots, x_n)\, dx_1, dx_2, \ldots, dx_n = 1. \qquad (5\text{-}57)$$

As with the individual probability density function described in section 5-3-2, the joint probability density function can be described in limit terms as

$$p(x_1, x_2, \ldots, x_n) = \lim_{T \to \infty} \left[\frac{t_n}{T(\delta x_1, \delta x_2, \ldots, \delta x_n)} \right]. \qquad (5\text{-}58)$$

It can be shown that if the sets of random variables are statistically independent then we can write for the joint properties:

$$p(x_1 x_2, \ldots, x_n) = p(x_1)\, p(x_2), \ldots, p(x_n)$$

and

$$(5\text{-}59)$$

$$P(x_1 x_2, \ldots, x_n) = P(x_1)P(x_2), \ldots, P(x_n).$$

Equation (5-59) permits the calculation of the joint functions from the product of their individual probability functions.

5-4 ANALYSIS METHODS

Consideration will now be given to practical methods of statistical analysis of random signals commencing with analog and hybrid methods.

The development of special-purpose analog devices for the routine statistical analysis of random signals has employed a combination of comparators and operational amplifiers with some decision logic [2]. The use of a general-purpose iterative analog or hybrid machine for statistical analysis has a number of advantages not least being the possibility of automatic processing of real-time signals [3,4,5]. A number of techniques for this purpose will be discussed in this and the following section.

5-4-1 ANALOG EVALUATION OF STATISTICAL QUANTITIES

The derivation of the parameters described in section 5-2 using analog methods is given below. The mean value is obtained by integration (see fig. 5-11). To derive this over a fixed time interval $T = t_2 - t_1$ mode control of an integrator over this period, is required to give

$$\bar{x} = \frac{1}{T} \int_{t_1}^{t_2} x(t).dt. \qquad (5\text{-}60)$$

This assumes that a specific mean value for a record of length T seconds is needed. More often with an analog signal the continuous mean value is required

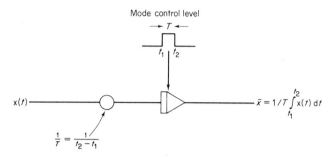

Fig. 5-11. Analog derivation of mean value.

and this mean value must take into account past values of the function $x(t)$. A running mean can be obtained using a low-pass filter if the time constant is large with respect to the longest period of interest in the signal. An alternative method, involving a weighting average, has been suggested by Otterman [6]. Using this method the values of earlier contributions to the continuous signal are weighted such that recent values count more than earlier ones. Thus the weighted mean value is given as

$$\bar{x}_m = \frac{\int_{-\infty}^{T} x(t).w(t).dt}{\int_{-\infty}^{T} w(t).dt}. \tag{5-61}$$

Many functions could be chosen for the weighting function $w(t)$, all of which would require to possess the property of becoming zero at an infinite negative time in the past. The one generally chosen due to its simplicity of simulation on the analog machine is an exponential and we can write

$$\bar{x}_m = \alpha. \exp(-\alpha T) \int_{-\infty}^{T} x(t) \exp(\alpha t).dt \tag{5-62}$$

This function has been defined as the Exponentially Mapped Past (E.M.P.) mean value. Implementation is shown in fig. 5-12 and will be recognised as a form of

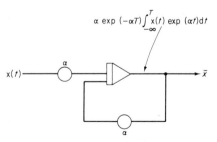

Fig. 5-12. Derivation of E.M.P. mean value.

216

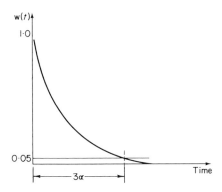

Fig. 5-13. E.M.P. weighting function W(t).

first-order lag. The effective time-constant of this circuit, that is the period of time that the past data is retained, can be shown to be $3/\alpha$. During this time interval there is a 95% decrease in the value of the weighting function (fig. 5-13).

The calculation of standard deviation S, directly from equation (5-9) requires the repeated calculation of the deviation from the mean. An improved method is to use the algebraically equivalent form for the variance:

$$\sigma_x^2 = x^2 - (\bar{x})^2 \tag{5-63}$$

from which the standard deviation for a given record length T, can be obtained from equation (5-1) and (5-3) as:

$$S = \sqrt{\left[\frac{1}{T}\int_0^T x^2.dt - \left(\frac{1}{T}\int_0^T x.dt\right)^2\right]}. \tag{5-64}$$

This is mechanised as shown in fig. 5-14. The mean-squared value and variance for the record can also be derived from this circuit. To use this for continuous

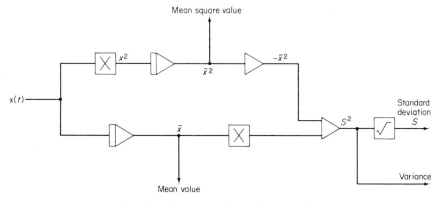

Fig. 5-14. Derivation of statistical quantities.

signals the integrator can be replaced by a first-order filter having a transfer function:

$$\frac{A}{1 + Tp}.$$

This will give the same result as the mechanisation of an E.M.P. standard deviation circuit from the variance equation developed using equations (5-6) and (5-62) namely:

$$\sigma_x^2 = \alpha \int_{-\infty}^{T} [x - \overline{x_m}]^2. \exp \alpha(t - T).dt. \tag{5-65}$$

This gives a weighted average form of the variance having the same conditions as regard the contributions of past signal values as the E.M.P. mean value expression. The implementation of equation (5-65), extended to include the

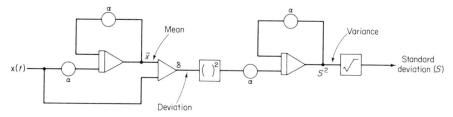

Fig. 5-15. E.M.P. derivation of statistical quantities.

square root operation necessary to extract the standard deviation, is given in fig. 5-15. Note that this requires less computing equipment than the circuit given in fig. 5-14, but does not permit the extraction of the mean-squared value directly.

5-4-2 PROBABILITY ANALYSIS

To obtain the probability, $P(A < x(t) \leqslant B)$ of a random signal, $x(t)$ using analog methods we can arrange for a Boolean function to be generated having the form:

$$b[x(t)] = \begin{matrix} 1 & \text{for } A < x(t) \leqslant B \\ 0 & \text{otherwise.} \end{matrix} \tag{5-66}$$

A simple arrangement using two comparators is given in fig. 5-16. The comparators provide a logical 1 output when, $x(t) < A(\text{or } B)$, and a logical 0 when $x(t) > A(\text{or } B)$. A truth table for the logic summation would be:

	B	A	$\overline{A}$	$C = \overline{A}.B$
$A < x(t) < B$	1	0	1	1
$A > x(t) < B$	1	1	0	0
$A < x(t) > B$	0	0	0	0

from which we see that a logical 1 is obtained for those periods of time that the signal lies between the defined limit A and B. The probability density function between a given level pair is defined as the fraction of the total time for which the signal lies between these levels divided by the amplitude window $(A-B)$ (equation 5-19). This gives a figure normalised to a discrete amplitude level mid-way between A and B. The probability function $C = \overline{A}B$ generated by fig.

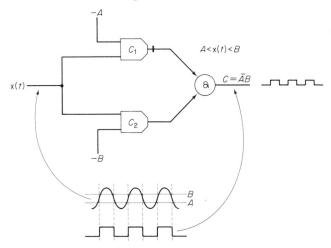

Fig. 5-16. Derivation of probability function.

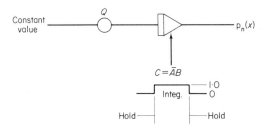

Fig. 5-17. Derivation of probability density function.

5-16 can be used to control the mode of an integrator to which a fixed voltage is applied, such that a linear integration is carried out when $C = 1$ and the integrator output held at its previously reached level when $C = 0$, (fig. 5-17). Potentiometer Q is scaled to average the accumulated time voltage by the product of record length and amplitude window.

To provide a full analysis over the amplitude excursion of the signal a number of controlled integrator circuits can be operated in parallel and the finally held values sampled or recorded at the end of the analysis period. Where many

219

discrete levels are required this can demand a large amount of equipment. If the signal can be stored initially and read out repeatedly to the analog computer, e.g. from a magnetic tape recording, then a single level interval detector can be used. In such a case the window centre value is changed for each repetition of the signal. An analysis configuration to achieve this from a loop or shuttle tape recorder is shown in fig. 5-18. The signal is applied to two comparators, C_1 and C_2 which compare the value of the signal at two levels x_j and $(x_j + \delta)$ where δ is the amplitude sampling interval and x_j corresponds to the level reached by

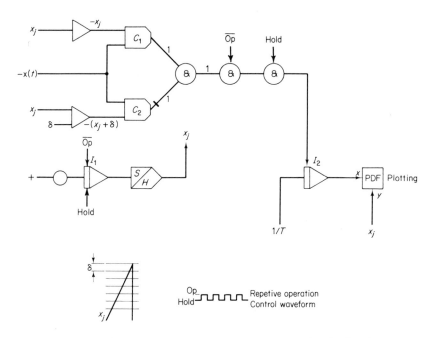

Fig. 5-18. An iterative method for deriving the probability density function.

the stair-step or ramp comparator input for the particular repetition of the input signal. The ramp generator is formed by integrator, I_1 which is controlled by repetitive operation/hold wave-form.

A logical 1 output is obtained from the sum of the comparator outputs if the signal lies between these two levels. This is used to operate the integrator, I_2 for a specified time. Some controlling logic is intervened to ensure that I_2 only integrates for the operating period. A steady input voltage proportional to the reciprocal of the record length is applied to I_2 so that the output represents a normalised value of the probability density function for the level pair. The complete density function is plotted as the controlling wave-form x_j, varies between its minimum and maximum value at a rate determined by the repetition of replay for the input signal.

Instead of using the comparator logic output to control the operation of an analog integrator it can be applied to a digital counter to record the number of times the signal lies between a given level pair. A series of counters will then be required, one for each difference level which can be read out at the end of the analysis period. Again the amount of equipment required will be large and the method inflexible.

An analog–digital quantising method is described in the next section which overcomes many of the disadvantages of these purely analog methods.

5-4-3 HYBRID METHODS

As will be noted in later chapters the application of hybrid methods to random signal processing offers considerable advantages, particularly in the reduction of computing time. The validity of hybrid methods rests on the amount of processing which can be carried out in a given time for a given cost. The availability of the original signal in analog form is not necessarily relevant to this. The improvements stem from the large amount of parallel operation that may be obtained in the analog section of the hardware. Digital operations are essentially sequential, permitting only one instruction to be obeyed at any given time. This slows down arithmetic operations (particularly integration), forming part of the analysis process. On the other hand a stored digital program controlling data transfer and carrying out updating and counting operations can contribute to the process in an area where purely analog methods perform badly.

A hybrid method of probability density analysis carried out in real-time from a continuous analog magnetic tape record is indicated by the flow diagram of fig. 5-19 using the hardware configuration shown in fig. 5-20. A maximum of 256 levels (8 bits) is adequate to define the probability density function and will introduce negligible quantising error (see chapter 4), and has therefore been adopted in this method. The analog signal will have been pre-processed to remove trends, limit the maximum excursion to the equivalent of 8 binary digits and to place a signal on a pedestal of slightly greater than half the maximum peak-to-peak excursion of the signal. This latter requirement is necessary to avoid the complication of sign bit allocation. The signal is applied to the converter and the first sample taken and converted to a digitised value designated by $x(n)$.

Following the completion of digitisation for this sample it is read into the digital computer as an address. Initially all 256 reserved locations in the core store are cleared and, at the address location, $x(n)$ a binary one is added to its content. The next sampled value, $x(n + 1)$, will generally define a different address to which a binary one is also added. The process continues with ones added at storage locations whose addresses correspond to discrete amplitude levels in the range 0 – 255. It is not necessary to use the first 256 storage locations in the core, however, since any 8 bits of the address word can be selected, either by program or simply by modifying the value of the analog pedestal level.

As shown in fig. 5-19, this continuous process of reading and updating values is halted in two ways namely:

(a) A storage location is reached when its value becomes equal to the word length of the store – an overflow state.

(b) The end of the analog record of interest is reached.

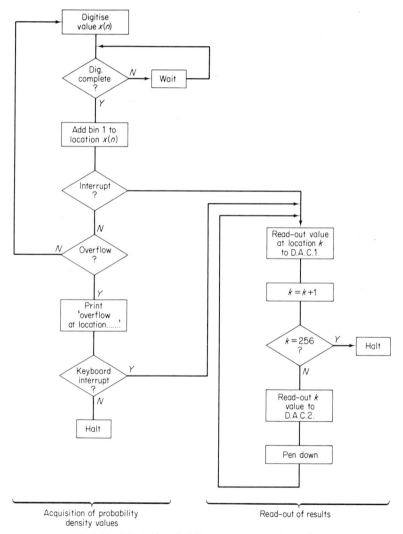

Acquisition of probability   Read–out of results
density values

Fig. 5-19. A hybrid probability density analysis routine.

Graphical printout in histogram form is obtained via a digital plotter using standard software routines or via digital–analog converters to a cheaper form of analog continuous X–Y plotter. This latter method also has the advantage of simpler program control than the digital case. The speed of data input can depend on the digital computer instruction loop time or the highest frequency, f contained in the analog signal (since the sampling interval must not be greater than $\frac{1}{2}f$). With many small process-control digital computers the instruction loop cycle time will set a limit on the speed of calculation. In order to indicate the amount of digitised signal that can be handled, using this method, we can assume an instruction loop time for read-in of values of 100 μsec and a typical highest frequency of interest of 5 kHz $(= f_s/2)$. It is reasonable to assume Gaussian

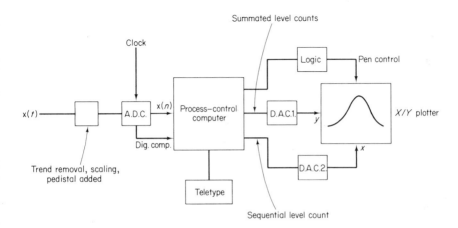

Fig. 5-20. Hardware configuration.

probability for the signal so that considering the level about zero (i.e. *peak-to-peak value*/256), then this will occur for 0·29% of its total time, and we can expect a maximum elapsed time before an overflow occurs of:

$$T = \frac{2^{16}}{2 \cdot 9 f_s} = \frac{12250}{f_s} \text{ sec.}$$

where f_s is the sampling rate in kHz. For the assumed value of $f_s \geqslant 10$ kHz then a record length of some 20 minutes can be accommodated. The limitation in speed-up factor is thus determined by the digital computer loop cycle time.

It can also be shown that the error calculated out to three standard deviations will only be of the order of 0·1% assuming a long record and an 8-bit quantisating level. The advantages of using hybrid methods for this type of calculation are therefore quite considerable.

223

5-4-4 DIGITAL PROGRAMS

Digital programming for the calculation of the statistical quantities described in this chapter is a fairly straightforward task. One unlooked for difficulty in the evaluation of probability density which can occur concerns the way in which a logical IF statement in Fortran may be implemented. This can result in the invariant selection of the lower quantising level for a value lying mid-way between two levels. The error generated is particularly noticeable with small numbers of values and a linearly varying signal (e.g. a sine wave). To avoid this a small amount of random noise (referred to as 'dither') can be added to the signal before this is tested by the program. This will ensure that these mid-way values are uniformly distributed to both adjacent quantisation levels.

5-5 LEVEL-CROSSING AND COUNTING

Other statistical quantities of interest are those associated with the number of times a signal crosses a given level per unit time, or the number of peaks counted in a given record. Evaluation of these quantities finds a use in frequency measurement, envelope detection, speech wave-form analysis, and fatigue studies of structures located in a random environment. The study of random noise also requires a knowledge of certain parameters which cannot be obtained from its behaviour in terms of autocorrelation or power spectral density measurement alone. It can be shown [7] that the inclusion of data derived from zero-mean crossing measurements can assist in determining a more complete definition of a

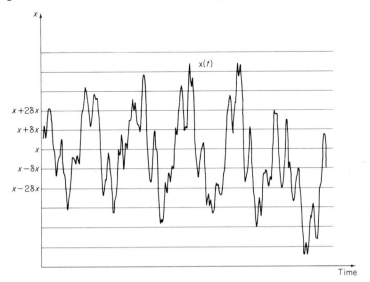

Fig. 5-21. Zero-crossing analysis.

particular random noise sample. It is possible to derive a value for level crossings per unit time from probability density considerations. Since this is useful in establishing definitions for the expected numbers of crossings this derivation will first be considered before discussing practical methods of estimation.

If we let $x(t)$ be a random signal (fig. 5-21), then the expected number of times that this crosses a given amplitude pair, x and $x + \delta x$ per unit time can be denoted by N. Since there are on average an equal number of crossings having positive and negative slopes, then the number of positive crossings will be $N_{+x} = N/2$. If we denote the time derivative of $x(t)$ as a velocity, $u(t)$, then we can write for the joint probability function of $x(t)$ and $u(t)$:

$$p(x, u) \, dx.du = \text{prob}(x < x(t) \leqslant x + \delta x \quad \text{and} \quad u < u(t) \leqslant u + \delta u) \qquad (5\text{-}67)$$

which gives the total time per unit period that $x(t)$ lies within an interval δx when its velocity is u (since δu is assumed to be very small). The number of crossings per unit time may be obtained by dividing this total time by the time taken by the signal to cross the interval, δx at a velocity u i.e.

$$N = \frac{p(x, u) \, dx.du}{T} \qquad (5\text{-}68)$$

where $\qquad\qquad T = dx/u,$

so that $\qquad\qquad N = u.p(x, u) \, du.$ $\qquad\qquad (5\text{-}69)$

Hence considering all possible velocities in the positive direction only we can write:

$$N_{+x} = \tfrac{1}{2} \int_{-\infty}^{\infty} u.p(x, u) \, du. \qquad (5\text{-}70)$$

The value for the zero level $x = 0$ is

$$N_0 = \int_{-\infty}^{\infty} u.p(0, u) \, du \qquad (5\text{-}71)$$

since both crossings are involved.

Rice [8] has also shown that a value of N for a Gaussian process can be obtained from the peak value of the autocorrelation function and its second derivative as

$$N_0 = \frac{1}{2\pi} \sqrt{\left\{ \frac{-R_x''(0)}{R_x(0)} \right\}} \qquad (5\text{-}72)$$

which can be obtained from the power spectral density function in the frequency domain. Thus the level-crossing and peak counting characteristics can be derived from probability considerations or from calculations in the time or frequency domains. It is generally, more convenient, however, to derive these quantities directly from measurements on the signal waveform, since the calculations of expressions, such as, equation (5-71) proves difficult in practise. Some of these direct methods will now be described.

A practical zero crossing detector using an operational amplifier is shown in fig. 5-22. The operation of this circuit makes use of the finite value of the threshold voltage required to render a diode conductive. If the signal lies within this threshold voltage the diodes exhibit their high impedance condition resulting in a large gain from the system. Voltages of either polarity outside this region will result one or other diode pair conducting and reducing the gain of the operational amplifier to a very small value. A hybrid method utilising this detector to determine interval times between given level crossings is shown in fig. 5-23. A digital number representing the given level is applied to a digital-to-

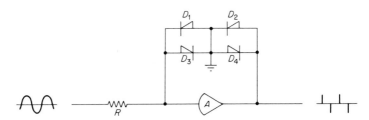

Fig. 5-22. Zero-crossing detector.

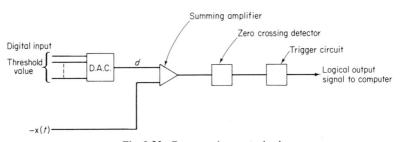

Fig. 5-23. Zero-crossing control values.

analog converter, which is summed with the signal having inverted sign. Zero crossing is detected and provides a logic level which is tested by the digital interrupt facility of the digital computer as part of an accounting loop. This method represents an increase in speed over purely digital means, since this loop need only contain testing and counter incrementation instructions, compared with the much larger instruction loop that would be required if transfer and data comparison are also included.

In addition to level-crossing detection, two forms of amplitude probability distribution are also used in the study of random signals. These are:

(*a*) Amplitude Probability Distribution, $p(A)$ which is defined as the percentage of time that a random signal lies above a series of n arbitrary levels taken sequentially.

226

(*b*) Peak Amplitude Probability Distribution, $p(P)$ which is a count of the number of peaks found above a series of n arbitrary levels, also considered sequentially to provide a distribution function.

These terms are illustrated by fig. 5-24 for a specific level x. The peak amplitude probability distribution permits the derivation of the envelope of a modulated signal since a new curve can be drawn from the value of the detected peaks.

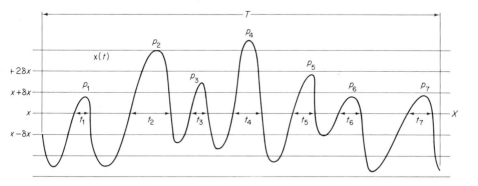

$$p(A) = \frac{100 \sum_{1}^{7} t}{T} = 43 \cdot 5 \% \quad (p(P) = 7)$$

Fig. 5-24. Measurement of amplitude and peak amplitude probability distribution.

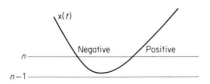

Fig. 5-25. Inflection between finite levels.

Any method that is devised for level detection will need to consider inflection between finite pairs of levels (fig. 5-25). If the signal is found to cross level n first with a negative derivative and again with a positive derivative without passing through the lower level $(n-1)$, then the second transfer must be counted as an additional peak for this level and all lower levels.

A simple analog method of measuring amplitude probability distribution is given in fig. 5-26. Detection of signal above a given level is detected by comparators C_1 to C_n, each of which produce a logical one when the signal lies above the compared level $x + x \, \delta x$. As long as the signal remains above this level gate, G_n is enabled and a series of clock pulses are provided to individual

227

counters, $N_1 - N_n$ at a rate several times higher than the highest frequency of interest. The final count divided by the total possible count in the period of measurement gives a measure of the amplitude probability for each discrete level

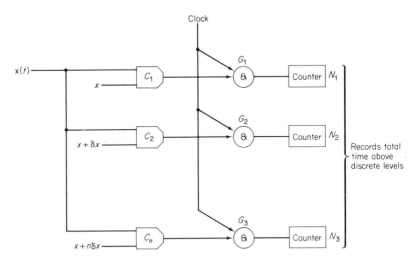

Fig. 5-26. Amplitude probability distribution measurement.

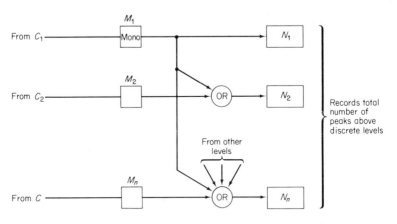

Fig. 5-27. Peak amplitude probability distribution measurement.

of x. Measurement of peak amplitude probability distribution requires in addition a method of detecting when the slope of the signal passing a given level is positive. This is equivalent to detecting when the logic output of the comparator shown in fig. 5-26 goes from 0 to 1, which can be obtained by a level-triggering mono-stable circuit preceding the counting circuit (fig. 5-27). The

228

count obtained will have to be translated down to all lower levels, as shown in the diagram, in order to provide the correct total count at all levels.

It is important during the calculation of peak amplitude probability distribution to set a minimum level below which the peaks will not be counted since in a practical case the noise level of a signal will result in numerous peaks, all occurring within the lowest band level. This is illustrated in fig. 5-28 which compares the peak probability distribution for a positive-going signal where in the first case the entire dynamic range of the signal is analysed, and in the second case the range of the signal is assumed to commence at a point just above the noise level.

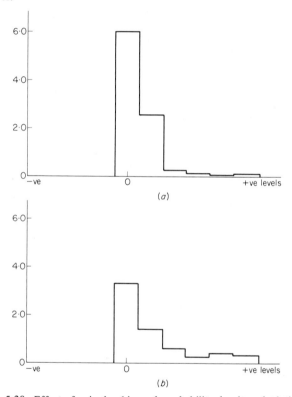

Fig. 5-28. Effect of noise level in peak probability density calculation.

BIBLIOGRAPHY

1. BENDAT, J. S. and PIERSOL, A. G. *Random Data: Analysis and Measurement Procedures.* John Wiley, New York, 1971.
2. GROUT, B. E. and KIMBER, G. A. A versatile electronic waveform analyser for determining density distribution of peaks, instantaneous values, periods and crossings. *R.A.E. Tech. Note,* ARM708 Nov. 1961.

3. RUBIN, A. I. Continuous data analysis with analog computers using statistical and regression techniques. *I.R.E. Trans.* (Elec. Comp.) EC12, Mar. 1963.
4. BRUBAKER, T. and KERN, G. A. Accurate amplitude distribution analyser combining analog and digital logic, *Rev. Sci. I.* **32**, Mar. 1961.
5. CAMERON, W. D. Hybrid computer techniques for determining probability distributions. Proc. Int. Symp. on analog and digital techniques in aeronautics, Liege, Belgium, 1963.
6. OTTERMAN, J. The properties and methods for computation of exponentially mapped past statistical variables. *I.R.E. Trans.* (Auto. Control). AC-5,7, 11-17, Jan. 1960.
7. BENDAT, J. S. Probability functions for random responses, prediction of peaks, fatigue damage and catastrophic failure. NASA, CR-33, Washington, April 1964.
8. RICE, S. O. Mathematical analysis of random noise. In *Selected Papers on Noise and Stochastic Processes,* Ed. N. Wax. Dover, New York, 1954.

Additional References

9. MILLER, I. and FREUND, J. E. *Probability and Statistics for Engineers.* Prentice Hall, 1965.
10. PARZEN, E. *Modern Probability Theory and its Applications,* John Wiley, New York, 1960.
11. DAVENPORT, W. B. and ROOT, W. I. *An Introduction to the Theory of Random Signals and Noise.* McGraw-Hill, New York, 1958.
12. KORN, G. A. *Random Process Simulation and Measurements.* McGraw-Hill, New York, 1966.
13. KAVANAGH, R. J. Automatic measurement and plotting of probability histograms using an analog computer with digital logic. *Simulation* **13**, 6, Dec. 1969.
14. ZEITLIN, R. A. *et al.* Combined analog-digital processing of cardiograms. *Ann. N.Y., Acad. Sci.* **115**, (2) July 1964.
15. BENDAT, J. S. Interpretation and application of statistical analysis for random physical phenomena. *I.R.E. Trans.* (Bio-med. Elec.). BME-9, 31-43, Jan. 1962.
16. COX, D. R. and LEWIS, P. A. W. *The Statistical Analysis of Series of Events.* Methuen, London, 1966.
17. KRONMAL, R. and TARTER, M. The estimation of probability densities and cumulatives by Fourier series methods. J.S.A., **63**, 925-52,.1965.
18. CRANDELL, S. H. Zero-crossings, peaks and other statistical measures of random responses. *Proc. Acoust. Soc. Amer.* Meeting, Seattle, Washington, 1962.
19. McFADDEN, J. A. The axis-crossing intervals of random functions. *I.R.E. Trans.* (Inf. Theory) IT-2, 146-50, Dec. 1956, and IT-4, 14-24, Mar. 1958.

Chapter 6

FOURIER SERIES AND ANALYSIS

6-1 INTRODUCTION

This chapter is concerned with the analysis of functions in the frequency domain, considered as continuous or discrete series. Where continuous functions are involved the information analysed may be termed the Signal wave-form and will generally be expressed as a function of time, $x(t)$, a typical example being the continuous signal obtained from a strain-gauge or pressure transducer. A discrete series, for example that obtained from a revolution counter, or the digitised form of a continuous signal, will be termed a Data series and expressed as x_i.

The analytical methods introduced by Fourier occupy a significant place in the solution of problems in physics and engineering. Originally derived for the analysis of problems in heat engineering, the properties of Fourier analysis are now widely applied in all branches of physics. These properties provide a method for determining the frequency components of a time-varying function. For example, a Fourier analysis of a signal derived from a time-varying displacement can provide the amplitudes of the frequencies present in the original process. The lowest frequency in this type of analysis is known as the fundamental frequency and the higher frequencies, which are integer multiples of the fundamental frequency, are known as the higher-order harmonics.

Where the function is periodic, that is, it repeats itself at regular intervals, then a Fourier series is used to obtain amplitudes of the fundamental and harmonic frequencies present. For this reason Fourier series analysis is often termed Harmonic Analysis.

Fourier analysis can also be extended to non-periodic functions. Here the period of analysis (but not the signal being analysed) is assumed to be infinitely long. This gives rise to the use of Fourier integral transformation where all frequencies are represented instead of just those which are integer products of the fundamental frequency.

These types of analysis can also be used relating to functions which are not time-varying. For example, the variation of thickness of rolled steel sheet as a

231

function of one or more of its spatial dimensions can be analysed to determine the characteristic periodicities present and so assist in location of the source of the thickness irregularities.

Finally, the methods of Fourier analysis can be applied to the analysis of discrete data series which results in a form of finite Fourier analysis which has important applications in digital computer calculations.

6-2 SERIAL INFORMATION

Information obtained as a result of a physical experiment or measurement often takes the form of a single or multiple time-dependent variable. Other bases are possible and will be discussed at a later stage. A common representation is a graphical one in which instantaneous values of a variable, (x), are plotted against uniform time increments. Other methods are valid if it is possible to reconstruct the original function from them. As an example, we could take the

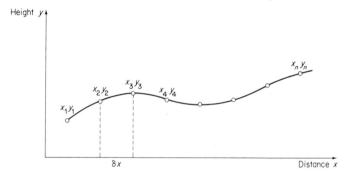

Fig. 6-1. Aircraft position in flight.

position of an aircraft in flight expressed as Cartesian coordinates relating to the height and distance along the flight path (fig. 6-1). The function could be represented (with some loss in accuracy) using the serial form:

$$f(xy) = x_1 y_1, x_2 y_2, x_3 y_3, \ldots, x_n y_n. \tag{6-1}$$

However, if we consider a uniform speed for the aircraft over the period of measurement T, then;

$$x = f(t)$$

and we can write, for the equal intervals of time between observed positions:

$$x_n - x_{n-1} = \delta x \equiv \delta t \tag{6-2}$$

so that elapsed time $T = n\delta t$, where n is the number of observed positions taken at equal intervals of time, δt, during the measurement period. Hence the height of the aircraft at time t can be expressed by the time series:

$$f_y(t) = f_y(n\delta t) = y_1 + (y_2 - y_1) + (y_3 - y_2), \ldots, (y_n - y_{n-1}). \tag{6-3}$$

232

We could equally well represent the observed position of the aircraft in terms of polar coordinates (fig. 6-2) taken from the origin $x_0 y_0$, i.e.

$$f(r\theta) = r_1\theta_1, r_2\theta_2, r_3\theta_3, \ldots, r_n\theta_n \qquad (6\text{-}4)$$

where

$$r_n = \frac{y_n}{\sin\theta_n} \text{ and } \theta_n = \arctan\left[\frac{y_n}{n\delta x}\right] \qquad (6\text{-}5)$$

so that, again assuming a uniform aircraft speed we can write

$$fy(t) = r_1 \sin\theta_1 + (r_2 \sin\theta_2 - r_1 \sin\theta_1)$$
$$+, \ldots, (r_n \sin\theta_n - r_{n-1} \sin\theta_{n-1}). \qquad (6\text{-}6)$$

this again expresses the height of the aircraft at any one of the discrete observational times but with entirely different data.

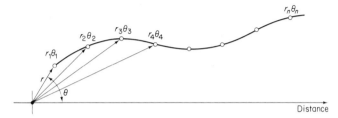

Fig. 6-2. Alternative coordinates for flight path.

Either series could be used to represent the flight path function of the aircraft in the two dimensions and the elements of one could be obtained, or reconstructed, from the elements of the other.

This is true generally for mathematical methods where the alternative form is known as a Transform (or Inverse Transform) of the other. We shall be considering two particular transforms in this chapter, a frequency series which is treated as a transform of a time series, and a time series which is considered as an inverse transform of a frequency series.

6-3 TIME SERIES

Functions of time can be expressed as a discrete series or as a continuous function. Mathematical methods of representing a given function are numerous and not all are amenable to calculation in order to extract meaningful information. A useful example, applicable over a wide range of variables found in the physical sciences is the representation of a function of time by means of a polynomial:

$$x(t) = a_0 + a_1 t + a_2 t^2 + a_3 t^3 +, \ldots, + a_n t^n \qquad (6\text{-}7)$$

233

for n discrete measurements in time. Coefficients a_0, a_1, a_2, etc., themselves give useful information about the function and a plot of their magnitude as a function of n will contain exactly the same information expressed in a different form as the original time series, $x(t)$.

6-4   FOURIER SERIES

Fourier series analysis applies to periodic functions only. A physical phenomena often produces a form of periodic data which may be characterised by the series:

$$x(t) = C_0 + C_1 \sin(\omega_0 t + \theta_1) + C_2 \sin(2\omega_0 t + \theta_2)$$

$$+, \ldots, + C_n \sin(n\omega_0 t + \theta_n) \qquad (6\text{-}8)$$

which may be considered as the sum of the fundamental frequency component, or first harmonic, $C_1 \sin(\omega_0 t + \theta_1)$, having a fundamental frequency of $f = \omega_0/2\pi$ together with its harmonics and a constant term C_0. The coefficients C_1, $\ldots$, C_n represent the peak amplitude excursion of the fundamental and harmonic components of the series. The angles $\theta_1, \ldots, \theta_n$ represent the phase relationship between the initial vector value for the fundamental and those of the harmonics at time t.

A feature of such a data series is that the value of the function, $x(t)$ exactly repeats itself at regular intervals, such that:

$$x(t) = x(t \pm nT) \qquad (6\text{-}9)$$
$$n = 0, 1, 2, 3, \ldots, \infty.$$

The interval required for one complete variation is known as the period T and the number of cycles per unit time is the fundamental frequency:

$$f_0 = \frac{\omega_0}{2\pi} = \frac{1}{T}. \qquad (6\text{-}10)$$

A more compact form for the general series given by equation (6-8) is obtained by substituting the relationship:

$$\sin(A + B) = \sin A \cdot \cos B + \cos A \cdot \sin B$$

so that we can write:

$$x(t) = C_0 + C_1 \sin\theta, \cos\omega_0 t + C_1 \cos\theta, \sin\omega_0 t$$

$$+, \ldots, + C_n \sin\theta_n \cdot \cos n\omega_0 t + C_n \cos\theta_n \cdot \sin n\omega_0 t \qquad (6\text{-}11)$$

but since $C_n \sin\theta_n$ = a constant = a_k and $C_n \cos\theta_n$ = a constant = b_k by letting $C_0 = a_0/2$, we can write:

$$x(t) = a_0/2 + a_1 \cos\omega_0 t + b_1 \sin\omega_0 t$$

$$+, \ldots, + a_n \cos n\omega_0 t + b_n \sin n\omega_0 t \qquad (6\text{-}12)$$

234

or

$$x(t) = \frac{a_0}{2} + \sum_{k=1}^{n} (a_k \cos k\omega_0 t + b_k \sin k\omega_0 t). \tag{6-13}$$

Equation (6-13) is known as the Fourier series for the function $x(t)$.

It is instructive to consider the synthesis of a time history, $x(t)$, using equation (6-13), from the sum of its constant term $a_0/2$ and a limited number of harmonic terms. If we consider the case where, $a_1 = a_2 = a_3 =, \ldots, a_n = 0, b_1 = 2/\pi, b_3 = \frac{2}{3}\pi, b_5 = \frac{2}{5}\pi$, and $b_2 = b_4 = b_6 = 0$, then the constituent cosinusoidal waveforms at their appropriate amplitude levels will be as shown in fig. 6-3.

These can be summed at unique times, t_1, t_2, etc., to form the continuous function shown in fig. 6-4, which represents the synthesis of a square wave from a limited number of harmonic terms. The converse operation of analysis will now be considered, in which amplitude coefficients for the harmonic terms will be obtained.

The instantaneous amplitudes for a_0, a_k and b_k may be derived theoretically by the use of certain integrals. It will be obvious that the area under a sinusoidal or cosinusoidal wave-form for a complete period $T = 1/f_0$ must be zero, i.e.

$$\frac{1}{T} \int_0^T \sin n\omega_0 t.dt = \frac{1}{T} \int_0^T \cos n\omega_0 t.dt = 0 \tag{6-14}$$

where n is an integer.

Also writing

$$\cos m\omega_0 t . \cos n\omega_0 t = \tfrac{1}{2} (\cos(m+n)\omega_0 t + \cos(m-n)\omega_0 t) \tag{6-15}$$

we see that, if m and n are unequal integers, then

$$\frac{1}{T} \int_0^T \cos m\omega_0 t . \cos n\omega_0 t.dt = 0. \tag{6-16}$$

Similarly

$$\frac{1}{T} \int_0^T \sin m\omega_0 t . \sin n\omega_0 t.dt = 0 \tag{6-17}$$

$$\frac{1}{T} \int_0^T \sin m\omega_0 t . \cos n\omega_0 t.dt = 0. \tag{6-18}$$

The only integrals of this type which have a finite value are:

$$\frac{1}{T} \int_0^T \sin^2 n\omega_0 t.dt = \frac{1}{T} \int_0^T \cos^2 n\omega_0 t.dt = \tfrac{1}{2}. \tag{6-19}$$

235

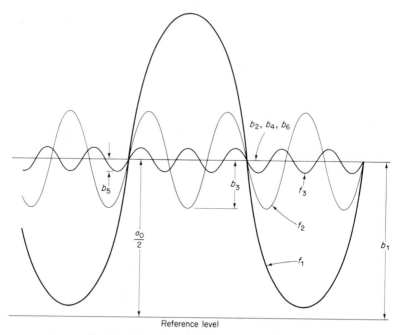

Fig. 6-3. Three harmonically related frequencies.

Equations (6-14) to (6-19) define a peculiar property of sinusoids known as Orthogonality, and can be summarised by stating that a finite value is obtained for the weighted average of the product of sine and sine or cosine and cosine if their frequencies and phase shifts are identical, and a zero result if they are not. Also, the weighted average of the product of sine and cosine will be zero irrespective of their frequencies.

This result is important and is the key to many analysis techniques used in signal processing.

It therefore follows from equation (6-12) that:

$$\int_0^T x(t).dt = \int_0^T \frac{a_0}{2}.dt = \frac{a_0}{2}.T$$

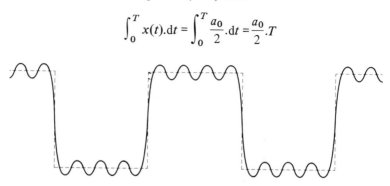

Fig. 6-4. Synthesis of a wave-form from the first five terms of a Fourier series.

236

since the integrals of all terms, other than $a_0/2$, will be zero (equation (6-14)). Therefore, the constant term:

$$\frac{a_0}{2} = \frac{1}{T} \int_0^T x(t).dt. \tag{6-20}$$

Also multiplying $x(t)$ by $\sin k \, \omega_0 t$, we have

$$\int_0^T x(t) \sin k\omega_0 t.dt = \int_0^T b_k \sin k\omega_0 t.dt = b_k T/2$$

since all terms other than the $\sin^2$ term will be zero (equations (6-16) to (6-19)). Therefore:

$$b_k = \frac{2}{T} \int_0^T x(t) \sin k\omega_0 t.dt. \tag{6-21}$$

Similarly

$$a_k = \frac{2}{T} \int_0^T x(t) \cos k\omega_0 t.dt. \tag{6-22}$$

These terms, a_k and b_k represent the amplitudes of the harmonics present in the original function, $x(t)$, and are known as the Fourier Coefficients.

6-5   FOURIER SPECTRUM

The values of the Fourier coefficients a_0, a_k, b_k, provide a further means of defining the function $x(t)$ as a series:

$$x(t) = f_a(a_0, a_1, a_2, \dots, a_n) + f_b(b_1, b_2, b_3, \dots, b_n) \tag{6-23}$$

where f_a and f_b denote a function of coefficients a and b.

A plot of these coefficients, representing the magnitude of the harmonic components of $x(t)$, gives information about the frequency content of the function and is known as the Fourier Spectrum. This is shown in fig. 6-5 as two separate relationships for a_k and b_k plotted against the common value of coefficients $\pm k$. It should be noted that whereas the cosine coefficients a_k exhibit an even symmetry about $k = 0$ the sine coefficients, b_k, show an odd symmetry.

Since the a_k, b_k coefficients have an orthogonal relationship a single vector sum can be derived to represent the modulus and phase of each of the harmonic amplitudes, viz.

$$C_k = \sqrt{(a_k^2 + b_k^2)} \tag{6-24}$$

and

$$\theta_k = \arctan\left[\frac{b_k}{a_k}\right]. \qquad (6\text{-}25)$$

When the Fourier coefficients are measured with a harmonic wave-analyser a_k and b_k are not indicated separately and the modulus and phase factor are measured in the form of equations (6-24) and (6-25). It is these latter two terms that are generally presented as the Fourier spectrum of a periodic function.

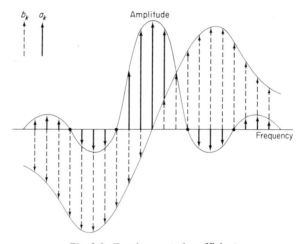

Fig. 6-5. Fourier spectral coefficients.

6-5-1 FOURIER ANALYSIS OF A RECTANGULAR FUNCTION

As an example of the derivation of the Fourier series and Fourier spectrum, let us consider a single repetition of the rectangular wave-form shown in fig. 6-6. For reasons given later, this must be made repeatable over a period $0 \to T$.

From equation (6-21) the sine coefficients are:

$$b_k = \frac{2}{T}\int_0^T x(t).\sin k\omega_0 t.dt.$$

We may note that $x(t)$ is zero outside the range $0 < t < T/2$. Also since $x(t)$ has a constant amplitude, h, within this range, we can write:

$$b_k = \frac{2}{T}\int_0^{T/2} h.\sin k\omega_0 t.dt.$$

Therefore,

$$b_k = \frac{2h}{T}\left[\frac{-\cos k\omega_0 t}{k\omega_0}\right]_0^{T/2}$$

238

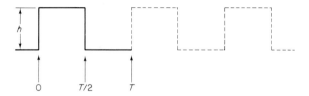

Fig. 6-6. A rectangular wave-form.

and since $\omega_0 = 2\pi/T$,

$$b_k = \frac{h}{\pi k}\left[1 - \cos k\pi\right].$$

Thus, for unit amplitude of the rectangular wave-form, ($h = 1$),

$$b_1 = 2/\pi; b_2 = 0; b_3 = 2/3\pi; b_4 = 0; b_5 = 2/5\pi; \text{etc.}$$

The cosine coefficients, a_k can readily be shown to be zero from equation (6-22), thus,

$$a_k = \frac{2h}{T}\int_0^{T/2}\cos k\omega_0 t.dt = \frac{2h}{T}\left[\frac{\sin k\omega_0 t}{k\omega_0}\right]_0^{T/2}$$

which is zero for all integer values of k.

The mean value coefficient $a_0/2$ is obtained from equation (6-20), as

$$\frac{a_0}{2} = \frac{1}{T}\int_0^{T/2}h.dt = \frac{h}{2}.$$

Hence the Fourier series for the rectangular wave-form, shown in fig. 6-6, is given as:

$$x(t) = \frac{h}{2} + \frac{2h}{\pi}\sin \omega_0 t + \frac{2h}{3\pi}\sin 3\omega_0 t + \frac{2h}{5\pi}\sin 5\omega_0 t +, \ldots, \frac{2h}{n\pi}\sin n\omega_0 t, \text{etc.}$$

$$(6-26)$$

To represent the rectangular wave-form fully, an infinite number of terms is needed. The result of plotting this function for the first five terms is shown in fig. 6-4. This indicates that $x(t)$ is a periodic function of time with a basic periodicity of T seconds equivalent to a fundamental repetition rate of $f = 1/T$ per second.

Thus, the Fourier series expansion over the integration interval T is identical to the expansion of the continuous function $x(t)$ taken over any period of time. The process of Fourier series transformation, therefore, implicitly assumes that the transformed time function repeats itself indefinitely outside the period, T, of

239

the observed function $x(t)$ at a rate equal to the fundamental frequency $f_0 = 1/T$. In other words, continuous Fourier analysis can only be carried out if the function is periodic or assumed to be periodic.

6-6   COMPLEX REPRESENTATION

Complex representation of Fourier series and the Fourier integral simplifies the notation and allows later development of the Fourier transform.

Referring to equation (6-13) we can expand the term:

$$(a_k \cos k\omega_0 t + b_k \sin k\omega_0 t)$$

by the use of

$$\cos k\omega_0 t = \tfrac{1}{2} [\exp (jk\omega_0 t) + \exp (-jk\omega_0 t)], \tag{6-27}$$

and

$$j \sin k\omega_0 t = \tfrac{1}{2} [\exp (jk\omega_0 t) - \exp (-jk\omega_0 t)], \tag{6-28}$$

thus:

$$(a_k \cos k\omega_0 t + b_k \sin k\omega_0 t) = \frac{a_k}{2} [\exp (jk\omega_0 t) + \exp (-jk\omega_0 t)]$$

$$+ \frac{b_k}{2j} [\exp (jk\omega_0 t) - \exp (-jk\omega_0 t)]$$

$$= A_k . \exp (jk\omega_0 t) + B_k \exp (-jk\omega_0 t) \tag{6-29}$$

where

$$A_k = \frac{a_k - jb_k}{2} \quad \text{and} \quad B_k = \frac{a_k + jb_k}{2}. \tag{6-30}$$

A_k and B_k represent the complex conjugate amplitude coefficients of the Fourier series and will now be expanded by the use of equations (6-21) and (6-22) to find the complex Fourier series. For the purposes of clarity, the variable of integration in these equations will be changed to p since we will need to retain t for time in the exponential relationship. This is permissable since the coefficients a_k and b_k are numbers represented by definite integrals. Therefore:

$$A_k = \frac{1}{T} \int_0^P x(p)[\cos k\omega_0 p - j \sin k\omega_0 p] \, dp$$

$$= \frac{1}{T} \int_0^P x(p) \exp (-jk\omega_0 p) \, dp \tag{6-31}$$

and similarly:

$$B_k = \frac{1}{T} \int_0^P x(p) \exp (jk\omega_0 p) \, dp \qquad (6\text{-}32)$$

Substituting equations (6-29), (6-31), (6-32) and (6-20) in equation (6-13) we obtain an expression for the complex Fourier series,

$$x(t) = \frac{1}{T} \int_0^P x(p) \, dp + \sum_{k=1}^{n} \left[\frac{1}{T} \int_0^P x(p) \exp (-jk\omega_0 p).dp \right] \exp (jk\omega_0 t)$$

$$+ \sum_{k=1}^{n} \left[\frac{1}{T} \int_0^P x(p) \exp (jk\omega_0 p).dp \right] \exp (-jk\omega_0 t). \qquad (6\text{-}33)$$

A simplification is possible since the expression represents a succession of terms: $\exp (jk\omega_0 p) \cdot \exp (-jk\omega_0 t)$ and $\exp (-jk\omega_0 p) \cdot \exp (jk\omega_0 t)$ both extending from $k = 1$ to $k = n$. The sign of k can be reversed in the second set of terms by summing from $k = -1$ to $k = -n$ so that the joint summation can be represented as a succession of terms: $\exp (jk\omega_0 p) \cdot \exp (-jk\omega_0 t)$ with k extending from $k = -n$ to $k = +n$. This now includes the term $k = 0$, when the summated terms reduce to:

$$\frac{1}{T} \int_0^P x(p).dp$$

which is equal to the constant term in equation (6-33) and therefore can be omitted from the resultant expression. Hence equation (6-33) can be written as:

$$x(t) = \sum_{k=-n}^{k=n} \frac{1}{T} \left[\int_0^P x(p) \exp (-jk\omega_0 p).dp \right] \exp (jk\omega_0 t). \qquad (6\text{-}34)$$

This represents the complex Fourier series for a time history, $x(t)$. We may note for later reference that the expression within the square brackets represents the modulus of the Fourier coefficients of the series.

6-7 THE FOURIER INTEGRAL TRANSFORM

The Fourier series transformation, described earlier as a means of analysis in the frequency domain, has two major limitations which prevent its application to many time series of practical interest. In the first case it assumes that the time function is of infinite duration, whereas practical data is often a transient having finite duration, and secondly the assumption is made that the data is periodic over an unlimited extent, whereas practical data is usually non-periodic, as well as being limited in duration.

It is possible in may cases to represent non-periodic data by means of a Fourier Integral Transform which will be defined in this section. This will be referred to as the Fourier Transform and with its use we may perform harmonic analysis, and obtain power spectra, correlation functions, filtered data and coherence functions. Consequently the Fourier transform plays a very important part in the analysis of vibration, shock and other random data.

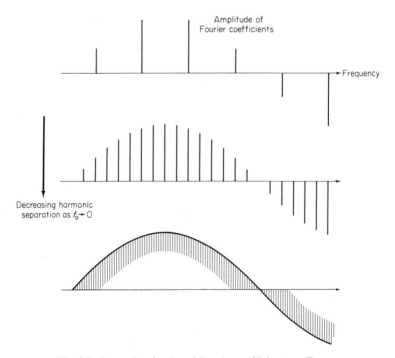

Fig. 6-7. Increasing density of Fourier coefficients, as $T \to \infty$.

Intuitively we realise that it will be necessary to extend the range of integration in equation (6-34) to infinity in order to include the complete transient time history, since we can no longer assume that the data will repeat itself outside the range of integration, previously limited to one fundamental period. Thus, we let $k, T \to \infty$ so that the fundamental frequency will tend towards zero, i.e. $f \to \delta f \to 0$. The Fourier coefficients will become continuous functions of frequency as the separation between the harmonics tends to zero and $k\omega_0 \to \omega$. This is illustrated in fig. 6-7 which shows repetitions of the same waveform expressed in terms of its Fourier coefficients with a decreasing order of time separation between them.

242

Thus equation (6-34) becomes:

$$x(t) = \int_{-\infty}^{\infty} df \left[\int_{-\infty}^{\infty} x(p) \exp(-j\omega p).dp \right] \exp(j\omega t)$$

$$= \int_{-\infty}^{\infty} \left[\int_{-\infty}^{\infty} x(p) \exp(-j\omega p).dp \right] \exp(j\omega t).df \qquad (6\text{-}35)$$

which is a complex form of the Fourier integral.

The term in brackets represents the amplitudes corresponding to the complex Fourier coefficients of a continuous time series. This is a frequency function and we can write:

$$X(f) = \int_{-\infty}^{\infty} x(p) \exp(-j\omega p).dp.$$

The variables in this definite integral can now be changed back to their original time form, viz.

$$X(f) = \int_{-\infty}^{\infty} x(t) \exp(-j\omega t).dt \qquad (6\text{-}36)$$

to give the Complex Fourier Transform, also known as the Complex Spectrum or, simply Fourier Transform of the time series, $x(t)$. The absolute value of this function yields the frequency amplitude content and the argument gives the phase content.

Equation (6-35) now becomes

$$x(t) = \int_{-\infty}^{\infty} X(f) \exp(j\omega t).df \qquad (6\text{-}37)$$

which is known as the Inverse Complex Fourier Transform, or Inverse Transform of the frequency series, $X(f)$. Equation (6-36) permits a simplification of the equation for the complex Fourier series given in (6-34), so that

$$x(t) = \sum_{-\infty}^{\infty} X(f) \exp(j\omega t). \qquad (6\text{-}38)$$

These forms of the Fourier transform find wide acceptance for practical applications [1] but for analytical purposes it is convenient to use the variable ω in place of f when it is necessary to add a scaling factor, $1/2\pi$ to the complex transform.

If the function, $x(t)$ is an even function, so that it is symmetric about the $t = 0$ axis, then the Fourier transform becomes:

$$X_c(f) = \int_{-\infty}^{\infty} x(t) \cos \omega t.dt \qquad (6\text{-}39)$$

which is known as the Fourier Cosine Transform, or Cospec (f).

243

For odd functions:

$$X_s(f) = \int_{-\infty}^{\infty} x(t) \sin \omega t.dt \qquad (6\text{-}40)$$

which is known as the Fourier Sine Transform, or Sinspec (f).

Equations (6-39) and (6-40) may be combined to obtain the Amplitude Spectrum, $A(f)$ and Phase Spectrum, $\theta(f)$ for a complex value $x(t)$ expressed as a continuous function rather than the discrete values implied by equations (6-24) and (6-25), viz.

$$A(f) = [(\text{cospec})^2 + (\text{sinspec})^2]^{\frac{1}{2}} \qquad (6\text{-}41)$$

and

$$\theta(f) = \arctan\left[\frac{\text{sinspec}}{\text{cospec}}\right]. \qquad (6\text{-}42)$$

6-8   ANALOG METHODS

The classical method of carrying out harmonic analysis is by the use of analog methods. Continuous integral transformation may be implemented easily on the analog computer and special-purpose wave-analysers are available which employ analog methods. Since the signals for analysis generally derive from transducers having a dynamic range well below the maximum attainable with analog computers there is no fundamental difficulty in obtaining the spectral accuracy required. The major problems are those associated with the analysis of very low or very high signal frequencies and with finite length signals.

A common method for the analysis of periodic signals is derived by consideration of the response of a linear undamped oscillator to a forcing function:

$$f(t) = p^2 x + \omega^2 x \qquad (6\text{-}43)$$

where p = an operator, d/dt. To obtain a solution to this equation we must first multiply equation (6-43) by $\sin \omega t$,

$$f(t) \sin \omega t = p^2 x \sin \omega t + \omega^2 x \sin \omega t. \qquad (6\text{-}44)$$

Integrating over the range 0 to t:

$$\int_0^t f(\tau) \sin \omega \tau. \, d\tau = px \sin \omega t - \omega x \cos \omega t. \qquad (6\text{-}45)$$

The variable of integration has been changed to τ to distinguish this from the range t.

244

Similarly, multiplying equation (6-43) by cos ωt, we obtain:

$$\int_0^t f(\tau) \cos \omega\tau.d\tau = px \cos \omega t + \omega x \sin \omega t. \tag{6-46}$$

Now multiplying equation (6-45) by cos ωt and equation (6-46) by sin ωt and subtracting gives:

$$- \omega x = \int_0^t f(\tau)[\sin \omega\tau. \cos \omega t - \cos \omega\tau. \sin \omega t] \, d\tau,$$

from which

$$x(t) = \frac{1}{\omega} \int_0^t f(\tau) \sin \omega(t - \tau).d\tau \tag{6-47}$$

which is the solution to the forcing function of equation (6-43). To mechanise this on the analog computer, we let

$$\omega = 2\pi k/T \tag{6-48}$$

where T is the fundamental period and k is the harmonic number. We can assume that $f(t)$ is a function of k/T, i.e.

$$f(t) = \frac{2\pi k}{T} y(t). \tag{6-49}$$

Substituting in equation (6-43)

$$\frac{2\pi k}{T} y(t) = p^2 x + \left(\frac{2\pi k}{T} \right)^2 x \tag{6-50}$$

and from equation (6-47)

$$x(t) = \int_0^t y(\tau) \sin \frac{2\pi k}{T} (t - \tau).d\tau, \tag{6-51}$$

from which the derivative is:

$$px(t) = \frac{2\pi k}{T} \int_0^t y(\tau) \cos \frac{2\pi k}{T} (t - \tau).d\tau. \tag{6-52}$$

If the upper limit of integration is set to the period of the forcing function T, then equations (6-51) and (6-52) reduce to

$$x(T) = \int_0^T y(\tau) \sin \left(\frac{2\pi k\tau}{T} \right).d\tau, \tag{6-53}$$

245

and

$$px(T) = \frac{2\pi k}{T} \int_0^T y(\tau) \cos\left(\frac{2\pi k\tau}{T}\right).d\tau. \tag{6-54}$$

These may be compared with the Fourier coefficient equations (6-21) and (6-22) from which we see that the values for a_k and b_k obtained there differ from $x(t)$ and $px(t)$ only by fixed scaling coefficients.

Re-arranging equation (6-50) to place the highest derivative on the left-hand side, we have:

$$p^2 x\left(\frac{T}{2\pi k}\right) = y(t) - \frac{2\pi kx}{T}. \tag{6-55}$$

This is mechanised by the analog solution shown in fig. 6-8 where equation (6-53) is represented at the output of the first integrator and equation (6-54) is

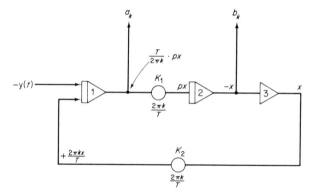

Fig. 6-8. Analog Fourier analysis.

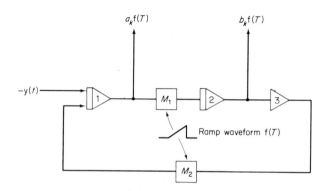

Fig. 6-9. Fourier analysis with automatic frequency updating.

represented at the output of the second integrator. For each setting of potentiometers K_1 and K_2, a single pair of a_k, b_k coefficient values can be obtained. A continuous spectrum can be obtained by replacing K_1 and K_2 by multipliers and controlling these by a ramp wave-form, (fig. 6-9). The duration of the ramp wave-form is made long compared with the fundamental period of the signal so that the frequency of analysis is caused to vary linearly with time, thus producing a range of coefficient values at the output of the two integrators.

6-8-1 FINITE BANDWIDTH METHOD

The method described above exhibits an infinitely narrow bandwidth and is thus limited to periodic signals, where discrete frequencies, consisting of the fundamental and harmonics of the function are searched for.

It is possible to introduce a bandwidth term into the forcing function method by replacing the integrators in figs. 6-8 and 6-9 by a particular weighted form known as the E.M.P. integrator. A general discussion of E.M.P. statistical variables was given in chapter 5, where it was seen that these variables are quantities derived from a set of observations in such a way that the recent observations contribute more strongly than the values taken in the past. The weighting function $w(t)$, chosen should therefore decrease with time, such that $w(t) \to 0$ as $t \to -\infty$. Many functions have this property and it is convenient in the case of analog computation to use an exponential function since the implementation of this will require very few computing elements. When applied to the Fourier transform generally the definition of equation (6-36) is expanded to include the weighting function, $w(t) = \exp\left[-\alpha(t - \tau)\right]$, i.e.

$$X(f) = \alpha \int_{-\infty}^{t} x(\tau).\exp\left[-\alpha(t - \tau)\right].\exp\left(-j\omega\tau\right).d\tau. \qquad (6\text{-}56)$$

This relates to an arbitrary moment, $-t$ in the past. Extension of the upper limit of integration to $+\infty$ is permissable if $x(\tau)$ is assumed to vanish for $\tau > t$, i.e. for future relative times of interest. Thus, we can write for equation (6-56):

$$X(f) = \alpha \int_{-\infty}^{+\infty} x(\tau).\exp\left[-\alpha(t - \tau)\right] u(t - \tau).\exp\left(-j\omega\tau\right).d\tau \qquad (6\text{-}57)$$

where $u(t - \tau)$ is a unit step function:

$$u(t - \tau) = 1 \text{ for } (t - \tau) \geqslant 0, u(t - \tau) = 0 \text{ for } (t - \tau) < 0.$$

It will be seen later that equation (6-57) is equivalent to a convolution operation on the Fourier transform and serves to give a bandwidth proportional to the value of the constant α, to the process of determination of the Fourier coefficients.

When applied to the forcing function method of harmonic analysis this involves merely the addition of feedback potentiometers K_3, K_4 to the

247

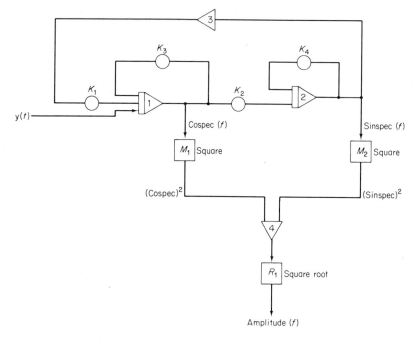

Fig. 6-10. Finite bandwidth Fourier analysis.

integrators shown in the oscillatory loop of fig. 6-10. These potentiometers are set to a value α determined by the bandwidth requirements. The output of integrator 1 in fig. 6-10 is the E.M.P. sinspec function:

$$x(t) = \int_{-\infty}^{t} y(\tau) . \exp\left[-\alpha(t - \tau)\right] . \sin \omega\tau . d\tau, \qquad (6\text{-}58)$$

and that of integrator 2 is the E.M.P. cospec function:

$$px(t) = \omega \int_{-\infty}^{t} y(\tau) . \exp\left[-\alpha(t - \tau)\right] . \cos \omega\tau . d\tau, \qquad (6\text{-}59)$$

(compare with equations (6-53) and (6-54)).

To obtain the amplitude spectrum it is necessary to square and add the outputs of the integrators and to find the square root of the sum as indicated in fig. 6-10.

This method of analysis requires that the signal be capable of repetition, to permit integration over the entire signal period for each value of the analysis frequency. With each repetition of the signal the potentiometers will need to be set to a new value to determine the next frequency band of analysis. An initial value, $A(f_\text{Q})$ is calculated, where f_Q is the lowest frequency of interest. This value is recorded or plotted and f_0 increased by δf. A new value of $A(f_\text{Q} + \delta f)$ is computed and the frequency increased in steps of δf until $f = f_\text{h}$ (the highest frequency of interest).

The process may be automated by recording the signal on magnetic tape, controlling the computer from the commencement of each pass of the tape and plotting $A(f)$ as a function of f on an X/Y plotter. A complete analog program for this automated procedure is given in fig. 6-11. Bi-polar amplifiers

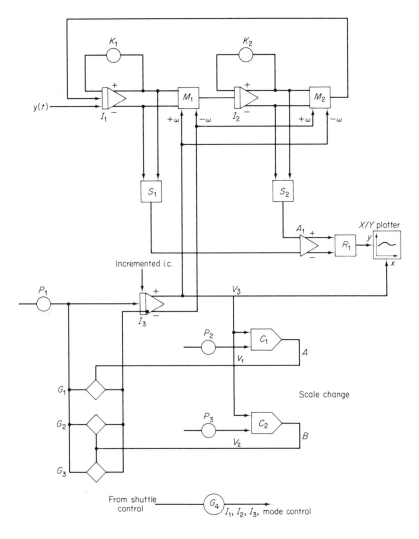

Fig. 6-11. Automated Fourier analysis using analog methods.

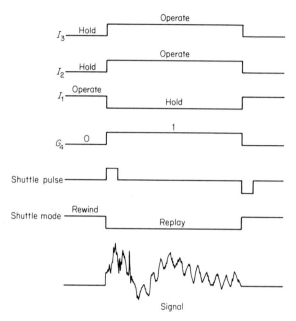

Fig. 6-12. Timing diagram for fig. 6-11.

are shown. In this example the input is derived from an analog tape record in which the shuttle facility of the transport is used. The updating logic shown in this diagram is controlled from pulses derived from this shuttle operation.

The forcing function oscillator consists of integrators I_1 and I_2 and multipliers M_1 and M_2. A potential proportional to the updated value of frequency, i.e. $(f_\varrho + (n-1)\delta f)$ is derived from integrator I_3. During a given repetition of the signal the integrators I_1 and I_2 are placed in the operate mode, whilst I_3 remains in the hold mode. A constant potential output of I_3, proportional to the frequency of the particular repetition, is maintained as input to the multipliers M_1 and M_2. At the end of the calculation the tape is rewound back to its commencing value, integrators I_1 and I_2 are placed in the hold position, and I_3 is allowed to operate so that its output can increment by a value proportional to frequency δf. At the commencement of the next signal repetition the output value reached is held and the integrators I_1 and I_2 caused to operate with their outputs multiplied by this new output value, proportional to $(f + n\delta f)$, which is held constant at the output of I_3.

The timing diagram of fig. 6-12 will make this process clear. The timing process is controlled automatically from the tape shuttle system. When the record is being replayed, an enabling level is applied to gate 4 permitting I_1 and I_2 to operate. Frequency updating is carried out during the rewind phase when a disabling (logic 0) output is delivered from the tape unit. The values reached by the integrators I_1 and I_2 at the end of each repetition are squared by S_1 and S_2, and summed in A_1. Finally a square root operation is carried out by R_1, prior to applying the output level to the plotter as the given value of Fourier amplitude $A(f_\varrho + n\delta f)$.

It is not necessary for the frequency incrementation to follow a continuous linear function and fig. 6-11 includes logic designed to change the scale of incrementation in accordance with the value of the frequency reached. The output of I_3, (V_3), is applied to both comparitors, C_1 and C_2. Fixed potential values, V_1 and V_2 are applied to C_1 and C_2 respectively. When $V_3 > V_1$, gate G_1 is enabled thereby decreasing the incremental value δf by a factor of 10. When $V_3 > V_2$, then a second gate G_2 is also enabled, reducing δf by a further factor of 10. Other scaling techniques are possible and will be described later in chapter 8 when the subject of power spectral density estimation is discussed.

This method of harmonic analysis has been used for the continuous evaluation of a frequency spectrum from a real time or an extremely lengthy recorded signal. It introduces a weighting factor into the time domain which favours recent signals and reduces all past signals existing beyond time $(t - 3/\alpha)$ to a negligible value. The resulting coefficient values obtained by equations (6-58) and (6-59) are thus dependent on both narrow frequency and time windows and are therefore very appropriate to the analysis of non-stationary data. This technique will be considered further in chapter 11, which deals with the analysis of non-stationary processes.

6-8-2 THE DIRECT FOURIER INTEGRAL METHOD

An analog method which overcomes the difficulties of adapting the forcing function technique to finite bandwidth frequency analysis is to derive the amplitude spectrum and phase of the signal directly by the use of the sine and cosine Fourier transform equations given in (6-39) and (6-40). This is shown diagrammatically in fig. 6-13.

Here the signal, $x(t)$ is simultaneously multiplied by $\cos(\omega t)$ and $\sin(\omega t)$ to form two products: $x(t) \cos(\omega t)$ and $x(t) \sin(\omega t)$.

If these products are filtered to remove all but the zero frequency component the resultants can be squared and added to produce a term proportional to the average power of the signal at a frequency $\omega/2\pi$. The square root of this will give the spectral amplitude for this frequency.

Thus, if we let $x(t) = A \sin(\omega_0 t + \theta)$,

and form

$$x(t) \sin \omega t = \tfrac{1}{2}A \cos \theta - \tfrac{1}{2}A \cos(2\omega t + \theta),$$

$$x(t) \cos \omega t = \tfrac{1}{2}A \sin \theta + \tfrac{1}{2}A \sin(2\omega t + \theta).$$

251

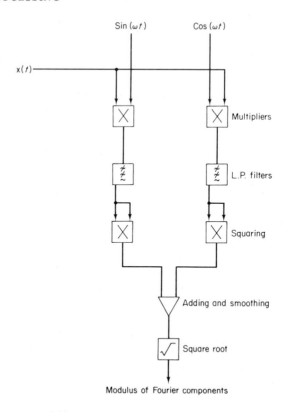

Fig. 6-13. Direct Fourier integral method of analysis.

Then, carrying out the filtering operation, $F(y)$ we have:

$$y_1(t) = F(x_t \sin \omega t) = \tfrac{1}{2}A \cos \theta$$
$$y_2(t) = F(x_t \cos \omega t) = \tfrac{1}{2}A \sin \theta,$$

so that

$$2[(y_1(t))^2 + (y_2(t))^2]^{\frac{1}{2}} = A$$

and phase angle:

$$\arctan \left[\frac{y_1(t)}{y_2(t)} \right] = \theta$$

as expected.

This is an important method since it permits the separate control of analysis frequency and bandwidth so that the bandwidth can be made constant over the entire frequency band or varied in a pre-determined manner.

252

Due to its flexibility the method is widely used for analog power spectral density analysis and will be considered in more detail in the next chapter when methods of implementation will be discussed. It is interesting to note that a discrete digital version of this technique has recently been applied to the analysis of non-stationary signals, when it is referred to as complex demodulation [2] for reasons which are apparent from the method of derivation for the controlling equations (6-39) and (6-40).

6-9 DISCRETE FOURIER SERIES

In the case of the analysis of digitised data, it is necessary to consider a finite version of the Fourier series developed earlier, and to derive a discrete form of the Fourier transform. The classical Fourier series described in section 6-4 is identical computationally to the discrete Fourier transform, although its theoretical derivation is quite different. We will consider its derivation following a consideration of the finite Fourier series.

If we consider the calculation of Fourier series and integrals from a discontinuous discrete data series, a number of limitations will be apparent which are not present when the time-history series is processed in continuous form. Broadly, these limitations stem from the need to preserve information in its conversion from a continuous to a discontinuous series, or where the analysis requirement is to treat this discrete information as if it were a continuous signal. Let us consider first the effect of these limitations on the derivation of a Fourier series suitable to express a finite ensemble of discrete data points.

Assuming a sample record of finite length T sec, where T may be termed the fundamental period of the data, let us consider the record to be divided into N equally spaced points, having adjacent points separated by a duration h sec. The series for $x(t)$ can be represented as a coefficient series having the form:

$$x(t) = \text{d.c. term} + \text{sine terms} + \text{cosine terms.}$$

in which only a limited number of frequency components are represented. This limit is determined by the Nyquist frequency, f_N, which is related to the sampling interval by $f_N = 1/2h$.

This may be understood by first noting that only a limited number of time points can exist at which data is able to be present.

These are

$$t = h.i (i = 1, 2, 3, \ldots, N)$$

where the total length of the record is $T = Nh$.

Consequently the frequency components must also be limited to a discrete number ($= N/2$)—a consequence of the Nyquist limitation given above:

$$f = fn \, [n = 1, 2, 3, \ldots, (N/2)].$$

253

Thus if we replace the fundamental frequency term, ω_0, in equation (6-13), by $2\pi/T$ and thus implicitly recognise that only a given number of discrete values of frequency are possible, we can write for the series:

$$x_i = \frac{a_0}{2} + \sum_{n=1}^{N/2} a_n \cdot \cos\left(\frac{2\pi nih}{T}\right) + \sum_{n=1}^{N/2} b_n \cdot \sin\left(\frac{2\pi nih}{T}\right) \tag{6-60}$$

or, since $h/T = 1/N$ and recognising that the mean value is a special case of the summation when $n = 0$, we can also write for x_i;

$$x_i = \sum_{n=0}^{N/2} a_n \cdot \cos\left(\frac{2\pi in}{N}\right) + \sum_{n=0}^{N/2} b_n \cdot \sin\left(\frac{2\pi in}{N}\right) \tag{6-61}$$

which is a definition for the discrete Fourier series.

The coefficients a_0, a_n and b_n are obtained in a similar manner to the derivation of equations (6-20), (6-21) and (6-22), as:

$$\frac{a_0}{2} = \frac{1}{N} \sum_{i=1}^{N} x_i \tag{6-62}$$

$$a_n = \frac{2}{N} \sum_{i=1}^{N} x_i \cos\left(\frac{2\pi in}{N}\right) \tag{6-63}$$

$$b_n = \frac{2}{N} \sum_{i=1}^{N} x_i \sin\left(\frac{2\pi in}{N}\right) \tag{6-64}$$

where $n = 1, 2, \ldots, N/2$.

6-10   THE DISCRETE FOURIER TRANSFORM

In order to derive a form of Fourier transform for discrete data from the continuous form given in equations (6-36) and (6-37) it is necessary to:
  (a) replace integrals by sums,
  (b) recognise that the limits of summation cannot be infinite.

We saw in the previous section that these limitations result in a discrete form of Fourier series representation which passes through all the sampled data values in the real discrete time-history, x_i. We can see that in order to use these equations we must let the spectral series be complex. This could then be written:

$$X_n = (a_n + jb_n), \tag{6-65}$$

which would represent both positive and negative frequencies.

254

Consequently equation (6-61) can be written as the complex Fourier series:

$$x_i = \sum_{n=-N/2}^{N/2} X_n \exp{(j2\pi in/N)} \quad (i = 1, 2, 3, \ldots, N). \quad (6\text{-}66)$$

Both x_i and X_n are periodic functions and an inverse representation is applicable as is indicated by equation (6-36). However, since the spectrum is given in complex form there are two spectral components generated for each real frequency. Consequently the summation of the pairs of components in the Fourier transform will result in a doubling of the amplitude of the spectral series produced. It will be seen later than this contravenes Parsevals' Theorem which states that the mean power of the signal must be equal to the sum of the powers contributed by each spectral component. The Fourier transform must therefore include a scaling factor of $1/N$ as shown below:

$$X_n = \frac{1}{N} \sum_{n=-N/2}^{N/2} x_i \exp{(-j2\pi in/N)}. \quad (6\text{-}67)$$

Equations (6-66) and 6-67) form a Fourier transform pair suitable for expressing a discrete data series.

Their equivalence can be proved by substituting (6-67) into (6-66) and applying the orthogonal relationship described in section 6-6 which, for complex frequencies, is given as

$$x_i = \sum_{n=-N/2}^{N/2} \exp{(j2\pi in/N)} = N \text{ if } n = 0, \pm N, \pm 2N \text{ etc.}$$
$$= 0 \text{ otherwise.} \quad (6\text{-}68)$$

It may be noted that the scaling factor, $1/N$ appears in the literature in either transform and is sometimes given as $1/\sqrt{N}$ in both transforms. Since the equations constitute a transform pair, these variants are equally valid providing consistency is maintained through their manipulation.

A simplified form of equations (6-66) and (6-67) can be obtained by noting that the series for x_i and X_n are symmetrical for positive and negative values of N and includes the zero value so that we can write:

$$X_n = \frac{1}{N} \sum_{i=0}^{N-1} x_i \exp{(-j2\pi in/N)} \quad (6\text{-}69)$$

$$x_i = \sum_{n=0}^{N-1} X_n \exp{(j2\pi in/N)} \quad [i, n = 0, 1, \ldots, (N-1)]. \quad (6\text{-}70)$$

255

These equations will be referred to as the Discrete Fourier Transform (D.F.T.) and the Inverse Fourier Transform (I.D.F.T.) respectively. The latter is, of course, the discrete Fourier series itself. A consequence of this notation when implemented on the digital computer using methods described below is that a sequential calculation of spectral values of X_n by a digital computer will be

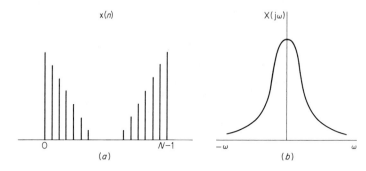

Fig. 6-14. Fourier spectrum and discrete Fourier spectrum.

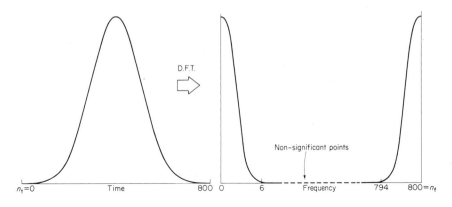

Fig. 6-15. Illustrating data reduction during transformation from time to frequency domain.

obtained as shown in fig. 6-14a compared with the more familiar representation of the continuous spectrum given in fig. 6-14b.

A major use of the discrete Fourier transform is the translation of a time series into an equivalent frequency series. We saw earlier that an N-point series in the time domain will transform into an $N/2$-point series in the frequency domain. It must not be expected however that the useful frequency range will be uniformly distributed over these $N/2$ points. For example if a rectangular waveform ($N = 1000$), consisting of 500 unit value samples bounded on either side by 250 zeros, is transformed we find that of the 500 frequency points obtained only about 10 will have a non-zero value. This is, of course, an argument for the expression of information in the frequency domain due to its data reduction properties. It can lead to an apparent loss of definition unless a much larger number of time samples are taken. An example is that of a Gaussian time function ($N = 800$) which we would expect to retain its function shape following transformation into the frequency domain. As shown in fig. 6-15 this will be expressed rather poorly by about six samples. A solution to this problem is interpolation by addition of zeros to the time history before transformation as discussed later in this chapter.

6-11 THE FAST FOURIER TRANSFORM

In the previous section expressions for Fourier transform pairs were described which are used for the calculation of a discrete time or frequency series. These discrete methods are applicable to digital computation and the remainder of this chapter will be concerned with practical methods of implementation on a digital machine.

The discrete complex Fourier transform is a 1:1 conversion of any sequence x_i, $i = 0, 1, 2, \ldots, N - 1$ consisting of N complex numbers, into another sequence defined by:

$$X_n = \sum_{i=0}^{N-1} x_i W^{in} \tag{6-71}$$

where $W = \exp(-2\pi j/N)$.

The factor $1/N$ will be omitted in this and later derivations in order to simplify notation. It serves as a scaling factor to the values of X_n obtained and needs to be considered only when numerical values are to be attributed to the frequency (or other) scale.

The direct calculation of equation (6-71) requires the solution of a matrix of terms, $[X_n] = [W^{in}] \cdot [x_i]$ which are all complex.

For example if there are eight samples in the time series ($N = 8$), then the matrix takes the form:

$$
\begin{vmatrix} X_0 \\ X_1 \\ X_2 \\ X_3 \\ X_4 \\ X_5 \\ X_6 \\ X_7 \end{vmatrix}
=
\begin{vmatrix}
W^0 & W^0 & W^0 & W^0 & W^0 & W^0 & W^0 & W^0 \\
W^0 & W^1 & W^2 & W^3 & W^4 & W^5 & W^6 & W^7 \\
W^0 & W^2 & W^4 & W^6 & W^8 & W^{10} & W^{12} & W^{14} \\
W^0 & W^3 & W^6 & W^9 & W^{12} & W^{15} & W^{18} & W^{21} \\
W^0 & W^4 & W^8 & W^{12} & W^{16} & W^{20} & W^{24} & W^{28} \\
W^0 & W^5 & W^{10} & W^{15} & W^{20} & W^{25} & W^{30} & W^{35} \\
W^0 & W^6 & W^{12} & W^{18} & W^{24} & W^{30} & W^{36} & W^{42} \\
W^0 & W^7 & W^{14} & W^{21} & W^{28} & W^{35} & W^{42} & W^{49}
\end{vmatrix}
\times
\begin{vmatrix} x_0 \\ x_1 \\ x_2 \\ x_3 \\ x_4 \\ x_5 \\ x_6 \\ x_7 \end{vmatrix}
$$

If matrix multiplication is carried out in this way we would have to calculate 64 complex products and 64 complex additions. In the general case N^2 complex multiplications and additions would be required. For large values of N this results in extremely lengthy calculation times on the digital computer.

An alternative method of obtaining the discrete Fourier transform, termed the Fast Fourier Transform (F.F.T.) method, was first described in a form suitable for machine calculation by Cooley and Tukey in 1965 [4]. This is a recursive algorithm, based on the factorisation of the above matrix into a series of 'sparce' matrices (i.e. simpler matrices having many zero terms). Using this technique much of the redundancy in the calculation of repeated products, required by equation (6-71), can be removed to enable a large reduction in calculation time to be realised.

This method is a rediscovery of the earlier work by Danielson and Lanczos [5] which found little application using the desk calculator methods of the time, and had to await the advent of the large-scale digital computer for its successful implementation.

Several other fast methods have since been published ([6], [7], [8], [9]) using essentially similar ideas to the original algorithm. They are all called fast Fourier transforms and will be distinguished from each other by the originators initials (e.g. $C-T$ Algorithm). The key to these methods lies in their exploitation of the possibilities for factorising the number of values of the series to be transformed. An explanation will first be given in terms of matrix factorisation and will be followed by the derivation of a general computing algorithm. Finally a third description will be given which provides a working algorithm suitable for high-level language implementation.

Commencing with the matrix description of the method we first note that the series representation of a variable is generally considered only in terms of a long series of coefficients. We can equally well describe the series in terms of r rows and s columns of a matrix, in which the total number of points is $N = rs$.

A Fourier transform of the elements of this matrix into an equivalent set of Fourier coefficients can then be arranged by carrying out s parallel Fourier transforms, each of r data points, on the individual columns and then summing the results. We have thus reduced the problem to a summation of a number of much shorter Fourier transforms instead of the calculation of one long transform. Quite apart from the saving in time (which will be shown below), this method has another major advantage in that a much smaller storage space is required to hold the intermediate calculations.

To illustrate the working of the algorithm in matrix terms a simple two-level factorising for the total number of terms, $N = rs$ will be considered first. Using this assumption we can express the N variables n and i in equation (6-71) as four subseries for $i_0, i_1, n_0,$ and n_1

$$n = n_1 r + n_0 \qquad (6\text{-}72)$$

and
$$i = i_1 s + i_0.\qquad(6\text{-}73)$$

where

$$\begin{aligned}
n_0 &= 0, 1, 2, \ldots, r-1,\\
n_1 &= 0, 1, 2, \ldots, s-1,\\
i_0 &= 0, 1, 2, \ldots, s-1,\\
i_1 &= 0, 1, 2, \ldots, r-1.
\end{aligned}$$

We can see that i and n each still contain N discrete values if we take some particular values and substitute these in the above. E.g. if $N = 20 = rs$, where $r = 5$ and $s = 4$, then

$$n_0 \text{ goes from 0 to 4,}$$
$$n_1 \text{ goes from 0 to 3,}$$
$$i_0 \text{ goes from 0 to 3,}$$
$$i_1 \text{ goes from 0 to 4,}$$

and a table for n will include all 20 values as required. This is shown in table 6-1

TABLE 6-1

$n_1 =$	0	1	2	3	n_0
$n =$	0	5	10	15	0
	1	6	11	16	1
	2	7	12	17	2
	3	8	13	18	3
	4	9	14	19	4

$$r = 5$$

derived from equation (6-72). Similarly for i from equation (6-73), shown in table 6-2.

TABLE 6-2

$i_1 =$	0	1	2	3	4	i_0
$i =$	0	4	8	12	16	0
	1	5	9	13	17	1
	2	6	10	14	18	2
	3	7	11	15	19	3

$$s = 4$$

Equation (6-71) can now be rearranged as two sums:

$$X_{(n_1,n_0)} = \sum_{i_0=0}^{i_0=s-1} \sum_{i_1=0}^{i_1=r-1} x_{(i_1,i_0)}\, W^{in}.\qquad(6\text{-}74)$$

This is a recursive formula in which we carry out complex multiplications and additions at every sth sample of the $N-1$ possible variables for i within the inner loop, and then advance the starting point for the first value of i in the

259

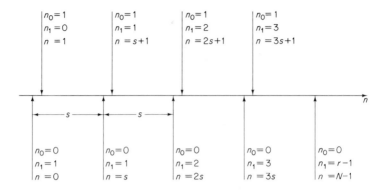

Fig. 6-16. Fast Fourier transform – simple partitioning.

outer loop. The inner loop procedure is then repeated from this new point (see fig. 6-16).

If we now expand W^{in} using equations (6-72) and (6-73) we have:

$$W^{in} = W^{(i_1 s + i_0)(n_1 r + n_0)}$$

$$= W^{i_1 n_1 rs} + W^{n_0 i_1 s} + W^{i_0 (n_0 + n_1 r)} \qquad (6\text{-}75)$$

but $i_1 n_1 r s = i_1 n_1 N$ and, since i_1, n_1 is an integer by definition, the exponent is a multiple of 2π and,

$$W^{i_1 n_1 N} = \exp\left(-j\frac{2\pi}{N}\right)^{i_1 n_1 N} = 1. \qquad (6\text{-}76)$$

We may also note for later use that:

$$W^{N/2} = \exp(-j2\pi/2) = -1. \qquad (6\text{-}77)$$

Equation (6-74) now becomes:

$$X_{(n_1 n_0)} = \sum_{i_0 = 0}^{i_0 = s-1} \cdot \sum_{i_1 = 0}^{i_1 = r-1} x_{(i_1, i_0)} W^{n_0 i_1 s} \cdot W^{i_0 (n_0 + n_1 r)}. \qquad (6\text{-}78)$$

The inner sum over i_1 depends only on i_0 and n_0 and can be defined as a new series:

$$x_{1(i_0, n_0)} = \sum_{i_1 = 0}^{i_1 = r-1} x_{(i_1, i_0)} W^{n_0 i_1 s} \qquad (6\text{-}79)$$

which we can recognise as the Fourier transform of a reduced array: $x_{(i_1, i_0)}$.

260

The original transform (equation (6-78)) can now be rewritten as:

$$X_{(n_1,n_0)} = \sum_{i_1=0}^{i_1=s-1} x_{1(i_0,n_0)} W^{i_0(n_0+n_1 r)}. \tag{6-80}$$

There are N elements in array x_1, but, unlike the original array x_i, we only require r operations to evaluate the transform, giving a total of Nr operations for array x_1 (remembering that these are complex operations). Similarly we require Ns operations in order to calculate x_1 from x_i. Consequently, this simple two-step algorithm requires a total of $T = N(r + s)$ complex operations.

We can see how this procedure may be extended by increasing the number of steps to p when the number of operations will be $T' = N(s_1 + s_2 + \ldots, s_p)$, where $N = s_1, s_2 \ldots, s_p$. If $s_1 = s_2 = s_3 \ldots, = s_p = s$ then $N = s^p$, so that s can be expressed as the radix of the number series and we can write $p = \log_s N$ from which

$$T' = s N \log_s N = N \left(\frac{s}{\log_2 s} \right) \log_2 N. \tag{6-81}$$

To determine the optimum radix for minimum T' we plot s against $s/(\log_2 s)$ From fig. 6-17 we see that the use of $s = 3$ is the most efficient, although for practical reasons in digital calculation a radix of two simplifies the algorithm. Using this radix an improvement in computing speed over direct evaluation of:

$$\frac{N^2}{2N \log_2 N} = \frac{N}{2 \log_2 N} \tag{6-82}$$

is obtained. For large values of N this saving in computing time can be quite considerable.

Although from fig. 6-17, the improvement in speed by choosing a radix other than two is marginal there are certain other advantages in the use of radices 4, 8, or even 16, which will be considered later in this chapter.

The improvements in computational speed obtained by F.F.T. methods have been extended to hardware logic design and a number of special purpose machines have been constructed to implement the F.F.T. in this way [10, 11, 12]. Experience in the use of these processors indicate that the increase in speed and reduction in running costs, compared with software implementation on a general purpose computer, can be very considerable [13]. The increase in speed enables the hardware F.F.T. processor to be used in a real-time environment such as radar or sonar signal processing and for realisation of real-time digital filtering. The second advantage, that of cost, is valuable for those applications where the bulk of data to be transformed would otherwise demand a dedicated digital computer for this purpose. Here the F.F.T. processor would act as a peripheral to the main computer and only create demands for peripheral transfers rather than central processor time.

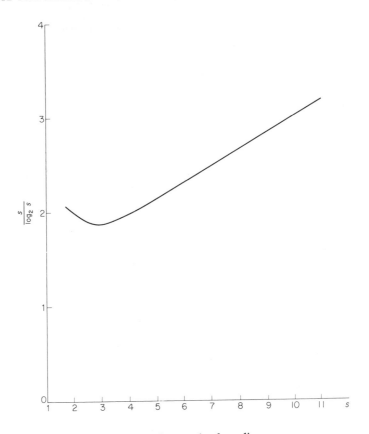

Fig. 6-17. Optimum value for radix s.

6-11-1 TRANSFORMS WHEN $N = 2^P$

A number of practical advantages arise from making the radix have a value of two or a power of two. In the case of $N = 2^P$ intermediate transforms are reduced to two elements and, from equations (6-76) and (6-77), two of the three exponentials shown in the expression of equation (6-75) will assume one of the two values $+1$ and -1, thus avoiding some of the need for complex arithmetic. Similar advantages are obtained for $N = 4^P$ when the exponentials obtained are ± 1 and $\pm j$. The implementation of the F.F.T., as described by Cooley and Tukey required that N be a power of two. This algorithm will be developed below by considering it as a general recursive device programmed by the digital computer. The original series is converted in stages, each stage providing an intermediate transform which supplies the input for subsequent stages, until the final series becomes the discrete transform of the original series.

With N limited to a power of 2 it is possible to express i and n in terms of the index, p, as a binary-weighted series:

$$i = i_{p-1} 2^{p-1} + i_{p-2} 2^{p-2} + , \ldots , + i_1 2 + i_0 \tag{6-83}$$

$$n = n_{p-1} 2^{p-1} + n_{p-2} 2^{p-2} + , \ldots , + n_1 2 + n_0. \tag{6-84}$$

In this way we can express all the N possible values of the indices in terms of a binary number and hence facilitate the consideration of storage in the digital computer. The input and output series of the Fourier transform, x_i and X_n are considered as being stored within the memory at location addresses defined by the binary representation of i and n. Using this convention, equation (6-71) can be written in terms of a factorised sum, exactly as carried out previously for $N = rs$, but with the factorising expressed in a binary-weighting form, i.e.

$$X_{(n_{p-1}, n_{p-2}, \ldots, n_0)} = \sum_{i_0=0}^{i_0=1} \sum_{i_1=0}^{i_1=1} , \ldots , \sum_{i_{p-1}=0}^{i_{p-1}=1}$$

$$x_{(i_{p-1}, i_{p-2}, \ldots, i_0)} \times W^{n(i_{p-1} 2^{p-1} + i_{p-2} 2^{p-2} + , \ldots , i_0)}. \tag{6-85}$$

That this is equivalent to equation (6-71) can be seen if we consider a simple case of:

$$N = 2^2 = 4_{10} = 1 \ 0 \ 0_2$$

$$\underset{i_2 \quad i_1 \quad i_0}{\diagup \quad \vert \quad \diagdown}$$

In the decimal case:

$$\sum_{i=0}^{i=4} x_{i(10)} = x_0 + x_1 + x_2 + x_3$$

and in the binary case:

$$\sum_{i_0=0}^{1} \sum_{i_1=0_2}^{1} \sum_{i=0}^{1} x_{i(2)} = x_{000} + x_{001} + x_{010} + x_{011} + x_{100}$$

which is seen to contain all the possible values of x_i in the range up to decimal 4, expressed in binary form.

If we now expand the first term of the exponential for equation (6-85) as we did in equation (6-75),

$$W^{n i_{p-1} 2^{p-1}} = W^{n_{p-1} 2^{p-1} i_{p-1} 2^{p-1}} + W^{n_{p-2} 2^{p-2} i_{p-1} 2^{p-1}}$$

$$+ , \ldots , + W^{n_1 2 i_{p-1} 2^{p-1}} + W^{n_0 i_{p-1} 2^{p-1}}, \tag{6-86}$$

263

we find that all the intermediate terms go to unity and:

$$W^{n_{i_{p-1}}\, 2^{p-1}} = W^{n_0 i_{p-1}\, 2^{p-1}}. \qquad (6\text{-}87)$$

This enables the innermost sum of equation (6-85) over $i_p - 1$ to be written as a shorter Fourier transform:

$$x_{1(n_0,\, i_{p-2}, \ldots,\, i_0)} = \sum_{i_{p-1}=0}^{i_{p-1}=1} x_{(i_{p-1},\, i_{p-2}, \ldots,\, i_0)}\, W^{n_0 i_{p-1}\, 2^{p-1}} \qquad (6\text{-}88)$$

which depends only on $n_0, i_{p-2}, \ldots, i_0$.

Unlike the complete Fourier transform, this sum consists of a set of N numbers only, each calculated from two of the original data points. Subsequent sums, proceeding outwards in equation (6-85), can be calculated using a generalised recursive expression for the exponential term:

$$W^{n_{i_{p-q}}\, 2^{p-q}} = W^{(n_{q-1}\, 2^{q-1} + n_{q-2}\, 2^{q-2+}, \ldots,\, n_0)\, i_{p-q}\, 2^{p-q}}.$$

$$(q = 1, 2, 3, \ldots, p) \qquad (6\text{-}89)$$

The successive sums are evaluated according to the equation:

$$x_{q\,(n_0, n_1, \ldots,\, n_{q-1},\, i_{p-q-1},\, i_{p-q-2}, \ldots,\, i_0)}$$

$$= \sum_{i_{p-q}=0}^{i_{p-q}=1} x_{q-1(n_0, n_1, \ldots,\, n_{q-2},\, i_{p-q}, \ldots,\, i_0)} \cdot$$

$$W^{(n_{q-1}\, 2^{q-1} + n_{q-2}\, 2^{q-2+}, \ldots,\, n_0)\, i_{p-q}\, 2^{p-q}}.$$

$$(q = 1, 2, 3, \ldots, p) \qquad (6\text{-}90)$$

which is the definition of the *C-T* Algorithm.

To apply this recursive formulae the initial set of data x_i is first made equal to x_0 ($q = 1$), thus:

$$x_i = x_{(i_{p-1},\, i_{p-2}, \ldots,\, i_0)} = x_0\, (i_{p-1},\, i_{p-2}, \ldots,\, i_0).$$

This leads to the derivation of succeeding arrays in x_q so that the final array will be that for X_n. Since we have represented the two sets of p arguments for x_q, namely $(n_0, \ldots, n_{q-1}, i_{p-q-1}, \ldots, i_0)$, as binary representations of their storage locations, this can be used to simplify the working of the algorithm and so reduce the storage requirements for the intermediate sums. This involves fetching values from these two storage locations, carrying out complex multiplication and addition in accordance with equation (6-90), and putting the results back into these same two locations. Over-writing the input array with output values thus occurs. This indexing scheme has one disadvantage in that the elements of the final array in X_n are stored in the incorrect order in the core memory. Thus for the last array calculated:

$$X_{(n_{p-1}, n_{p-2}, \ldots, n_0)} = x_{p(n_0, \ldots, n_{p-2}, n_{p-1})}. \tag{6-91}$$

This shows that, in order to find X_n, the order of the bits in the binary representation of n must be reversed so as to obtain the index of the memory location where X_n may be found from the x_p array. From equation (6-90) we see that for each element of x_q, one complex multiply/add operation is required. Also, a further complex multiply/add operation is required for the recursive generation of the complex exponential. Thus, the total number of operations is $T' = 2Np$. But, since there are $N = 2^p$ elements in each array x_q where $q = 1, 2, 3, \ldots, p$, the improvement in speed over direct evaluation is $T/T' = N/(2 \log_2 N)$, which is the result obtained earlier for the general case.

An alternative form of the algorithm is obtained if the roles of the indices i and n are interchanged when carrying out the expansion of the exponential term in equation (6-86). This is known as the Sande-Tukey version [6] and leads to the following recursion:

$$x'_q (n_0, n_1, \ldots, n_{q-1}, i_{p-q-1}, i_{p-q-2}, \ldots, i_0)$$

$$= \sum_{i_{p-q}=0}^{i_{p-q}=1} x_{q-1}(n_0, n_1, \ldots, n_{q-2}, i_{p-q}, \ldots, i_0)$$

$$\cdot W^{(i_{p-q} 2^{p-q} +, \ldots, i_0) n_{q-1} 2^{q-1}}, \tag{6-92}$$

which is the definition of the $S\text{-}T$ algorithm.

The two algorithms are operationally very similar although some advantages in speed, particularly for in-place array manipulation, are claimed for the $S\text{-}T$ algorithm, due to its simpler exponential structure.

6-11-2 PROGRAMMING THE ALGORITHM

Although the recursive algorithms described previously, can be implemented directly, the alternative derivation given below, is preferred since this will avoid the unnecessary calculation of some complex exponentials. This derivation also assumes that N is a power of two.

Referring to the definition of the discrete Fourier transform given by equation (6-71), if we consider the series x_i to be divided into two interleaved series:

$$x_{(2i)} = y_i$$
$$x_{(2i+1)} = z_i \quad [i = 0, 1, \ldots, ((N/2) - 1)] \tag{6-93}$$

where y_i consists of the even-numbered samples and z_i the odd-numbered samples, the discrete Fourier transforms of these two series can now be written (omitting the scaling factor $2/N$) as

$$Y_n = \sum_{i=0}^{(N/2)-1} y_i W^{2in} \tag{6-94}$$

$$Z_n = \sum_{i=0}^{(N/2)-1} z_i W^{2in}. \tag{6-95}$$

Note that in now represents a half-length series so that $W_{N/2} = \exp[(-2\pi j)/(N/2)] = W^2{}_N$. The transformed series Y_n and Z_n are displaced by one sampling interval so that to obtain the discrete Fourier transform from the complete N-point series using equations (6-94) and (6-95) we write:

$$X_n = \sum_{i=0}^{(N/2)-1} [y_i W^{n(2i)} + z_i W^{n(2i+1)}]$$

$$= \sum_{i=0}^{(N/2)-1} y_i W^{2in} + W^n . \sum_{i=0}^{(N/2)-1} z_i W^{2in} \tag{6-96}$$

since W^n is a constant for a given value of n. Thus from equations (6-94) and (6-95):

$$X_n = Y_n + W^n Z_n. \tag{6-97}$$

But n is limited to $(N/2) - 1$ samples so that to complete the sequence for X_n we require to find the terms from $N/2$ to $N - 1$. We note from the theory of the discrete Fourier transform that for $n > N/2$ the transforms Y_n and Z_n will repeat periodically the values obtained with $0 < n < N/2$ so that $(n + N/2)$ can be substituted for n in the phase shift term only, viz

$$X_{(n+N/2)} = Y_n + W^{n+N/2} Z_n. \tag{6-98}$$

But

$$W^{n+N/2} = \exp(-j2\pi n/N - j\pi),$$

which, from equation (6-77),

$$= -\exp(-j2\pi n/N) = -W^n. \tag{6-99}$$

Hence

$$X_{(n+N/2)} = Y_n - W^n Z_n. \tag{6-100}$$

Equations (6-97) and (6-100) enable the first and last $N/2$ points to be evaluated from the separate transforms formed from $x_{(2i)}$ and $x_{(2i+1)}$.

266

Neglecting addition and subtraction as taking negligible computing time compared with multiplication, we see that this result requires $N + 2\,(N/2)^2$ multiplications compared with N^2 for direct evaluation. For large values of N this represents a reduction by almost half of the computing time required.

The process of dividing the series by two and then obtaining the discrete Fourier transform from the transforms of each half of the series can be repeated to obtain further reductions in computing time. In the limit, providing that the series is capable of division by two, a two-point transform is obtained from two single point transforms. The discrete Fourier transform of a single point is, however, the point itself, thus:

$$X_n = \sum_{n=0}^{n=0} x_i W^{in} = x_i. \qquad (i = 0)$$

so that the method reduces to a simple series of additions and multiplication of an exponential factor, W^n. In this limiting case the improvement in computing

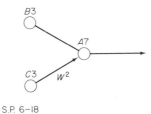

S.P. 6-18

Fig. 6-18. Principle of the signal flow graph.

speed is, $N/(2 \log_2 N)$ as obtained previously. This version of the algorithm is calleed the Successive Doubling Method and, as described above, has been termed Decimation in Time [3].

A similar technique divides the sequence for x_i directly into two non-interleaved sequences:

$$x_i = y_i$$

$$x_i + N/2 = z_i \qquad (i = 0, 1, \ldots, (N/2 - 1))$$

and interleaves the values of the final sequences obtained for X_n [6]. This is called Decimation in Frequency.

A description of the successive doubling method which illustrates the formation of a suitable computer algorithm makes use of what are known as Signal Flow Graphs. A Signal Flow Graph consists of a series of nodes or points, each representing a variable as the sum of other variables originating from the left of the diagram and connected together by means of straight lines. These additions may be weighted by a number appearing at the side of an indicating arrow shown on these connecting lines. Thus, from fig. 6-18 the variable $A7$ is

N=4

0	}
2	}
1	}
3	}

N=8

0	0	0
1	2	4
2	4	2
3	6	6
4	1	1
5	3	5
6	5	3
7	7	7

N=16

0	0	0	0
1	2	4	8
3	4	8	4
4	6	12	12
5	8	2	2
6	10	6	10
7	12	10	6
8	14	14	14
9	1	1	1
10	3	5	9
11	5	9	5
12	7	13	13
13	9	3	3
14	11	7	11
15	13	11	7
	15	15	15

Fig. 6-19. Shuffling of data.

derived from variables originating at nodes $B3$ and $C3$, with the latter weighted by a variable, W^2 so that we can write:

$$A7 = B3 + W^2 C3.$$

The complete decimation down to pairs of single-value transforms will lead to a shuffling of the data into apparently unrelated pairs of values. This shuffling is due to the repeated regrouping of the position of the intermediate results as indicated in fig. 6-19 for several small values of number sequences. If the process

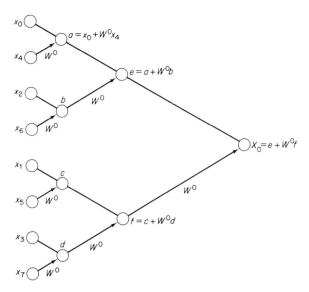

Fig. 6-20. Extraction of the first transform sample.

commences with shuffled data then it becomes possible to carry out computation using a minimum of array storage space. Assuming this to have been carried out (a mechanism for doing this will be suggested later) then the procedure is shown in the signal flow graph of fig. 6-20 for the extraction of the first converted sample, X_0 from a sequence of eight values of x_i. This is seen to produce a summation which, when multiplied by the scaling factor $1/N$ gives:

$$X_n = \tfrac{1}{8} \sum_{i=0}^{i=7} x_i W^{in} = \tfrac{1}{8} \sum_{i=0}^{i=7} x_i \quad \text{(for } n = 0)$$

which we know to be correct.

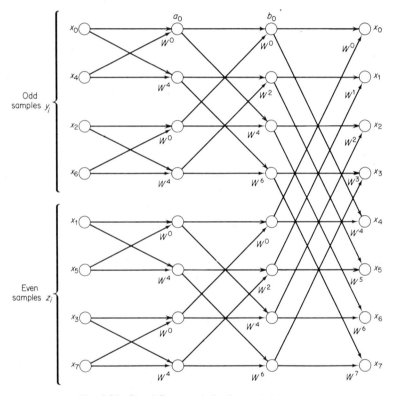

Fig. 6-21. Signal flow graph for 8 sampled data values.

A complete flow graph for an eight point sequence is given in fig. 6-21. From this we see that it is not necessary to allocate separate storage for the intermediate arrays of values, a_0, b_0, etc. Storage is only required sufficient to accommodate the initial number of data points, and a similar area for the calculation of product values (since each calculated value is needed twice). The first set of intermediate transforms can be calculated sequentially and placed in

269

the locations previously occupied by the first set of values, which are now no longer required. The second set can then be calculated and returned to the locations occupied by the first set, and so on until the final transform is formed. The procedure is called 'in-place' computation. The complex exponentials required can be calculated in advance and stored in a table, or computed during the calculation of the intermediate transforms. Standard sine/cosine subroutines are generally employed although the use of a recursive method, such as that suggested by Singleton [14], can lead to faster computation. We may also note from fig. 6-21 that advantage can be taken of the relationship given by equation (6-99) to reduce the number of complex exponentials that need to be generated or stored.

The data shuffling required in this version of the algorithm corresponds to the bit-reversal noted earlier. If the value of i is written in binary form, then the reversed order of the digits, retranslated back to decimal form, will give the correct order for the location of the input-sample. Note that this initial data shuffling can also be carried out 'in-place' as described previously.

It is possible to re-arrange the flow graph to produce a bit-reversed transform from correctly-ordered data (as is the case for the original C-T algorithm). In this case the powers of W, needed in the computation, are required in bit-reversed order [3]. A version can be obtained in which both input and transformed output data can be in the correct natural order. However, in this case the computation can no longer be carried out 'in-place' and at least double the array storage is necessary [15].

Machine-dependant routines, using logic functions for bit-reversal, are necessary with some Fortran compilers. With ASA Fortran IV this is not necessary and a two-part procedure can be implemented as shown in (fig. 6-22). A re-ordering section shuffles the location of the data into a bit reversed order and a second section carries out the F.F.T. calculation using repeated calls to the exponential subroutine from the main loop.

This program applies only to the calculation of the discrete Fourier transform. Calculation of the inverse transform can, of course, be obtained by a similar method since the two transforms (as defined) differ in form only by a scaling factor ($= 1/N$). A second program is not necessary, however, as a reversal of the order for the data to be inverse-transformed will enable this to be entered correctly into the discrete transform program to produce an inverse transform.

6-11-3 TRANSFORMATION OF REAL DATA

The basic F.F.T. algorithms described above have been developed for complex data series. When applied to practical time history analysis real data will be available and hence transforms of a real data series will be required. We can, of course, set the imaginary parts to zero and insert the relevant data samples in the real part of the synthesised complex sequence. However, this will be wasteful

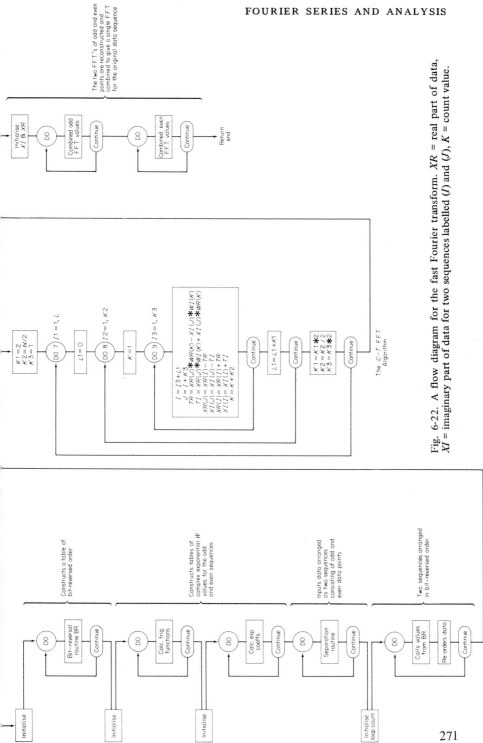

Fig. 6-22. A flow diagram for the fast Fourier transform. XR = real part of data, XI = imaginary part of data for two sequences labelled (I) and (J), K = count value.

271

both of computer time and storage space. A more efficient procedure makes use of the conjugate symmetry of a real variable.

If a time history series x_i is real then:

and
$$x_i = {}^*x_i$$
$$X_n = {}^*X_{-n} = {}^*X_{(N-n)}. \qquad (6\text{-}101)$$

This latter relationship follows from the periodic nature of the exponent W in equation (6-71)

$$W^{in} = W^{(n+N)i} = W^{(i+N)n}. \qquad (6\text{-}102)$$

Hence

$$x_i = x_{(kN+i)}$$
$$X_n = X_{(kN+n)}. \qquad (k = 0, 1, \ldots, \text{etc.})$$

Therefore

$$x_{-i} = x_{(N-i)}$$
$$X_{-n} = X_{(N-n)}$$
$${}^*X_{-n} = {}^*X_{(N-n)}$$

This means that if x_i is real then we only need half the data points in order to specify the frequency spectrum (fig. 6-23).

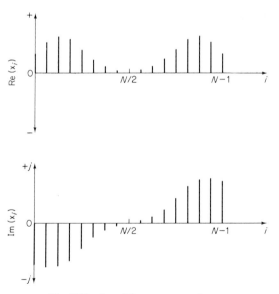

Fig. 6-23. A real frequency spectrum

If we consider the real data series to be separated into two interleaved series, as defined in equation (6-93), defining y_i as the real part, and z_i as the imaginary part of a complex function x_i, we can write: $x_i = (y_i + jz_i)$. The discrete Fourier transform of this will give (neglecting scaling fact $1/N$):

$$X_n = \sum_{i=0}^{(N/2)-1} (y_i + jz_i)W^{2in}$$

$$= \sum_{i=0}^{(N/2)-1} y_i W^{2in} + j \sum_{i=0}^{(N/2)-1} z_i W^{2in}$$

which we can represent as:

$$X_n = Y_n + j Z_n. \tag{6-103}$$

But from equation (6-97)

$$X_n = Y_n + W^n Z_n. \quad [n = 0, 1, \ldots, (\tfrac{1}{2}N - 1)]$$

Each of these transforms will contain $\tfrac{1}{2}N$ points, and since the original series y_i and z_i were real, then from equation (6-101) only $\tfrac{1}{4}N$ points will be required to describe them.

Hence

$$Y_{(\frac{1}{4}N - m)} = {}^*Y \tag{6-104}$$

and

$$Z_{(\frac{1}{4}N - m)} = {}^*Z_{(\frac{1}{4}N + m)} \quad [m = 0, 1, \ldots, (\tfrac{1}{4}N - 1)]$$

Also

$$^*X_n = Y_n - j Z_n \tag{6-105}$$

so that by adding and subtracting equation (6-103) and (6-105) we can obtain:

$$X_n + {}^*X_n = 2Y_{(\frac{1}{4}N - m)} \tag{6-106}$$

and

$$X_n - {}^*X_n = 2j Z_{(\frac{1}{4}N - m)}.$$

This enables Y_n and Z_n to be obtained from our original transform in X_n and substituted in equation (6-97) to obtain X_n for the first $\tfrac{1}{4}N$ points.

The second $\tfrac{1}{4}N$ points are obtained from:

$$X_{(\frac{1}{2}N - m)} = Y_n - W^n Z_n. \tag{6-107}$$

A similar procedure can be carried out to evaluate simultaneously the Fourier transforms of two equal length sets of real data. Thus if we let the two series be x_1 having a Fourier transform X_1 and x_2 having a Fourier transform X_2, then substituting these arrays to form the real and imaginary parts of a complex series x_i (note that an interleaved series is not involved here): $x_i = x_1 + j\,x_2$ and we obtain the transform:

$$X_n = X_1 + jX_2. \tag{6-108}$$

From equation (6-101) we can replace n by $N - n$ so that taking the complex conjugate of both sides we obtain:

$$*X_{(N-n)} = X_1 - jX_2. \tag{6-109}$$

Solving equations (6-108) and (6-109) for X_1 and X_2 gives:

$$\left.\begin{aligned}
X_1 &= \tfrac{1}{2}[X_n + *X_{(N-n)}] \\
X_2 &= \tfrac{1}{2}(X_n - *X_{(N-n)}). \quad [n = 0, 1, \ldots, (\tfrac{1}{2}N - 1)]
\end{aligned}\right\} \tag{6-110}$$

Since X_1 and X_2 are real the symmetry property of equation (6-101) applies and only half of each array need be computed and stored.

Thus only the same amount of storage will be required to evaluate X_1 and X_2 taken together as would be needed to obtain a single complex transform, X_n. Computation time will be only slightly increased by the necessity to form sums and differences using equation (6-110)

6-11-4   FURTHER PROGRAMMING CONSIDERATIONS

Whilst considerable saving in memory storage is obtained by the use of 'in-place' computation, the evaluation of the F.F.T. by any of the methods discussed above still requires considerable storage for any reasonable signal length. Thus with many types of machine to obtain a transform of N complex samples, $2N$ storage locations are required, plus a further $2N$ working locations. Consequently using a $N = 2^p$ algorithm and allowing for programming and systems allocation we cannot hope for much more than a transformation of a 4096 sample array using a 32K word memory.

Several schemes have been proposed in order to make use of backing storage e.g. magnetic tape or disc, and to compensate for the slow transfer-rate of these peripherals by making a minimal number of transfer requests to core storage.

Straight-forward use of overlay (virtual memory) software is slow due to block transfer of data, although by some rearrangement of calculation sequence a doubling of sample size is possible before a time penalty occurs [16]. Suggested methods involving minimal transfer requests fall into two categories:

(a) those using the successive doubling algorithm,
(b) those using simultaneous serial reading/writing from a number of backing-store files.

An example of the former is the system proposed by Gentleman and Sande [6]. Here the sequence is divided into a number of smaller sequences by a process of decimation in time and each sequence stored separately on a disc or magnetic tape file. The size of these sub-sequences is such that in-core transformation is possible and the transformed sub-sequence can be returned to the original file. Combination of the transforms, including matrix multiplication by factors of W^n, is carried out using equation (6-97). Note that complete blocks of data can be transferred in this way, thus reducing considerably the calls to the slow peripheral.

The second method uses a parallel reading and writing process sequentially through a number of files and a repeat of this p times, where the number of samples processed, $N = 2^p$. This is illustrated in fig. 6-24 for a two file sequence.

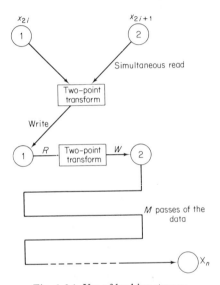

Fig. 6-24. Use of backing storage.

The data is shuffled into a series of values, x_i. These are placed on two files in interleaved order, i.e. data x_{2i} in one file and x_{2i+1} in the second file. On the first pass of the data into the working core area two samples one from each file are simultaneously read and combined with the W^n factors to produce an intermediate transform which is written back onto one of the files in its original location. The process is repeated with the second, third pair etc., until all $N/2$ pairs of points have been transformed. On the second pass of the data into working core area (following disc re-traversal or tape rewind) the second set of transforms is formed from pairs of points taken from one file and transferred to the other, successive passes of data from each of the files transfer the data backwards and forwards until the final transform is contained on one of the files. This is, of course, the method of 'in-place' computation extended for use with backing storage files. Computational speed comparable with in-core programs can be obtained but the input/output times increased by a factor of two or more.

Singleton [16] has suggested various ways of improving the input/output speed for this method by the use of alternate computation and permutation of the transforms on to both files before further computation, and also by the use of auxiliary files.

Some areas of application for the F.F.T. are described in succeeding chapters. One particular use to overcome the data reduction limitation of the transform mentioned in section 6-10, is that of ideal band-limited interpolation. Given a sampled time function having N points, a band-limited interpolated function can be obtained having rN points (where r need not be an integer). An F.F.T. of the N point function is obtained and $(r-1)N$ zeros are inserted in the middle of the transformed function, preserving the symmetry of the waveform. An inverse transform of this new series will now yield an interpolated time series having rN points. This is illustrated in fig. 6-25 for a Gaussian function.

Cooley and Tukey's original paper indicated that the F.F.T. algorithm was most efficient for radix 3 and that only a small reduction in speed was obtained

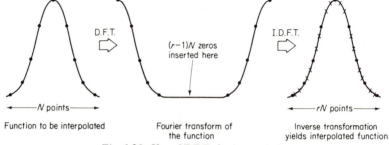

Fig. 6-25. Use of F.F.T. for interpolation.

by the use of radix 2 or 4. Where radix 2 algorithms are employed the use of tables for sines in preference to calculation by sine and cosine subroutines can result in a worthwhile reduction in execution time. Only one quadrant of sines need be stored so that the penalty of increased program size need not be very great. If higher based algorithms are used then several sine–cosine pairs can be obtained from the basic sine values by the use of trignometric relations, e.g. sin $2x = \sin x \cos x$ etc.

Symmetries in the calculation of the exponential terms can also be used. Thus multiplication by W^0 or $W^{N/2}$ will not demand actual machine multiplication. Neither will other integer powers of $N/4$. Thus:

$$W^N = W^{-N} = W^{NC} = W^{-NC} = +1$$
$$W^{\frac{1}{2}N} = W^{-\frac{1}{2}N} = W^{\frac{1}{2}NC} = W^{-\frac{1}{2}NC} = -1$$
$$W^{\frac{1}{4}N} = W^{\frac{1}{4}NC} = -j$$
$$W^{-\frac{1}{4}N} = W^{-\frac{1}{4}NC} = +j$$

where C is integer.

It can also be seen from:

$$(a + jb)(2^{-\frac{1}{2}} + j2^{-\frac{1}{2}}) = 2^{-\frac{1}{2}}(a - b) + j2^{-\frac{1}{2}}(a + b)$$

that any power of W producing a phase shift of $45°$ will result in only two multiplications being required instead of four. These symmetries have been exploited [6] in developing algorithms of radix 4 which are found to be more efficient than either radix 3 or 2. Later developments have shown that a mixed radix of 4 and 2 or 8 and 4 can increase the speed of computation still further [8, 17]. The technique used is to carry out as many of the iterations as possible with the higher radix and then complete the transform using the lower radix. The main disadvantage of these and other high-based systems is the increased complexity of the logic and resultant increase in program size.

6-11-5 ACCURACY

For fairly evenly distributed data, the method of calculataion used in the evaluation of the F.F.T. is such that pairs of values of roughly similar magnitude are added together. This helps to minimise the errors introduced by round-off or truncation and shifting carried out in a floating-point calculation. A further fundamental consideration is that, due to the economy of the F.F.T. algorithm, there are fewer arithmetic operations to be carried out so that the F.F.T. is actually more accurate in terms of round-off error than other D.F.T. methods.

The mean-square errors introduced by direct evaluation of the Fourier transform can be shown to be proportional to N, from Parseval's Theorem, since:

$$\sum_{i=0}^{N-1} |x_i|^2 = N \sum_{n=0}^{N-1} |X_n|^2 \tag{6-111}$$

Several derivations of the comparable error using the F.F.T. have been made [18, 19] and for a radix 2 evaluation this is found to be proportional to $\log_2 N$ so that the method can be seen to give a reduction in mean-square error similar to the reduction in calculation time. It is worth noting that a significant reduction in error is obtained if rounding takes the place of truncation in the transform process [6].

Smaller machines do not have floating-point hardware for addition, subtraction, multiplication, or division so that fixed-point arithmetic is often used for the transform algorithm. The special features of this implementation are discussed by Singleton [14] who comments on the care needed to avoid overflow and underflow with fixed-point operations. Accuracy problems are particularly acute for the small machine where the word length is limited (e.g. 16 bits). It has been shown that in such a case the upper bound of the mean square error is proportional to $\sqrt{N}$ whilst the lower bound is similar to that obtained in the floating-point case.

BIBLIOGRAPHY

1. CAMPBELL, G. A. and FOSTER. R. M. Fourier integrals for practical applications. *Bell Syst. Tech. J.* 7, 639, 1928.
2. GRANGER, C. W. J. and HATANEKA, M. *Spectral Analysis of Economic Time Series.* Princeton University Press, 1964.
3. COCHRAN, W. T. *et al.* What is the fast Fourier transform? *I.E.E.E. Trans.* (Audio and Electroacoust.) AU-15, 2, 45, 1967.
4. COOLEY, J. W. and TUKEY, J. W. An algorithm for the machine calculation of complex Fourier series. *Math. Comp.* 19, 297, 1965.
5. DANIELSON, G. C. and LANCZOS, C. Some improvements in practical Fourier analysis and their application to X-ray scattering from liquids. *J. Franklin, I.,* 233, 365, 1942.
6. GENTLEMAN, W. M. and SANDE, G. Fast Fourier transforms for fun and profit. *A.F.I.P.S. Proc. Fall Joint Comp. Conf.* 29, 563, 1966.
7. SINGLETON, R. C. ALGOL Procedures for the fast Fourier transform. *Comm. A.C.M.* 11, 773, 1968.
8. BERGLAND, G. D. A fast Fourier transform algorithm using base 8 iterations. *Math. Comp.* 22, 275, 1968.
9. ANDREWS, H. A high-speed algorithm for the computer generation of Fourier transforms. *I.E.E.E. Trans.* (Comp.) C-17, 373-5, 1968.
10. BERGLAND, G. D. Fast Fourier transform hardware implementation—A survey. *I.E.E.E. Trans.* (Audio and Electroacoust.) AU-17, 2, 109, 1969.
11. SMITH, R. A. A fast Fourier processor. *Bell Tel. Labs. Rep.* Whippany, N.J., 1967.
12. SHIVELY, R.R. A digital processor to generate spectra in real time. *I.E.E.E. Trans.* (Comp.) C-17, 485, 1968.
13. BERGLAND, G. D. Fast Fourier transform hardware implementations—an overview. *I.E.E.E. Trans.* (Audio and Electroacoust.) AU-17, 2, 183, 1969.
14. SINGLETON, R. C. On computing the fast Fourier transform. *Comm. A.C.M.* 10, 647, 1967.
15. GLASSMAN, J. A. A generalisation of the fast Fourier transform. *I.E.E.E. Trans.* (Comp.) C-19, 2, 165, 1970.
16. SINGLETON, R. C. A method for computing the fast Fourier transform with auxiliary memory and limited high-speed storage. *I.E.E.E. Trans.* (Audio and Electroacoust.) AU-15, 2, 91, 1967.
17. SINGLETON, R. C. An algorithm for computing the mixed radix fast Fourier transform. *I.E.E.E. Trans.* (Audio and Electroacoust.) AU-17, 2, 93, 1969.
18. WEINSTEIN, C. Roundoff noise in floating point fast Fourier transform computation. *I.E.E.E. Trans.* (Audio and Electroacoust.) AU-17, 3, 1969.
19. COOLEY, J. W., LEWIS, A. W. and WELCH, P. D. The fast Fourier transform and its applications. *I.E.E.E. Trans.* (Ed.) E-12, 1, 27, 1969.
20. WELCH, P. A fixed-point fast Fourier transform error analysis. *I.E.E.E. Trans.* (Audio and Electroacoustics) AU-17, 2, 151, 1969.

Additional references

21. WIENER, N. Generalised harmonic analysis. *Acta. Math. Stockh.* 55, 117, 1930.
22. PAPOULIS, A. *The Fourier Integral and its Application.* McGraw-Hill, New York, 1962.

23. SHANNON, C. E. A mathematical theory of communication. *Bell Syst. Tech J.* 27, 623-56, 1948.
24. LANBER, R. A comparison of Fourier analysis methods using an analog computer. *Proc. Fourth A.I.C.A. Conf.* Brighton, 1964.
25. COOPER, G. and BROOM, P. Fourier spectrum analysis by analog methods. *Instrs. Control Syst.*, 35, 1962.
26. BROCK, J. J. Analog cross-spectral density analysis—applications and limitations. Lecture series, Applications and methods of random data analysis, (Conf.) Southampton University, July, 1969.
27. GOOD, I. J. The interaction algorithm and practical Fourier series. *J. Roy. Stat. Soc.*, B, 20, 361, 1960.
28. McCOWAN, D. W. Finite Fourier transform theory and its applications to the computation of convolutions, correlations and spectra. *Earth Sci. Div., Tel. Ind., Virginia U.S.A. Tech. Memo,* 8-66, 1966.
29. PEASE, M. C. An adaptation of the fast Fourier transform for parallel processing. *J. A.C.M.* 15, 2, 1968.
30. COOLEY, J. W., LEWIS, A. W. and WELCH, P. D. Historical notes on the fast Fourier transforms. *Proc. I.E.E.E.* 55, 1675, 1967.
31. COOLEY, J. W. The fast Fourier transform algorithm, programming and accuracy considerations in the calculations of cosine, sine and Laplace transforms. Lecture series, Applications and methods of random data analysis, Southampton University, July, 1969.
32. ROBINSON, E. A. *Multichannel Time Series Analysis with Digital Computer Programs.* Holden-Day Inc., San Francisco, 1967.
33. GROGINSKY, H. L. and WORKS, G. A. A pipeline fast Fourier transform. *I.E.E.E., Trans,* (Comp.) C-19, 11, Nov. 1970.
34. CORINTHIOS, M. J. The design of a class of fast Fourier transform computers. *I.E.E.E. Trans.* (Comp.) C-20, 6, 617-22, June 1971.
35. PEASE, M. C. Organisation of a large-scale Fourier processor. *J. A.C.M.* 16, 474-82, July 1969.
36. GOLD, B. The F.D.P., a fast programmable signal processor. *I.E.E.E. Trans.* (Comp.) C-20, 1, Jan. 1971.

Chapter 7

DIGITAL FILTERING

7-1 INTRODUCTION

The three-point smoothing routine referred to in chapter 1 forms an example of a digital filter. This averaging process can be extended to take into account a much larger number of data samples to increase the effectiveness of the process. The process of digital filtering thus represents an operation on a discrete series of input values such that the output series produced is dependent on the input series and the modifying coefficients defining the filter characteristics. It is necessary to assume that the process is a linear one so that the principle of superposition applies to the input and output series.

The principal use of the digital filter or numerical filter is to smooth a data series in the time or frequency domain. It will show the characteristics of its analog counterpart, i.e. low-pass, band-pass, high-pass etc., but can also have some properties not possible for the analog form. As an example we can consider the filter to operate in other than real time so that its response to a unit impulse may not be zero for $t < 0$. It may also be arranged to have zero phase characteristics and accept data in a reverse order. This is because, unlike the analog filter, it does not have to be physically realisable. Digital filters are especially economical for very low frequency operation in the region of 0·01 to 1 Hz where the size of analog elements would be prohibitive and where the simulation of filter characteristics by means of cascaded integrators proves extremely difficult. Realisation of filter characteristics is also somewhat easier with digital filters due to the effective isolation of each resonant filter element so that elaborate transforms, equivalent to complex analog networks are not required.

We can approach the specification of a digital filter via either the time domain or the frequency domain.

The impulse response function in its discrete form as a number sequence can be used to specify the filter in the time domain. The sequence may be used directly to form a convolution of the signal with the response function, or indirectly via a transformation in the frequency domain. The advent of the fast

Fourier transform algorithm has played an important role in the indirect filtering operation from the time series which often proves more economical than the direct method.

Frequency specification in terms of amplitude and phase characteristics expresses the 'classical' method of considering continuous analog filters, and may also be used with discrete digital filters. Since a wealth of design information on the transfer function characteristics of continuous filters is available, synthesis methods have been developed to permit conversion of these characteristics into discrete forms so that the design of a digital filter can proceed.

The definition of a transfer function in terms of its polynomials was used in chapter 3 to derive continuous (analog) filter representations. Modern network synthesis is based on a quite different representation, that of poles and zeros. Before we consider a response function approach to digital filter design we must first understand the relationship between these two forms of filter representation.

7-1-1 POLE AND ZERO LOCATIONS OF TRANSFER FUNCTIONS

As described earlier, a linear system, such as a filter, can be characterised by a complex transfer function expressed in Laplace form as:

$$H(s) = \frac{Y(s)}{X(s)} = \frac{a_0 s^n + a_1 s^{n-1} + a_2 s^{n-2}, \ldots, a_n}{b_0 s^m + b_1 s^{m-1} + b_2 s^{m-2}, \ldots, b_m}. \tag{7-1}$$

This can be factored into roots of the numerator and denominator polynomials,

$$H(s) = K \frac{(s - a_1)(s - a_2), \ldots, (s - a_n)}{(s - b_1)(s - b_2), \ldots, (s - b_m)} \tag{7-2}$$

where a_1, a_2 etc., designate the roots of $Y(s)$ and b_1, b_2, etc., designate the roots of $X(s)$. When the complex frequency $s = j\omega$ assumes any value, $a_1, a_2,$ ..., a_n the system response is zero. When it assumes any value, $b_1, b_2, \ldots, b_m$ the response is infinite. These values of s are the zeros and poles of the transfer function, $H(s)$ respectively.

The substitution of $s = j\omega$ in equation (7-2) will enable the frequency behaviour of the filter to be determined. It enables filter performance to be completely specified by means of a complex frequency diagram (s-plane), in terms of real and imaginary values and a constant filter gain, K. Figure 7-1 shows a typical pole–zero configuration for a continuous stable filter. Poles are represented by crosses and zeros by circles. It can be shown that the criterion for stability is that all the poles must lie in the left-hand half of the s-plane. Also since the filter impulse response, as representative of a physical system, is a real function of time, then any unreal poles or zeros must exist in complex conjugate pairs. The number of poles must be exactly equal to the number of zeros,

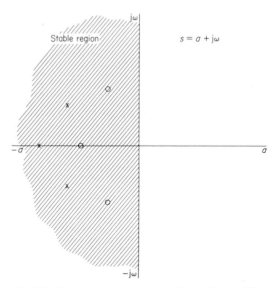

Fig. 7-1. Pole-zero representation of a continuous filter.

although it is possible for certain zeros to be placed at infinity (e.g. if s is very large).

This representation in terms of poles and zeros can yield the frequency and phase characteristics of the filter either by graphical methods from the s-plane, or by substitution of the complex frequency for s in the factored form of the transfer function.

7-1-2 DISCRETE FILTER REPRESENTATION

Equation (7-1) is a form of differential equation applicable to the representation of continuous filters. In the discrete case we use the linear difference equation:

$$y_i = \sum_{k=0}^{P} b_k y_{i-k} + \sum_{k=0}^{M} a_k x_{i-k} \quad (i = 1, 2, \ldots, N) \tag{7-3}$$

where P and M are positive integers and a_k and b_k are real constants. When $M = 0$ then the filter is autoregressive and implies a form of feedback analogous to the analog active filter. This is known as the recursive filter. For $P = 0$ then equation (7-3) reduces to a non-recursive form, which is analogous to the analog passive filter and is hence inherently stable.

The two classes of filter have quite different properties. The non-recursive filter represents the summation of a limited number of input terms and thus has a finite memory. It has excellent phase characteristics but requires a large number of terms to obtain a relatively sharp attenuation characteristic.

The recursive filter represents the summation of both input and output terms

282

so that it can be considered as having an infinite memory. It requires relatively few terms for a similar attenuation characteristic, but will possess poorer phase performance.

The appropriate transform for difference equations is the z-transform [1], which performs the same role with difference equations as the Laplace transform carries out for differential equations. The use of the z-transform permits the specification of a digital filter from the continuous transfer function directly in terms of delays, multipliers, and adders which is the correct form for hardware filter implementation [2]. Its use for 'software' filters permits a convenient form of expression for the difference equations derived from the general form of equation (7-3), particularly when recursive filters are considered. It is of less value when used in the implementation of software non-recursive filters since these are inherently stable, and generally imply the use of Fourier transform methods.

In summary we can recognise three fundamental techniques used in digital filtering. These are:

(*a*) convolution (direct filtering),
(*b*) Fourier transformation (indirect filtering),
(*c*) autoregression (use of difference equations).

The first two techniques are generally confined to the non-recursive filter and the third to the recursive filter.

Essentially the synthesis of a digital filter follows the form familiar with analog filters. That is, we need to choose the ideal desired characteristic in the frequency domain, determine an acceptable approximation for this and finally synthesise the filter by the calculation of the filter weights. This last requirement is, of course, equivalent to the calculation of the analog filter lumped constants (L, C, R etc.). With the recursive filters to be discussed later then the synthesis used has close parallels to that of analog filter specification using cascaded integrators rather than the complex impedance network which uses a single active element.

Before filter synthesis is considered in detail the behaviour of filters in the time and frequency domain will be described.

7-1-3 THE IMPULSE RESPONSE FUNCTION

Considering first the filter operation in the time domain. A linear network system operating as a filter can be represented simply by a four-terminal black box, as shown in fig. 7-2. The characteristics of the system can be determined

Fig. 7-2. A black box representation of a linear network system.

precisely by applying to its input terminals a known function and measuring the function appearing at its output terminals. Let this input be a Dirac Impulse Function $\delta(t)$, (fig. 7-3) defined as:

$$\delta(t) = \lim_{\substack{H \to \infty \\ \Delta t \to 0}} f(t) \tag{7-4}$$

where the integral of $f(t)$ is always unity, i.e.

$$\int_{-\infty}^{\infty} f(t) \, dt = \lim_{\substack{H \to \infty \\ \Delta t \to 0}} H\Delta t = 1. \tag{7-5}$$

The total response of the system to this unit impulse is determined by the system characteristics, expressed as a weighting function $h(t)$, which in this case can be termed the Impulse Response Function. $h(t)$ uniquely defines the system and will always follow the same form for a Dirac input function. Thus for a second order damped system corresponding to a narrow-band resonant filter the impulse response would be the decaying wave-form shown in fig. 7-4 which characterises the performance of the system.

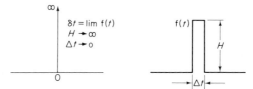

Fig. 7-3. The Dirac impulse function.

This argument can be extended to any arbitrary wave-form $x(t)$ applied to the system by using the principle of superposition. Referring to fig. 7-5, we assume that the driving wave-form consists of a series of narrow samples, each representing a delta function multiplied by a coefficient proportional to the height of the curve. This is a direct consequence of equation (7-4) since the integral of a function multiplied by δt must be equal to the function itself at $t = 0$. Thus we can define the output wave-form as:

$$x(t) = \int_{0}^{t} x(\tau).\delta \, (t - \tau).d\tau \tag{7-6}$$

where $x(\tau)$, represents the wave-form coefficient at $t = \tau$ and, since the unit impulse response of a physically realisable linear system is zero for negative values of τ, then we can replace the infinite lower limit of equation (7-5) by zero. The response function to this arbitrary input wave-form, $x(t)$ will be the total sum of the input response functions for each individual unit impulse thus:

$$y(t) = \int_{-\infty}^{\infty} x(\tau).h(t - \tau).d\tau. \tag{7-7}$$

284

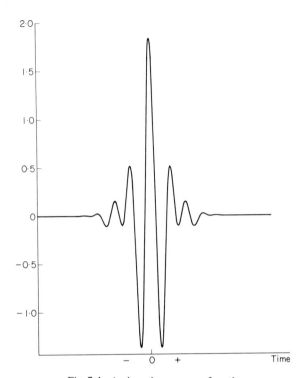

Fig. 7-4. An impulse response function.

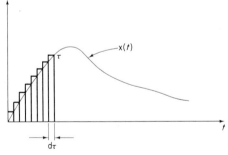

Fig. 7-5. Synthesis of a driving wave-form.

We can consider equation (7-7) to be obtained from the product of the instantaneous value of $x(t)$, namely $x(\tau)$ and its weighting function, $h(t - \tau)$, together with current values of impulse response functions initiated earlier. This process is illustrated in fig. 7-6 for positive time only. The output function $y(t)$ expresses the convolution product $x(t)$ and $h(t)$ and can be written:

$$y(t) = x(t)*h(t). \qquad (7\text{-}8)$$

285

Equation (7-7) represents a convolution integral and may also be written:

$$y(t) = \int_{-\infty}^{\infty} h(\tau)x(t - \tau).d\tau \qquad (7\text{-}9)$$

The weighting function of time, $h(\tau)$ describes the system characteristics in the time domain and, as included in equation (7-9), would consist of a continuous function. A discrete form of equation (7-9) will be considered in section (7-2), where it is used to implement the non-recursive digital filter.

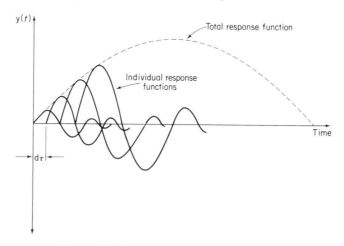

Fig. 7-6. Synthesis of output response function.

7-1-4 THE FREQUENCY RESPONSE FUNCTION

An alternative, and perhaps more familiar way of describing a linear network system such as a filter, is via a frequency domain representation. If we again consider the black box representation of fig. 7-2, but this time consider the input to be an arbitrary frequency function $X(\omega)$, where $X(\omega)$ is the Fourier transform of $x(t)$:

$$X(\omega) = \int_{-\infty}^{\infty} x(t) \exp(-j\omega t).dt \qquad (7\text{-}10)$$

then the output frequency response will be determined by the system transfer function $H(\omega)$ or frequency response function, where (ω) will be a complex quantity. Since the transfer function also uniquely defines the system in the frequency domain, we would expect a relationship to be obtained between this and the impulse response function, $h(t)$ in the time domain. They are, in fact, related by a Fourier transform pair, viz.

$$H(\omega) = \frac{1}{2\pi} \int_{-\infty}^{\infty} h(t) \exp(-j\omega t).dt \qquad (7\text{-}11)$$

and

$$h(t) = \int_{-\infty}^{\infty} H(\omega) \exp{(j\omega t)}.d\omega. \qquad (7\text{-}12)$$

The output of the system $y(t)$ can be obtained by taking the Fourier transform of equation (7-7) viz.

$$Y(\omega) = \frac{1}{2\pi} \int_{-\infty}^{\infty} \exp{(-j\omega t)}.dt \int_{-\infty}^{\infty} x(\tau)h(t - \tau).d\tau$$

If we let the variable of integration be, $p = t - \tau$, then:

$$Y(\omega) = \frac{1}{2\pi} \int_{-\infty}^{\infty} \exp{[-j\omega(p + \tau)]}.dp \int_{-\infty}^{\infty} x(\tau)h(p).d\tau$$

$$= \left[\int_{-\infty}^{\infty} x(\tau) \exp{(-j\omega\tau)}.d\tau \right]\left[\frac{1}{2\pi} \int_{-\infty}^{\infty} h(p) \exp{(-j\omega p)}.dp \right].$$

We recognise the bracketed terms as the Fourier transforms of the input function, $X(\omega)$ and the transfer function, $H(\omega)$ respectively which enables the important relationship between them to be stated as:

$$Y(\omega) = X(\omega).H(\omega) \qquad (7\text{-}13)$$

This may directly be compared with equation (7-8). We may note here the significant result that a product in the frequency domain transforms to a convolution in the time domain.

This relationship provides a method of obtaining filtering action using the Fourier transform of equation (7-12), thus.

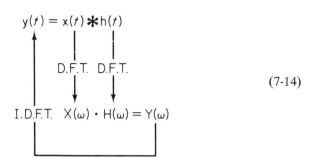

$$ (7\text{-}14) $$

This method is particularly valuable with digital computation, since it proves more economical in computing time to obtain a convolution of two time series by the product of their transforms and subsequent inverse transformation, than the direct product method of evaluating summations.

It should be noted that the transfer function defined by equation (7-11) is a complex quantity, $H(\omega) = a(\omega) + jb(\omega)$, where $a(\omega)$ represents the real part and

287

$b(\omega)$ the imaginary part. Consequently the amplitude characteristic of a smoothing filter defined from $H(\omega)$ is given as

$$|H(\omega)| = [a^2(\omega) + b^2(\omega)]^{\frac{1}{2}} \qquad (7\text{-}15)$$

and the phase characteristic by

$$\phi(\omega) = \arctan\,[b(\omega)/a(\omega)]. \qquad (7\text{-}16)$$

7-2 THE NON-RECURSIVE DIGITAL FILTER

Ideal filter representations in the frequency domain are shown in fig. 7-7. The approximation problem in filter design (analog or digital) is to decide upon a suitable criterion or set of criteria which will permit a satisfactory approximation to these ideal characteristics to be realised. Whatever criterion is selected several basic limitations to the filter model will be found. With digital filter

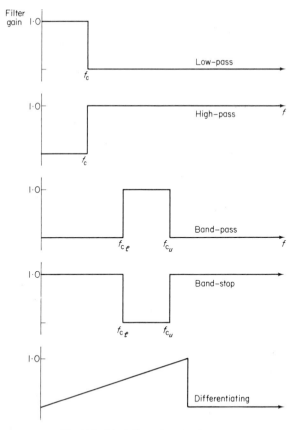

Fig. 7-7. Ideal filter characteristics.

design the most common limitation is that for a given number of filter weights a steep change in slope is likely to be accompanied with considerable frequency ripple in the realised filter shape corresponding to large amplitude side-lobes present in its impulse response function. The non-recursive digital filter consists of the summation of the products of M weighting coefficients h_k and the present and past N samples of the signal wave-form,

$$y_i = \sum_{k=0}^{M-1} h_k.x_{i-k} \quad (i = 1, 2, \ldots, N; k = 0, \ldots, M-1) \quad (7\text{-}17)$$

which will be recognised as the discrete sample equivalent to the continuous convolution equation of (7-9). The phase characteristics of such a filter are linear

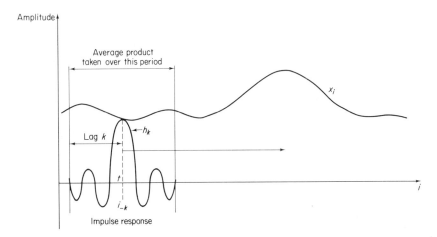

Fig. 7-8. Evaluation of a moving time average.

and under some circumstances can be made constant at all frequencies. The problems associated with frequency aliasing will be present in filter design and in order to avoid these it is necessary to ensure that the signal sampling rate is at least twice that of the highest frequency of interest, f_s applied to the filter.

The numerical filter itself consists of a set of fixed weights, h_k and the process of filtering consists of the determination of the sum of products $h_k . x_{i-k}$ from the set of filter weights, at each sampled data value. This is illustrated in fig. 7-8 where the set of filter weights, h_k, form an impulse response function over a time interval smaller than the time period of the signal, x_i. For convenience the function h_k is assumed to move along the time axis from left to right. At each point along the time axis the average product of $h_k . x_i$ is determined so that the entire process can be considered as the production of a 'moving average' of the data series as the filter is moved along it. Note here that the discrete impulse response function, defined by h_k, does not have to be zero

289

for $\tau < 0$ as with a real physical system. We will see later that the determination of filter weights via the Fourier transform commences with a two-sided response in the frequency domain and results in an impulse response function, or series of filter weights, symmetrically disposed about $\tau = 0$.

The moving average value produced by the process of convolution described above will be dependent on the frequency content of the data time history. As an example we may consider the representation of a low-pass filter where the filtered output will be unchanged (i.e. the convolved value, $\Sigma\, h_k \cdot x_{i-k}$ will be unity) for frequencies below cut-off frequency. Above this frequency the time history will contain several alternations and the incremental convolved values, $h_k * x_{i-k}$ will have alternative signs and will therefore tend to cancel during the summation process. The non-recursive filter may thus be considered as a direct equivalent of the analog filter, but with the operation considered in terms of the time domain rather than the frequency domain. The frequency characteristic of the filter may be derived from equation (7-12), replacing the filter transfer function by its discrete Fourier transform

$$Y(\omega) = X(\omega) \sum_{k=-M}^{M} h_k \exp\left(-j\omega kT\right) \tag{7-18}$$

where T is the constant time interval between successive samples. We can write for the transfer function of the filter:

$$H(\omega) = \frac{Y(\omega)}{X(\omega)} = \sum_{k=-M}^{M} h_k \exp\left(-j\omega kT\right). \tag{7-19}$$

This is a complex number and can be expressed as:

$$H(\omega) = h_k \cdot \cos\theta(k) + j\, h_k \cdot \sin\theta(k) \tag{7-20}$$

where the filter gain is $|h_k|$ and the phase shift is $\theta = \omega T$. If we now let $h_k = h_{-k}$ for all the values of k from $k = 0$ to $k = M$ (equivalent to a phase shift of zero or π radians) then the transfer function will become a real number for any value of k and we can write:

$$H(\omega) = h_0 + 2. \sum_{k=1}^{M} h_k \cdot \cos\omega kT \tag{7-21}$$

which defines a cosine non-recursive filter having zero phase shift. Similarly, for $h_{-k} = -h_k$ (equivalent to a phase shift of $+\frac{1}{2}\pi$ radians), then $H(\omega)$ will become purely imaginary and will define the sine non-recursive filter.

$$\frac{H(\omega)}{j} = 2. \sum_{k=1}^{M} h_k \cdot \sin\omega kT. \tag{7-22}$$

This possibility of obtaining zero or fixed phase shift at all frequencies is a particularly valuable feature of digital filters.

290

The discrete Fourier transform used in the derivation of equations (7-21) and (7-22) represents a finite approximation to the infinite integral of equation (7-11). As a consequence the Fourier series will not converge smoothly and errors will result from the truncation of the infinite series. Methods of minimising this error by modification to the Fourier series will be considered in section 7-2-2 when the derivation of weights for the non-recursive filter is discussed.

7-2-1 THE USE OF THE FAST FOURIER TRANSFORM

The direct method of evaluating the integral given by equation (7-17) proves uneconomic when large numbers of sample values are considered so that the indirect method is preferred. This makes use of the equivalence of a product in the frequency domain and a convolution in the time domain, given by equation (7-14). Here the fast Fourier tranform can be used to reduce substantially the amount of computation time required.

To carry out convolution in this way the signal and the impulse response function are transformed into the frequency domain, their products formed, and the new series inversely transformed back into the time domain to provide a filtered version of the original signal. Unfortunately due to the need for handling complex numbers at all stages of these operations and the need to avoid circular-convolution, the length of data that can be handled in the main storage of a digital computer is limited. For a 32K word processor, for example, using Fortran as the source language, the limitations of the method can result in only 1024 points being convolved. This is due partly to the need to store a complex variable in 2- or 4-digital words and the Cooley–Tukey F.F.T. algorithm which requires that the length of the data be a power of 2. Additionally to avoid circular convolution, zeros, equal in number to the length of the signal, need to be inserted in the signal series.

Fortunately certain techniques are available which permit the convolution of a long data series by a process of segmentation and continuous convolution which uses less in-core storage. These are known as the 'select-save' [3] and 'overlap add' [4] methods. The first of these is described here.

If we consider the signal series, x_i subject to a filter series of length, h_k where $i = 0, 1, \ldots, (N - 1); k = 0, 1, \ldots, (M - 1)$; and $M \ll N$ then equation (7-17) (repeated below),

$$y_i = \sum_{k=0}^{M-1} h_k x_{i-k}$$

represents a difference equation describing the non-recursive filter. A procedure will now be evolved which considers the repeated convolution of a modified filter series h_l with the total signal series x_i divided into a number of smaller series x_m, each of identical length to the modified filter sereis h_l. The addition of these fractional convolutions will be shown to be equivalent to the complete convolution given by equation (7-17). If we consider a signal series; x_m where $m = 0, 1, \ldots, (L - 1)$, and $L < N$, and a modified filter series, h_l where $l = 0, 1, \ldots, (L - 1)$. Then if

$$X_n = \text{D.F.T. of } x_m = \frac{1}{L} \sum_{m=0}^{L-1} x_m \exp(-j2\pi mn/L) \qquad (7\text{-}23)$$

and

$$H_n = \text{D.F.T. of } h_l = \frac{1}{L} \sum_{l=0}^{L-1} h_l \exp(-j2\pi ln/L) \qquad (7\text{-}24)$$

we can write the convolved product of x_m and x_l as the I.D.F.T. of $X_n . H_n$, i.e.

$$y_p = \sum_{n=0}^{L-1} X_n H_n \exp(j2\pi pn/L) \qquad (7\text{-}25)$$

substituting for X_n and H_n and re-arranging the summations:

$$y_p = \frac{1}{L^2} \sum_{m=0}^{L-1} \sum_{l=0}^{L-1} x_m h_l \sum_{n=0}^{L-1} \exp[k2\pi n/L(p - l - m)] \qquad (7\text{-}26)$$

but

$$\sum_{n=0}^{L-1} \exp[j2\pi n/L \ (p - l - m)] = L \ \text{ if } (p - l - m = qL) = 0 \text{ otherwise}$$

where q is an integer. This follows from the principle of orthogonality. Now, since p, l, m vary in the range 0 to $L - 1$ then q can only be 0 and -1, so that: $p - l - m = 0$ and $l = p - m$, $\quad p - l - m = -L$ and $l = L + p - m$ are the two possible ranges for the variable, h_l and we can substitute these in equation (7-26) and sum over the two sets of values for l, i.e.

$$y_p = \frac{1}{L} \sum_{m=0}^{p} x_m . h_{p-m} + \frac{1}{L} \sum_{m=p+1}^{L-1} x_m . h_{L+p-m}. \qquad (7\text{-}27)$$

As we shall see later when considering correlation this form of expression represents a 'circular' convolution which gives the sum of two functions acting in time opposition so that a particular value obtained at time t is added to a value also present at a time separated by $L/2$ sampling periods from t. To separate the two functions we can consider h_l to consist of the original filter series h_k, containing $M - 1$ terms, plus $L - M$ zeros, so that for $L - 1 > p \geqslant M$ then the

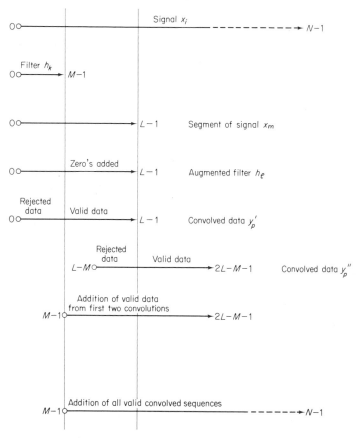

Fig. 7-9. The select-save method of continuous convolution.

second summation in equation (7-27) will be zero and the convolution reduced to:

$$y'_p = \frac{1}{L} \sum_{m=0}^{p} x_m . h_{p-m} \qquad (7\text{-}28)$$

which will be seen to be equivalent to equation (7-17) for the smaller series of $L - M$ terms, taken from the $N - 1$ terms of the total series, if we neglect the scaling term $1/L$. To obtain this result we must therefore reject the first $M - 1$ values of the series in y_p as productive of erroneous circular-convolved data.

This process is shown diagrammatically in fig. 7-9. The lines indicate length of data sequences and terminate with an arrow followed by the number of terms for that sequence. Thus the complete signal x_i to be filtered is shown as a line of length $N - 1$ terms and the filter weights h_k by a much shorter line of $M - 1$ terms. A section of the signal, x_m is chosen of length $L - 1$ terms and this is convolved with an augmented filter h_k of the same length, consisting of $M - 1$ terms of the original filter plus $L - M$ zero terms.

The results of this convolution, y'_p for the first section of the signal are seen to consist of $M - 1$ terms of invalid data which are rejected, and the remaining $L - M$ terms which form the first acceptable fraction of the convolved data. The selected values for x_m forming the next section to be convolved, are taken from terms x_{L-M} to x_{2L-M-1} and a second convolution, y''_p is carried out in the same way. After the rejection of the first $M - 1$ terms this second fraction of the convolved data is added to the first as shown. Continuing in this way and adding the fractional convolutions along a time axis gives a total convolution series equivalent to that obtainable from the direct application of equation (7-17). The missing $M - 1$ terms at the beginning of the signal are included by simply prefixing k zeros to the original data series, x_i. A flow diagram is given in fig. 7-10 which indicate the various stages in the continuous convolution process.

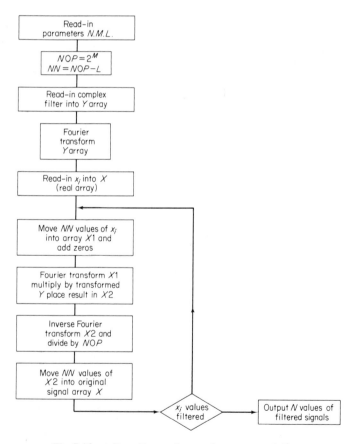

Fig. 7-10. A flow diagram for continuous convolution.

294

It will have become apparent from the recurrent operation of this technique, that an optimum ratio for L/M must exist for a given value of L, since the choice is a compromise between a small number of lengthy convolutions or a larger number of smaller and hence faster convolutions. This is indicated in fig. 7-11, derived from the calculations given by Helms in his paper.

The 'overlap-add' method is essentially similar. In this case $M - 1$ zeros are added to the end of a section of the signal, x_m so that when this is convolved

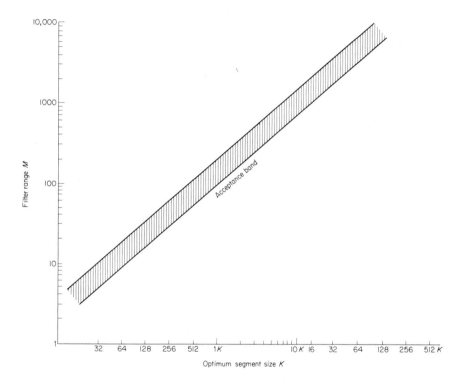

Fig. 7-11. Choice of convolution segment.

with the filter the result contains no circularly-convolved section. Consequently this may be overlapped by $M - 1$ values with the result of the next partial convolution and added to it along the time axis. The computing times required for the two methods are very similar and there seems little to choose between them.

It will have been noted that a scaling factor, $1/L$ is included in the resultant convolution. As stated earlier the averaging term, $1/L$ for the F.F.T. can appear in either the direct transform or the inverse transform. Since the process of obtaining a convolution by this method involves the inverse transformation of the product of two transforms, we can arrive at a situation in which the convolved value can be averaged by either $1/L$ or $1/L^2$, dependent on where we choose to include the averaging term. If this is included in the transform, as shown in equation (6-69), then the final output filter values will need to be multiplied by L to achieve a zero insertion loss.

7-2-2 CALCULATION OF FILTER WEIGHTS

Synthesis procedures for non-recursive filters can derive from any representation of the desired filter performance. If the required transfer function is known analytically, this can be directly transformed to the impulse response by way of the inverse Fourier transform. Alternatively the required frequency characteristics can be stated in graphical form, samples taken from this, and transformed to the time domain to form a series of filter weights. In either case an infinite series of non-zero Fourier series coefficients are required to represent the transformed frequency characteristic exactly. Since this is not possible an approximation to this must be obtained by truncation of the Fourier series. The nature of the approximations are well known and lead to the Gibbs phenomenon [5] which manifests itself as a fixed percentage overshoot and ripple before and after the truncation discontinuity.

In a practical case a modified Fourier series is used in which a weighting function is applied to the raw Fourier series. This time-limited weighting function is called a Window. A major requirement for this is that its defining impulse response function must contain most of its energy in the main lobe, and have relatively small side lobes. A window used for digital filter work is that of a cosine bell, having the form:

$$W_k = \tfrac{1}{2}\left(1 + \cos\frac{2\pi k}{N}\right) \quad (-\tfrac{1}{2}N \leqslant k < \tfrac{1}{2}N) \tag{7-29}$$

which contains 90% of its energy in the main lobe and peak amplitude of the side-lobes reduced to about 2% of the fundamental peak. The effect of this in modifying a rectangular frequency function, (corresponding to truncation of a low-pass filter impulse response function) is shown in fig. 7-12. This improves the ripple in the pass-band at the expense of a slower fall-off at the band edges. An improved performance is obtained by the use of the Dolph–Chebychev function [6]:

$$W_k = \frac{\cos\left(P \arcos\left(E \cos \pi k/P\right)\right)}{\cosh\left(P \operatorname{arcosh} E\right)} \quad (k = 0, 1, \ldots, P - 1) \tag{7-30}$$

where E is a function of frequency ripple amplitude. This has been shown [7] to result in a maximum ratio of main lobe width and side lobe ripple for a given number of terms and allows the numerical specification of ripple and resolution.

The choice of a suitable smoothing window to minimise the effects of truncation of the Fourier transform will be discussed in more detail in the next chapter.

Before describing a general procedure for evaluating the filter weights it will be helpful to consider the equivalence between convolution and transformed products derived earlier. It has been shown (section 7-1-2) that convolution in the time domain transforms to the products of transformation in the frequency domain, i.e.

$$y(t) = h(t) * x(t) \Longleftrightarrow Y(\omega) = H(\omega).X(\omega) \qquad (7\text{-}31)$$

Where the quantities are expressed in discrete form the domain identification is lost, since we are manipulating a series of discrete numbers rather than a time or frequency-dependent continuous function. Hence it is equally valid to state:

$$y_i = h_k.x_i \Longleftrightarrow Y_i(\omega) = H_k(\omega) * X_i(\omega) \qquad (7\text{-}32)$$

providing the equations are dimensionally correct.

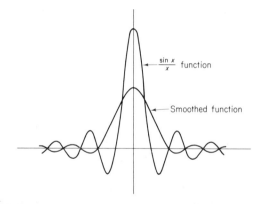

Fig. 7-12. Effect of smoothing the impulse response by means of a cosine bell.

A procedure for the derivation of the required filter weights via the F.F.T. can be stated as follows:

1. The required response in the positive frequency domain is drawn and repeated in reversed order over the negative frequency region, fig. 7-13. The sharpest discontinuity is noted and its frequency duration, f_d is divided into the Nyquist frequency $(\frac{1}{2}f_s)$. For an overshoot of less than 0·3% peak in the achieved filter response the minimum number of filter weights n is given by Kuo and Kaiser [8] as:

$$n \geqslant \frac{8(\frac{1}{2}f_s)}{f_d}. \qquad (7\text{-}33)$$

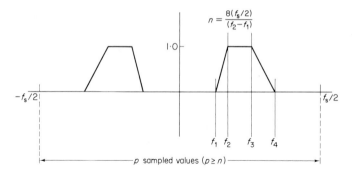

Fig. 7-13. Representation of desired frequency response.

2. Samples of the desired frequency response are taken at intervals of:

$$\frac{f_s k}{p} \qquad \text{for } 0 \leqslant k < \tfrac{1}{2}p$$

$$f_s\left(\frac{k}{p} - 1\right) \qquad \text{for } \tfrac{1}{2}p \leqslant k < p - 1$$

where $\qquad k = 0, 1, \ldots, p - 1$, and $p \geqslant n$. $\qquad$ (7-34)

The amplitude values of the samples are taken as 1·0 within the pass-band to obtain correct scaling.

By sampling the desired frequency response at a number of frequencies, p, larger than the required number of filter weights it is possible to obtain fairly good approximations to the first n Fourier coefficients.

3. The samples are transformed, using an F.F.T. routine and stored as a complex array.

4. A smoothing window of n points is derived and multiplied point by point with the p transformed samples, such that the n significant products are retained and truncation to n values takes place.

5. The series of n weights form the filter coefficient series, h_k required for the filter.

To test the filter action and to compare this with the desired frequency response h_k can be transformed to form the filter transfer function using equation (7-19).

This procedure for obtaining filter weights is shown in fig. 7-14. Note that the transformation from the required filter frequency response characteristic to the time domain can be obtained with either a D.F.T. or I.D.F.T. since H_k is real. The choice of intermediate amplitude values between 0 and 1.0 in the frequency transition periods is only critical if the filter operates near the Nyquist limit. For

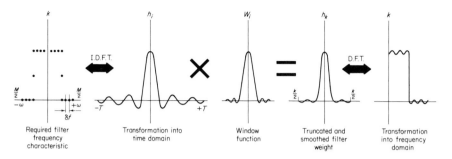

Fig. 7-14. Derivation of filter weights from a sampled frequency representation.

many uncritical situations a linear set of amplitude values would be chosen, as indicated in fig. 7-14.

An alternative method of filter definition is given by Ormsby [9] who found an improvement in filter performance for a given number of filter weights by introducing a specified value of slope into the filter characteristic (shown in fig. 7-15 for the low-pass filter) where the transfer function is:

$$H(\omega) \begin{cases} = 1 & \text{for } 0 \leqslant \omega < \omega_c \\ = \left[\dfrac{\omega - \omega_d}{\omega_c - \omega_d}\right]^n & \text{for } \omega_c \leqslant \omega \leqslant \omega_d \\ = 0 & \text{for } \omega_d \leqslant \omega \end{cases} \qquad (7\text{-}35)$$

and n defines the rate of cut-off for the filter in units of 6 db/octave. The Ormsby filters are characterised by their ability to pass polynomials without change whilst reducing their noise content and, due to the symmetry, they have zero phase shift. A disadvantage is a slow fall-off in gain and the amount of ripple included in the pass-band of the realised filter.

For the low-pass filter it can be shown that a given set of filter weights can be applied for other cut-off frequencies and sampling intervals than the original

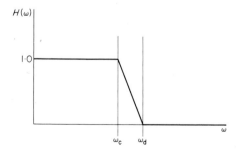

Fig. 7-15. The Ormsby low-pass filter shape.

299

designed values, providing we keep their product constant. Advantage may be taken of this by obtaining a long series of weights for a low value of f_c and small sampling interval, t, and using this as data for a simple decimation program. Filter weights for a given cut-off frequency F_c and sampling interval, T, are obtained by abstracting every Rth point (including the first value) where:

$$R = \frac{F_c T}{f_c t} \qquad (R_{min.} > R \geqslant 1) \qquad (7\text{-}36)$$

and $R_{min.}$ is a minimum value dependent on the attenuation characteristic and product $f_c t$ for the original filter, derived from equation (7-33).

It will be necessary to scale the abstracted weights to produce a filter having zero insertion loss. Thus if the input series x_i have unit value then from equation (7-17):

$$y_i = \sum_{k=0}^{M-1} h_k = 1. \qquad (7\text{-}37)$$

Hence for a given (unscaled) series of filter weights h_k we can derive a new (scaled) series:

$$h_k = \frac{h'k}{H} \qquad (7\text{-}38)$$

where H represents the summation of the unscaled series. This can be expressed simply by saying that for zero insertion loss a normalisation to unity of the area under the impulse response function is required.

7-3   THE RECURSIVE DIGITAL FILTER

A disadvantage of the non-recursive filter is that a large number of filter weights is necessary to approximate the desired function. We may for example, require 200 filter weights to obtain a satisfactory approximation to the ideal shape for a given filter shown in fig. 7-7. This means 200 multiplications/additions per data point, which can lead to lengthy computational time in addition to demanding substantial memory storage.

Recursive filters will reduce both of these requirements by an order of magnitude. They will, however, be found to exhibit a more complex structure in which the output includes the weighted sum of both input and output terms. In this respect they are similar to analog active filters where electrical feedback takes the place of recursion. Due to the economy in filter series length a fast acting filter is obtained which makes it valuable in real-time applications. Whilst implementation of the recursive filter can be carried out using the fast Fourier transform [10] this does not confer such great advantages as with the non-recursive case and the following discussion is limited to direct evaluation of the difference equation.

The recursive filters described below will, in general, have a non-zero phase characteristics. In many applications this is unimportant but where this is likely to affect adversely the transient response to the data then the following procedure can be adopted. A filter is designed to give half the attenuation slope required. The data is passed through the filter twice, once in the usual way, and a second time with the data points arranged in reverse order. The phase changes will be found to cancel and a real value of transfer function obtained. A slight correction will be required for cut-off frequency.

A linear difference equation for the recursive filter was given as equation (7-3), and is known as the compound recursive filter. Its transfer function may be derived in a similar manner to that described earlier for the non-recursive filter as:

$$H(\omega) = \frac{Y(\omega)}{X(\omega)} = \frac{\sum_{k=0}^{M} a_k . \exp(-j\omega kT)}{1 - \sum_{k=0}^{P} b_k \exp(-j\omega kT)} \tag{7-39}$$

which has been shown by Holz and Leondes [11] to be a rational function of sines and cosines having a polynomial form, and therefore equivalent to the general form for the continuous analog active filter given by equation (7-1).

A simpler form of the recursive filter is given as:

$$y_i = Cx_i + \sum_{k=0}^{P} b_k . y_{i-k} \tag{7-40}$$

for which its Fourier transform yields a transfer function:

$$H(\omega) = \frac{Y(\omega)}{X(\omega)} = \frac{C}{1 - \sum_{k=0}^{P} b_k . \exp(-j\omega kT)}. \tag{7-41}$$

This type of relation is somewhat easier to implement since the poles of the transfer function are sufficient to define the filter weights.

A simple example of this is the first order difference low-pass filter given by:

$$y_i = x_i + Ky_{i-1}. \quad (i = 1, 2, \ldots, N) \tag{7-42}$$

The transfer function can be obtained by letting $y_i = A \sin i\omega t$, where $\omega = 2\pi f$ and $f = 1/2T$. Thus

$$y_i = KA \sin i\omega(t - T) + x_i$$

$$= KA \sin i\omega t . \cos i\omega T - KA \cos i\omega t . \sin i\omega T + x_i.$$

But $\sin i\omega T = \sin i(2\pi T/2T) = \sin i\pi = 0$, since i is an integer. Therefore $y_i = Ky_i \cos i\omega T + x_i$ and

$$H(f) = y_i/x_i = \frac{1}{1 - (K \cos i\omega T)}. \qquad (7\text{-}43)$$

K can be related to filter cut-off frequency by defining f_c to be at the half-power point:

$$|H(f)|^2 = \tfrac{1}{2} = \frac{1}{1 + (K^2 \cos^2 i\omega T) - (2K \cos i\omega T)}$$

but $\cos^2 i\omega T = \cos^2 [i(2\pi T/2T)] = \cos i\pi = 1$ since i is an integer, so that

$$K^2 - 2K \cos \omega_c T = 1$$

where $\omega_c = i\omega$ and $K = \cos \omega_c T \pm \sqrt{(\cos^2 \omega_c T + 1)}$. This can be substituted in the recursive expression given in equation (7-42) to enable a simple algorithm to be implemented for the low-pass filter. Note that only one multiplication and one addition is required to realise a single output filter point.

A number of design techniques have been developed for the recursive filter and whilst some methods enable the design to be carried out completely in the frequency domain [12], a considerable simplification ensues if the z-transform is used. A brief introduction to the z-transform is given in the next section.

7-3-1 THE z-TRANSFORM
One way of defining the z-transform is from the Laplace transform and, since filter theory for continuous analog filters is generally expressed in Laplace form, we can consider the z-transform as a logical extension of this for a discrete series.

A sampled series, x_i can be considered to be the product of a continuous signal, $x(t)$ and a set of uniformly spaced unit impulses (see chapter 9). Thus the Laplace transform, $H(s)$ of a sampled series can be expressed as:

$$H(s) = \sum_0^\infty [a_0\delta(t) + a_1\delta(t - T) + a_2\delta(t - 2T) + , \ldots,] \exp(-sT) \quad (7\text{-}44)$$

where $a_0, a_1, a_2, \ldots$, represent the sample amplitudes and T is the sampling interval. Hence

$$H(s) = \sum_{k=0}^\infty a_k \exp(-ksT). \qquad (7\text{-}45)$$

This may be expressed in simplified terms to facilitate the algebraic manipulation by replacing $\exp(sT)$ by z, thus: $\exp(ST) = z$, or

$$\exp(-sT) = z^{-1} \qquad (7\text{-}46)$$

and by replacing $H(s)$ by $H(z)$ so that:

$$H(z) = \sum_{k=0}^{\infty} a_k z^{-k} \tag{7-47}$$

$H(z)$ is thus by definition the z-transform of x_i and represents a power series in z^{-1} with coefficients, a_k representing the amplitude of successive samples of x_i in the time domain.

Tables of z-transforms for sampled data series can be obtained similar to the Laplace transforms derived for a continuous signal by using the transformation given by equation (7-46). It is convenient to use the transform for z^{-1} rather than z since the multiplicative factor of z^{-1} is equivalent to a delay of T, or a delay of one sampling period.

Some properties of the z-transform will now be derived by consideration of a number of typical functions for x_i.

(i) Step function: $x_i = \begin{cases} 1 & : n \geqslant 0 \\ 0 & : n < 0. \end{cases}$

From equation (7-47)

$$H(z) = z^0 + z^{-1} + z^{-2}, \ldots, z^{-n}.$$

If $z^{-1} < 1$ then the series converges to:

$$H(z) = \frac{1}{1 - z^{-1}} \tag{7-48}$$

(ii) $x_i = a^n$.

$$H(z) = \sum_{n=0}^{\infty} a^n z^{-n} = \sum_{n=0}^{\infty} (az^{-1})^n$$

which converges to:

$$H(z) = \frac{1}{1 - az^{-1}}. \tag{7-49}$$

(iii) A sampled data series: $x_i = x[nT] = \exp(-anT)$.

$$H(z) = \sum_{n=0}^{\infty} \exp(-anT)z^{-n} = \frac{1}{1 - \exp(-aT)z^{-1}}. \tag{7-50}$$

(iv) $x_i = na^n$.

$$H(z) = \sum_{n=0}^{\infty} na^n z^{-n} = z \sum_{n=0}^{\infty} na^n z^{-(n+1)}$$

$$= z \sum_{n=0}^{\infty} a^n \frac{dz^{-n}}{dz}$$

$$H(z) = z \left[\frac{d}{dz} \sum_{n=0}^{\infty} a^n n^{-n} \right]$$

303

so that substituting $H(z)$ from equation (7-49) and differentiating:

$$H(z) = \frac{az}{(1 - az^{-1})^2},$$ (7-51)

from this example we can obtain the general relationship:

$$H[(z)nx(t)] = z\left[\frac{dH(z)[x(t)]}{dz}\right].$$ (7-52)

(v) A sampled data series delayed by K units: $x_i = x[(n - k)T]$.

$$H(z) = \sum_{n=0}^{\infty} z[(n - k)T]z^{-n} = z^{-k} \sum_{n=0}^{\infty} x[(n - k)T]z^{-(n-k)}$$

which, from equation (7-47),

$$H(z) = z[x(n - k)T] = z^{-k}[x(nT)].$$ (7-53)

This is the delay property or shifting theorem of the z-theorem. Each delay by one sampling interval corresponds to a multiplication by z^{-1} in the z-domain.

Thus $H(z) = z^{-2}$ corresponds to a sample taken with a delay of two sampling units, i.e. x_{i-2}. Similarly $H(z) = z^3$ corresponds to a forward shift in time by three sampling intervals, i.e. x_{i+3}.

In general the z-transform possesses equivalent properties to those of the Laplace transform. The z-transfer function, $H(z)$ is equal to the ratio of the output and input functions expressed as z transforms. Multiplication of functions by z is equivalent to convolution of functions of time. Finally a complex function can be expressed graphically on a z-plane in much the same way as it can be described in terms of the s-plane. The two representations are related in a manner which will be discussed in section 7-3-2.

An important application of the z-transform which we shall be using in this chapter is in the representation of difference equations. Continuous time functions can be regarded as the solutions of differential equations in which integration, represented as $1/s$ in the s-domain, plays an essential part. Similarly, for sampled time functions, these are the solutions of difference equations in which the essential element is that of unit delay, represented by z^{-1} in the z-domain. Both can be represented by block schematic form, as shown in fig. 7-16.

A transfer function for the recursive filter was given in equation (7-39). This can be expressed in z-form as:

$$H(z) = \frac{\displaystyle\sum_{k=0}^{M} a_k z^{-k}}{1 - \displaystyle\sum_{k=0}^{P} b_k z^{-k}}.$$ (7-54)

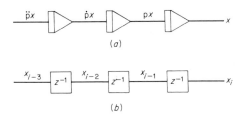

Fig. 7-16. Schematic representation of integration and delay.

The filter is realisable when the filter parameters, a_k and b_k are real and constant and where the poles of $H(z)$ lie inside the unit circle on the z-plane (see later). Equation (7-54) can also be expressed in product form as a general function of z corresponding to equation (7-2).

$$H(z) = \frac{Y(z)}{X(z)} = K \frac{(z - a_1)(z - a_2), \ldots, (z - a_n)}{(z - b_1)(z - b_2), \ldots, (z - b_m)} \qquad (7\text{-}55)$$

where $a_1, a_2, \ldots, a_n$ are zeros, and $b_1, b_2, \ldots, b_m$ are simple poles.

The polynomials in z represented by $Y(z)$ and $X(z)$ have real coefficients since a_n and b_m are all either real or occur in complex conjugate pairs. Since a multiplication in z implies a time shift of one sampling interval the z-transfer function can be converted easily into a difference equation. An example will be given using a transfer function having one zero and two complex poles, viz.

$$H(z) = \frac{Y(z)}{X(z)} = \frac{z - a}{(z - \alpha - j\beta)(z - \alpha + j\beta)} \qquad (7\text{-}56)$$

where a is a zero in the z-plane, and $\alpha \pm j\beta = b_1, b_2$ are complex poles in the z-plane.

This can be re-arranged as:

$$Y(z)[z^2 - 2\alpha z + \alpha^2 + \beta^2] = X(z)[z - a]$$

and divided by z^2 to give:

$$Y(z)[1 - 2\alpha z^{-1} + (\alpha^2 + \beta^2)z^{-2}] = X(z)[z^{-1} - az^{-2}]. \qquad (7\text{-}57)$$

Replacing the present input and output sampled values by x_i and y_i and invoking the shifting theorem then equation (7-57) can be written:

$$y_i - 2\alpha y_{i-1} + (\alpha^2 + \beta^2)y_{i-2} = x_{i-1} - ax_{i-2}$$

or

$$y_i = x_{i-1} - ax_{i-2} + 2\alpha y_{i-1} - (\alpha^2 + \beta^2)y_{i-2}. \qquad (7\text{-}58)$$

305

This recurrence relationship or difference equation allows a new value for the current output value, y_i, to be obtained from past input and past output values, weighted by given z-plane zero and pole coefficients.

7-3-2 MAPPING THE s-PLANE INTO THE z-PLANE

The Laplace transform of a single pole continuous function, $x(t) = a_i \exp(s_i t)$ is by definition:

$$X(s) = \frac{a_i}{(s - s_i)}. \tag{7-59}$$

A sampled data series, $x_i = x(nT)$ may be formed from this continuous series, $x(t)$ by sampling at uniform intervals, separated by T. This can be regarded as a delta series modulated by an amplitude function which, in this case, is an exponential term. Thus:

$$x(nT) = \sum_{n=0}^{\infty} a_i \exp(s_i t)\delta(t - nT)$$

$$= a_i[\delta(t) + \exp(s_i t)\delta(t - T) + , \ldots , + \exp(s_i t)\delta(t - nT)]. \tag{7-60}$$

This represents a series of delayed impulses in which the coefficients, $a_i \exp(s_i T)$, a_i, $\exp(2s_i T)$ etc., are constants.

The Laplace transform of a unit impulse is given as $\exp(-snT)$, so that the transform of the sampled data series may be represented as:

$$X(s) = \sum_{n=0}^{\infty} a_i \exp(s_i - s)T = a_i[1 + \exp(s_i - s)T + \exp(s_i - s)2T + , \ldots ,$$

$$+ \exp(s_i - s)nT] \tag{7-61}$$

which is a geometrical progression and converges to:

$$X(s) = \frac{a_i}{1 - \exp(s_i - s)T} \tag{7-62}$$

This may be compared with the Laplace transforms of a continuous function given by equation (7-59).

The poles of equation (7·62) occur where $\exp(s_i - s)T = 1$. Thus $(s_i - s)T = 0$ and $s = s_i$ which correspond to the single pole of the continuous function. However we can also find unit value for $\exp(s_i - s)T$ when this is equal to $\exp(\pm j2\pi n)$ where $n = 0, 1, 2, \ldots$, so that: $(s_i - s)T = \pm j2\pi n$, and

$$s = s_i \pm j\frac{2\pi n}{T} = s_i + jn\omega_s \tag{7-63}$$

where ω_s is the sampling frequency, $2\pi f_s$.

Hence the effect of sampling on a continuous form of a transfer function, $H(s)$ will take the form of folding or repetition of the frequency characteristic so that instead of $H(s)$ we obtain a sampled version,

$$H^*(s) = \sum_{n=-\infty}^{\infty} H(s + jn\omega_s) \qquad (7\text{-}64)$$

which is equivalent to:

$$H(z) = T \sum_{n=0}^{\infty} h(nT)z^{-n}. \qquad (7\text{-}65)$$

Thus the pole–zero pattern of the original signal will be repeated at intervals of $2\pi/T$ over the s-plane shown in fig. 7-17. This is a result of the sampling theory

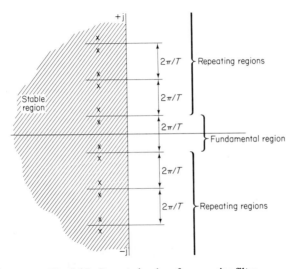

Fig. 7-17. Repeated poles of a recursive filter.

discussed in chapter 4 and has similar consequences. It is essential to ensure that the sampling rate for the data to be filtered is adequate to match the particular filter design if stability is to be preserved.

For wide-band filters where $\omega_c \simeq \omega_s/2$ then folding errors can occur and a solution proposed by Kaiser [8] is to precede the desired wide-band filter $H(s)$ by a low-pass filter, $G(s)$ having a high attenuation slope to give a cascaded filter;

$$H_c(s) = H(s).G(s). \qquad (7\text{-}66)$$

This may be transformed into z-transfer form and implemented. The resultant z-transfer function, $H_c(z)$, will however be complex and contain considerably more terms than the simpler realisation for $H(s)$.

The correspondence between the s-plane and the z-plane can be seen if we write:

$$z = \exp (sT) = \exp (\alpha + j\omega)T \qquad (7\text{-}67)$$

so that:

$$|z| = \exp (\alpha T) \text{ and } \angle z = \omega T$$

and a point s_i in the s-plane will transform to a point z_i in the z-plane (fig. 7-18).

Since the path along the imaginary axis for the s-plane where the poles are situated corresponds to $\alpha = 0$ then $|z| = 1$ and the angle $\angle z$ varies between $\pm \pi$ radians. We thus infer that the $j\omega$ axis in the s-plane maps into a unit radius circle in the z-plane. The left-hand half of the s-plane maps inside the unit circle

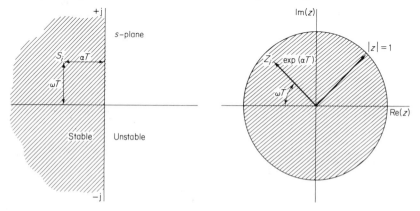

Fig. 7-18. Relationship between the s-plane and the z-plane.

of the z-plane. This is the region of stability for a system where the uncancelled poles must lie within the unit circle. The effect of containing the entire stable region of the s-plane within the unit circle of the z-plane means that all the repeated poles of the sampled function will now result in a single pole within the unit circle in the z-plane. This has the disadvantage that undesirable aliasing is not readily apparent from the z-plane representation.

The z-transform offers considerable manipulational convenience due to the elimination of exponential terms and normalisation with respect to the sampling interval. It also possesses the ability to directly translate a polynomial in z to a difference equation by the use of the shifting theorem. In the next section we will consider synthesis methods using this s- to z-plane transformation.

7-3-3 GENERAL METHODS

Synthesis methods for recursive filters operate either from considerations of the pole–zero requirements or indirectly from the transfer function of the desired continuous filter.

An example of the first of these is the direct synthesis from the impulse transfer function into a z-transform realisation and its implementation as a difference equation or set of difference equations derived from this realisation.

Two examples of the indirect method are:

(a) Frequency domain synthesis from squared magnitude transfer functions. This is similar to a method of design for analog filters, in which the poles and zeros are first determined to define the filter characteristics. From these the filter weights are obtained directly and the recursive expression, given by equation (7-3), is implemented.

(b) Frequency domain synthesis from the continuous analog filter, this time considered in the s-plane and known as the bilinear transform. This method overcomes the folding disadvantage of the standard z-transform by mapping the entire complex plane into a single horizontal strip in the s-plane corresponding to the fundamental region shown in fig. 7-17. It is an indirect method in which this band-limited s-plane realisation is converted into the z-plane and from this a difference equation derived which can be implemented as a recursive algorithm.

(c) Simulation of the filter characteristics in the frequency domain by means of a frequency sampling technique [13]. Here the process of sampling is used to reconstruct the filtered signal from its elemental time responses.

A brief discussion of these methods is given below and followed by a fuller discussion of one of these, namely the bilinear transform method, which is attractive due to its simple realisation and economy.

The direct synthesis from the impulse transfer function is a method of design from the known characteristics of continuous systems in terms of its poles and zeros, and is sometimes referred to as the impulse invariance method. The transfer function is given and a sampled version of the impulse response function obtained, which can be converted into a difference equation and used to simulate the recursive digital filter [8]. Thie method will be explained in terms of a single real pole transfer function:

$$H(s) = \frac{1}{s+a} \tag{7-68}$$

where the single pole at $s = -a$ is known. The Laplace transform of a discrete series of unit impulses $x_i = x(nT)$ has been shown from equation (7-62) for a pole $s_i = -a$ to be:

$$X(s) = \frac{k}{1 - \exp(-a - s)T} \tag{7-69}$$

or

$$H(z) = \frac{k}{1 - \exp(-aT)z^{-1}}. \tag{7-70}$$

The filter can be realised from the recurrence relation:

$$y_i = kx_i + \exp(-aT)y_{i-1}. \tag{7-71}$$

To determine k we equate the gains of the continuous and discrete transfer function at zero frequency. Thus with $s = j\omega = 0$ then equation (7-68) became $1/a$ and that of equation (7-70) becomes $k/[1 - \exp(-aT)]$, so that:

$$k = \frac{1 - \exp(-aT)}{a}.$$ (7-72)

To use this method the continuous transfer function is factorised into one or more elementary transfer functions such as that given by equation (7-68). Each function is converted into its corresponding discrete version and the polynomial in z converted into a difference equation using the shifting theorem. The programming of the recursive algorithm is simplified if the exponential term, $\exp(-aT)$ is calculated outside the recursive loop since this only needs to be calculated once after the sampling rate has been determined. The impulse method has been applied to the calculation of equi-ripple Lerner filters [14] where the impulse response function is given in terms of a number of pole pairs. For filter-bank simulation used for spectral density evaluation the four-pole Lerner filter has an advantage that adjacent filters can share poles and thus reduce the computation required [15].

The non-recursive filters and a number of recursive filters discussed in the previous sections are characterised by symmetrical weighting functions and zero phase shift. Their characteristics in other directions are however far from optimum. A class of filters having asymmetrical weighting functions can be defined by transfer functions having the form:

$$H(s) = \frac{1}{\displaystyle\prod_{i=1}^{m} (s_i + s)}$$ (7-73)

These filters have m poles at finite values of z and zeros at infinity. They can be designed to have optimum amplitude characteristics but will exhibit pronounced phase shift in the region of cut-off frequency.

The most important of these are the Butterworth and Chebychev filters. The Butterworth filter has a modulus squared frequency response function of the form:

$$|H(j\omega)|^2 = \frac{1}{1 + (\omega/\omega_c)^{2n}}.$$ (7-74)

The pole positions for a Butterworth filter will be found to lie equally spaced on a circle of radius, ω. For stability the poles must be contained in the left-hand half of the s-plane. Thus the first pole occurs at an angle of $\pi/2n$ and subsequent poles at angles of π/n, with the final pole at an angle of $\pi/2n$ (fig. 7-19a). The transfer function can be obtained by consideration of the pole locations. For example in the case of a second-order transfer function ($n = 2$):

$$H(s) = \frac{1}{1 + P_a} \cdot \frac{1}{1 + P_b}$$ (7-75)

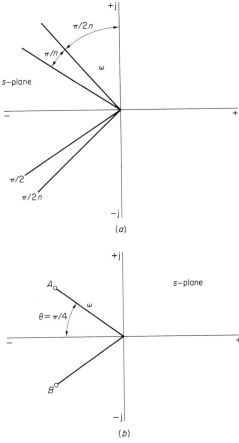

Fig. 7-19. Pole-zero representation for a Butterworth filter.

The poles are located at $\pi/4$ radius with respect to the real axis (fig. 7-19b). Their coordinates are:

$$\left.\begin{array}{l} -\omega \cos \pi/4 + j\omega \sin \pi/4 \text{ for pole } A \\ \omega \cos \pi/4 + j\omega \sin \pi/4 \text{ for pole } B \end{array}\right\} \qquad (7\text{-}76)$$

Hence, substituting in (7-75):

$$H(s) = \cfrac{1}{\left(1 - \dfrac{2^{\frac{1}{2}}\omega}{2} + j\dfrac{2^{\frac{1}{2}}\omega}{2}\right)\left(1 + \dfrac{2^{\frac{1}{2}}\omega}{2} + j\dfrac{2^{\frac{1}{2}}\omega}{2}\right)}$$

$$= \frac{1}{1 + 2^{\frac{1}{2}}j\omega + (j\omega)^2}$$

311

and in normalised form, referred to cut-off frequency ω_c, with $j\omega$ replaced by s

$$H(s) = \frac{1}{1 + 2^{\frac{1}{2}}(s/\omega_c) + (s/\omega_c)^2} \qquad (7\text{-}77)$$

which is the required transfer function.

A digital transfer function closely resembling that of the Butterworth low-pass filter has been proposed by Holz and Leondes [11], viz.

$$|H(j\,\omega)|^2 = \frac{1}{1 + \left[\dfrac{\tan \frac{1}{2}\omega T}{\tan \frac{1}{2}\omega_c T}\right]^{2n}} \qquad (7\text{-}78)$$

where ω_c is the cut-off frequency, defined as the frequency at which the amplitude squared value for $H(\omega)$ is reduced by half from its value at $\omega = 0$ (-3db). This filter will give a maximally flat characteristic (see chapter 3).

A similar form is obtained for the Chebychev filter.

$$|H(j\omega)|^2 = \frac{1}{1 + E^2 V_n^2 \left[\dfrac{\tan \frac{1}{2}\omega T}{\tan \frac{1}{2}\omega_c T}\right]} \qquad (7\text{-}79)$$

where V_n is an nth order Chebychev polynomial and E is related to the fractional pass-band ripple:

$$R = 1 - \frac{1}{(1 - E^2)^{\frac{1}{2}}}. \qquad (7\text{-}80)$$

The phase characteristics for both filters are non-linear but their amplitude characteristics are optimum for a given number of filter weights. A better attenuation characteristic outside the pass-band is achieved by the Chebychev filter at the expense of a controlled amount of constant amplitude ripple in the pass-band given by equation (7-80). Both Butterworth and Chebychev filters exhibit rather lengthy settling time (for a second order Butterworth low-pass filter this can be several periods at the cut-off frequency). An improvement can be obtained with Elliptic filters at the expense of a finite ripple content in both pass-band and stop band [2].

Equation (7-78) can be rewritten in z-transform notation by letting $z = \exp(j\omega T)$ and substituting

$$\tan(\tfrac{1}{2}\omega T) = j\,\frac{\exp(j\frac{1}{2}\omega T) - \exp(-j\frac{1}{2}\omega T)}{\exp(j\frac{1}{2}\omega T) + \exp(-j\frac{1}{2}\omega T)}$$

so that,

$$|H(z)|^2 = \frac{k^n}{k^n + (-1)^n \left[\dfrac{z-1}{z+1}\right]^{2n}} \qquad (7\text{-}81)$$

where $k = \tan^2(\tfrac{1}{2}\omega_c T)$.

The poles of the function are found by substituting $s = (z-1)/(z+1)$ in equation (7-81) and will be found to be uniformly spaced around a circle of radius, $\tan(\tfrac{1}{2}\omega_c T)$ in the s-plane, which thus corresponds with that found for equation (7-74). A similar mapping of the poles for the Chebychev filter will be found to lie on an ellipse in the s-plane and can be determined by a similar substitution of $s = (z-1)/(z+1)$.

The recursive realisation of these filters can be obtained by multiplying out the numerator and denominator polynomials in $H(z)$ and applying the shifting theorem. An alternative realisation has been proposed by Otnes [16] where the filter weights are determined by conformal mapping, retaining the results in terms of a polynomial of complex frequency terms. In this case the coefficients for the recursive equation are evaluated in terms of a natural frequency and damping ratio. A suitable computational procedure for this is described by Enochson and Otnes [12] for a number of filter types. These realisations can lead to difficulties in implementation on the digital computer if high-order filters are attempted. This will be discussed further in section 7-3-9.

The effects of sampling in the time domain has been considered earlier (chapter 4). If a continuous signal, band-limited to $\pm B$ Hz, is sampled, the original signal can be reconstructed by passing the sampled signal through a filter having a $\sin x/x$ shape for its impulse response function. Since each sampled value can be considered as exciting the filter to produce a $\sin x/x$ response we can visualise the simulation of the complete filter response as comprising the summation of each individual sample response in the frequency domain (fig. 7-20). This principle is used in the frequency sampling filter [13]. The signal is applied to a comb filter which is designed to have m zeros equally spaced around the unit circle. The filter is followed by a lossless resonator, defined by two complex conjugate poles which lie directly on the unit circle. If the angle of the resonator pole is made equal to one of the zeros of the comb filter then an output at the fundamental resonant frequency of the resonator is obtained. If this pole-zero cancellation is not obtained then little output results. By applying the comb filter output to m resonators connected in parallel, each resonant at a different discrete frequency related to the 'teeth' frequencies of the comb, then the output response of each resonant filter can be added as a scalar quantity. This is shown in fig. 7-21. The impulse response of the resonators is arranged to have a finite period, mT, to limit its contribution, considered in the time domain, following excitation. The resonators are followed by attenuators set to provide a gain at each resonator frequency corresponding to the required filter

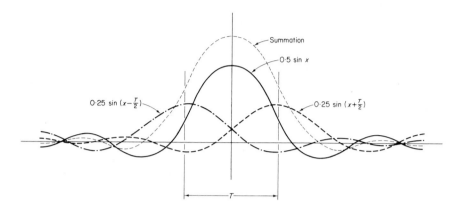

Fig. 7-20. Synthesis method used in frequency sampling filters.

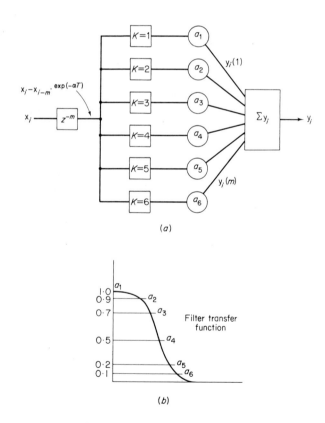

Fig. 7-21. Method of frequency sampling filtering.

response. Summation of the attenuated resonator responses gives a synthesised output corresponding to the filtered signal.

The designed filter has a linear phase characteristic and allows very simple specification of response in the frequency domain (simply the attenuation required at each frequency sampled value). However, since $m - 1$ resonators are required then a substantial amount of memory is required in the computer to retain the intermediate results before attenuation and addition are carried out to form the final output. These storage requirements approach those needed for non-recursive filters. Realisation of the comb filters response is obtained from the difference equation:

$$y_i = x_i - x_{i-m} \cdot \exp(-\alpha T) \tag{7-82}$$

which requires simply a series of m delays. The loss-less sinusoidal resonator is easier to achieve with a digital recursion than its analog counterpart and can be represented as:

$$y_i = x_i + y_{i-1}[\exp(j\omega k - \alpha)T] \tag{7-83}$$

where $k = 1, 2, \ldots, m - 1$.

From the preceding brief description the filter synthesis will be seen to have similarities with Fourier synthesis methods and shares a common requirement with them of requiring much greater digital storage than the simpler recursive techniques. Advantage can be taken however of the complex arithmetic capabilities of many computers which renders the method easy to program in Fortran.

7-3-4 THE BILINEAR z-TRANSFORM METHOD

The two advantages of this method are:

1. The transformation is purely algebraic in form and so is applicable to continuous analog transfer functions.

2. Aliasing errors, possible with the direct z-transform method, are removed.

It is an indirect method since the requirements of an equivalent analog filter are first postulated and then transformed into a discrete form, from which the filter weights can be derived. Essentially the transformation is from the integrating operation $1/s$ to the z-transform.

If we consider a continuous signal, $x(t)$ to be operated upon by an ideal integrator having a transfer function, $H(s) = 1/s$ then the output, $y(t)$ can be represented by $x(t)*h(t)$ where * represents a convolution operation with the impulse response function for $H(s)$. An integrator having a transfer function $H(s) = 1/s$ is of course, ideal. Its impulse response function will be a step function: $h(t) = 1$ for $t \geqslant 0$; $= 0$ for $t < 0$ and we can define the results of such integration on a signal, $x(t)$ as the convolution operation:

$$y(\tau) = \int_0^t x(\tau).h(t - \tau).d\tau \tag{7-84}$$

since $h(t)$ becomes zero for $\tau > t$ and unity for $\tau \leqslant t$. This can be expressed in discrete form by considering t_1 and t_2 to be two consecutive sampling times separated by the sampling interval T where:

$$x(t) = x(nT) \text{ and } n = 0, 1, 2, \ldots, \tag{7-85}$$

so that the continuous integral defined by equation (7-84) and having limits t_1 and t_2 approximates to:

$$\tfrac{1}{2}T[x(nT) + x(n-1)T] = y(nT) - y(n-1)T$$

and we can write for the z-transfer function:

$$H(z) = \frac{y(z)}{x(z)} = \tfrac{1}{2}T\left[\frac{1 + z^{-1}}{1 - z^{-1}}\right]. \tag{7-86}$$

This represents a trapezoidal approximation to integration using the s-operator expressed as a function of z. Thus the transfer function, $H(s)$ for continuous system can be replaced by the z-transfer function $H(z)$ using the relationships:

$$\frac{1}{s} \rightarrow \tfrac{1}{2}T\left[\frac{z + 1}{z - 1}\right]$$

$$s \rightarrow \frac{2}{T}\left[\frac{z - 1}{z + 1}\right], \tag{7-87}$$

giving

$$H(z) \equiv H(s)\Bigg|_{s = \frac{2}{T}\left[\frac{z - 1}{z + 1}\right]} \tag{7-88}$$

This transform is known as the bilinear z-transform [17]. It maps the imaginary axis of the s-plane into a unit circle of the z-plane such that the left-hand side of the s-plane corresponds to the interior of the circle (fig. 7-18). Thus repeating poles of a discrete transfer function resulting from aliasing effects are coalesced into single unique positions within the circle (assuming a stable filter) removing the possibility for aliasing errors present with the direct z-transform method [18]. If the transformation is carried out precisely as a unique 1:1 relationship the filter will be stable and of the same order as the original Laplace transfer function. The frequencies of the two relationship will, however, be different so that the replacement of s shown in equation (7-88) will result in a digital filter having a cut-off angular frequency, ω_d, related to the continuous frequency ω_a, by:

$$\omega_a = \frac{2}{T}\tan\left[\frac{\omega_d T}{2}\right]. \tag{7-89}$$

This has the effect of compressing the complete continuous frequency characteristics into a limited digital filter frequency range of $0 < \omega T < \pi$ as indicated in

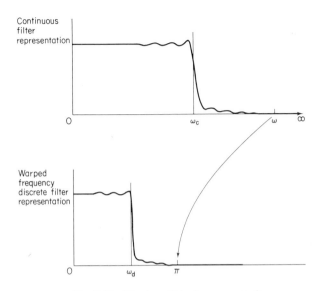

Fig. 7-22. Warping of the frequency scale.

fig. 7-22. In the practical application of equations (7-88) and (7-89) we will be calculating the ratio s/ω_a so that the $2/T$ term may be dropped and we can write:

$$s = \frac{z - 1}{z + 1} \tag{7-90}$$

$$\omega_a = \tan \left[\tfrac{1}{2}\omega_d T\right]. \tag{7-91}$$

This is the form in which these terms often appear in the literature.

The method of design using the bilinear transform can be summarised as follows:

1. A new set of frequencies for the continuous transfer function, $H(s)$ is calculated from the desired digital frequencies using equation (7-91).

2. A suitable continuous transfer function, $H(s)$, is chosen to give the required filter performance using the frequencies derived from step (1).

3. The operator s in $H(s)$ is replaced by a function in z and the new transfer function, $H(z)$ expressed as a ratio of polynomials in z.

4. $H(z)$ is converted into a difference equation by the application of the shifting theorem and the algebraic equation rearranged to give a recursive equation for the output sample.

5. The recursive equation, considered as an algorithm, is programmed for the digital computer.

This procedure will now be described for a number of the filter types.

317

A low-pass Butterworth filter will be taken as the filter model. The transfer functions for the first-order, $(n = 1)$ and second-order $(n = 2)$ have been given previously as:

$$H(s) = \frac{1}{1 + s/\omega_c} \tag{7-92}$$

and,

$$H(s) = \frac{1}{1 + 2^{\frac{1}{2}}(s/\omega_c) + (s/\omega_c)^2} \tag{7-93}$$

These transfer functions will be used as a base for design, commencing with the low-pass case.

(i) Low-pass filter design
A difference equation will be developed for a first-order $(n = 1)$ and a second-order $(n = 2)$ transfer function, using equations (7-92), (7-93), (7-90) and (7-91).

1st-order case. Given a cut-off frequency, f_c Hz, the required equivalent continuous cut-off angular frequency is from (7-91):

$$\omega_a = \tan(\pi f_c T). \tag{7-94}$$

The z-transform function is from (7-92) and (7-90)

$$H(z) = \frac{Y(z)}{X(z)} = \frac{1}{1 + \left[\dfrac{z-1}{z+1} \cdot \dfrac{1}{\omega_a}\right]}$$

$$= \frac{z+1}{z\left[1 + \dfrac{1}{\omega_a}\right] + \left[1 - \dfrac{1}{\omega_a}\right]}.$$

Multiplying by z^{-1} and equating input and output terms:

$$Y(z)\left[\left(1 + \frac{1}{\omega_a}\right) + z^{-1}\left(1 - \frac{1}{\omega_a}\right)\right] = X(z)[z^{-1} + 1] \tag{7-95}$$

and applying the shifting theorem gives:

$$y_i = Gx_i + Gx_{i-1} - Hy_{i-1} \tag{7-96}$$

where

$$G = \frac{1}{1 + \cot(\pi f_c T)}$$

and

$$H = \frac{1 - \cot(\pi f_c T)}{1 + \cot(\pi f_c T)}.$$

The filter gain is $A = 1/G$.

2nd-order case. The z-transfer function is from (7-93) and (7-90):

$$H(z) = \frac{1}{1 + \left[\dfrac{2^{\frac{1}{2}}}{\omega_a} \cdot \dfrac{z-1}{z+1}\right] + \left[\dfrac{z-1}{z+1}\right]^2 \cdot \dfrac{1}{\omega_a^2}}.$$

Substituting for ω_a from equation (7-91) and invoking the shifting theorem, we can write the difference equation for the second order case as:

$$y_i = Cx_i + 2Cx_{i-1} + Cx_{i-2} - Dy_{i-1} - Ey_{i-2} \qquad (7\text{-}97)$$

where

$$C = \frac{1}{1 + [2^{\frac{1}{2}} \cdot \cot(\pi f_c T)] + [\cot(\pi f_c T)]^2},$$

$$D = 2[1 - [\cot(\pi f_c T)]^2],$$

and

$$E = 1 - [2^{\frac{1}{2}} \cdot \cot(\pi f_c T)] + [\cot(\pi f_c T)]^2.$$

The filter gain is $1/C$.

Transformation of the continuous low-pass filter transfer functions to the equivalent high-pass, band-pass, and band-stop forms can be obtained through well-known conformal mapping techniques [19]. A list of these is given in table 7-1. Use is made of the low-pass/high-pass transformation as described below.

TABLE 7-1 s-plane transformations of
low-pass continuous filters

Required Filter	Replace s/ω_c by:
Low-pass	s/ω_c
High-pass	ω_c/s
Band-pass	$\dfrac{s^2 + \omega_l \omega_u}{s(\omega_u - \omega_l)}$
Band-stop	$\dfrac{s(\omega_u - \omega_l)}{s^2 + \omega_l \omega_u}$

ω_c = cut-off angular frequency,
ω_l = lower cut-off angular frequency,
ω_u = upper cut-off angular frequency.

319

For the band-pass and band-stop transformation more economical algorithms are obtained if the transformation is carried out directly in the z-plane and these will be used in the following derivations.

(ii) High-pass filter design
1st-order case. From table 7-1 and equations (7-92), (7-90), and (7-91), the z-transfer function is

$$H(z) = \frac{z - 1}{z(1 + \omega_a) + (\omega_a - 1)}$$

from which the difference equation is obtained as:

$$y_i = G'x_i - G'x_{i-1} - H'y_{i-1} \tag{7-98}$$

where

$$G' = \frac{1}{1 + \tan(\pi f_c T)},$$

and

$$H' = \frac{1 - \tan(\pi f_c T)}{1 + \tan(\pi f_c T)}.$$

The filter gain is $1/G'$
2nd-order case. The z-transform function is obtained as

$$H(z) = \frac{(z - 1)^2}{(z - 1)^2 + 2^{\frac{1}{2}}\omega_a(z^2 - 1) + \omega_a^2(z + 1)^2}$$

from which the difference equation is:

$$y_i = C'x_i - 2C'x_{i-1} + C'x_{i-2} + D'y_{i-1} - E'y_{i-2} \tag{7-99}$$

where

$$C' = \frac{1}{1 + 2^{\frac{1}{2}}\tan(\pi f_c T) + [\tan(\pi f_c T)]^2},$$

$$D' = 2[1 - (\tan(\pi f_c T))^2],$$

and

$$E' = 1 - 2^{\frac{1}{2}}(\tan(\pi f_c T) + [\tan(\pi f_c T)]^2.$$

The filter gain is $1/C'$

We can note from the similarity of equations (7-96) and (7-98) also (7-97) and (7-99) that only two algorithms are needed to evaluate first and second-order low-pass and high-pass filters. To convert from a low-pass to a high-pass difference equation we need only to:

(a) change the sign of the odd-numbered weights for both x_i and y_i,
(b) substitute $\tan(\omega f_c T)$ for $\cot(\omega f_c T)$ in the evaluation of the filter weights.

It can also be shown that if only the signs of the weights are changed for the low-pass filter then it will behave as a high-pass filter having a cut-off frequency of $((1/2T) - f_c)$.

(iii) Band-pass filter design

The transformation from the s-plane to a cylindrical surface resulting from the extension of the unit circle in an amplitude domain has been described by Constantinides [20]. This results in a new transformation which goes directly from the Laplace transfer function of a low-pass filter to the discrete band-pass case given by:

$$s = \frac{z^2 - 2z\alpha + 1}{z^2 - 1} \tag{7-100}$$

where

$$\alpha = \frac{\cos \pi T(f_u + f_l)}{\cos \pi T(f_u - f_l)} = \cos \omega_0 T. \tag{7-101}$$

ω_0 represents the warped band centre frequency which is derived from the difference of the warped values of f_u and f_l and is different from:

$$\omega_a = \tan \pi T(f_u - f_l) \tag{7-102}$$

which represents the equivalent mean frequency of the continuous filter and will not be central.

These relationships will now be used to derive difference equations for band-pass filter.

1st-order case. Substituting (7-100) in (7-92) gives the z-transfer function:

$$H(z) = \frac{1}{1 + \frac{(z^2 - 2z\alpha + 1)}{z^2 - 1} . 1/\omega_a}$$

Carrying out algebraic manipulation and dividing by z^2 gives:

$$H(z) = \frac{\omega_a - \omega_a(z^{-2})}{(\omega_a + 1) - 2\alpha z^{-1} + (1 - \omega_a)z^{-2}} = \frac{Y(z)}{X(z)}$$

from which the difference equation may be obtained as:

$$y_i = Jx_i - Jx_{i-2} + Ky_{i-1} + Ly_{i-2} \tag{7-103}$$

where

$$J = \frac{1}{1 + 1/\omega_a} = \frac{1}{1 + \cot \pi T(f_u - f_l)},$$

$$K = \frac{2\alpha}{\omega_a + 1} = \frac{2 \cos \pi T(f_u + f_l)}{[1 + \tan \pi T(f_u - f_l)] \cos \pi T(f_u - f_l)},$$

321

and

$$L = \frac{(\omega_a - 1)}{(\omega_a + 1)} = \frac{\tan \pi T(f_u - f_1) - 1}{\tan \pi T(f_u - f_1) + 1}.$$

The filter gain is $1/J$.

2nd-order case. Similarly for the second-order case:

$$H(z) = \left[\frac{1 + 2^{\frac{1}{2}}}{\omega_a}\left(\frac{z^2 - 2z\alpha + 1}{z^2 - 1}\right) + \left(\frac{z^2 - 2z\alpha + 1}{\omega_a(z^2 - 1)}\right)^2\right]^{-1}.$$

After algebraic manipulation and dividing by z^4 we obtain:

$$H(z) = \frac{\omega_a^2(z^{-4} - 2z^{-2} + 1)}{(\omega_a^2 + 2^{\frac{1}{2}}\omega_a + 1) + z^{-1}(-2.2^{\frac{1}{2}}\omega_a\alpha - 4) + z^{-2}(-2\omega_a^2 + 4\alpha^2 + 2)} \\ + z^{-3}(2.2^{\frac{1}{2}}\omega_a\alpha - 4\alpha) + z^{-4}(\omega_a^2 - 2^{\frac{1}{2}}\omega_a + 1)$$

from which the difference equation can be obtained as:

$$y_i = Mx_i - 2Mx_{i-2} + Mx_{i-4} - Oy_{i-1} - Py_{i-2} - Qy_{i-3} - Ry_{i-4}$$

$$(7\text{-}104)$$

where $M = \omega_a^2/N$, $N = \omega_a^2 + 2^{\frac{1}{2}}\omega_a + 1$, $O = (-2.2^{\frac{1}{2}}\omega_a\alpha - 4\alpha)/N$, $P = (-2\omega_a^2 + 4\alpha^2 + 2)/N$, $Q = (2.2^{\frac{1}{2}}\omega_a\alpha - 4\alpha)/N$, $R = (\omega_a^2 - 2^{\frac{1}{2}}\omega_a + 1)/N$, and ω_a and α are defined by equations (7-101) and (7-102). The filter gain is $1/M$.

(iv) Band-stop filter design

Similar equations apply for the band-stop filter. The warped analog frequency, ω_a, is now:

$$\omega_a = \cot \pi T(f_u - f_1) \qquad (7\text{-}105)$$

and the transformation relationship is:

$$s = \frac{z^2 - 1}{z^2 - 2z\alpha + 1} \qquad (7\text{-}106)$$

1st-order case. From equations (7-92) and (7-106):

$$H(z) = \frac{1}{1 + \left(\dfrac{z^2 - 1}{z^2 - 2z\alpha + 1}\right) \cdot \dfrac{1}{\omega_a}}.$$

Rearranging and dividing by z^2 gives:

$$H(z) = \frac{\omega_a - 2z^{-1}.\alpha\omega_a + \omega_a.z^{-2}}{(\omega_a + 1) - z^{-1}.2\alpha\omega_a + z^{-2}(\omega_a - 1)} = \frac{Y(z)}{X(z)}$$

from which the difference equation can be obtained as

$$y_i = J'x_i - K'x_{i-1} + J'x_{i-2} + K'y_{i-1} - L'y_{i-2} \qquad (7\text{-}107)$$

where:

$$J' = \frac{1}{1 + 1/\omega_a} = \frac{1}{1 + \tan \pi T.(f_u - f_l)},$$

$$K' = \frac{2\alpha\omega_a}{\omega_a + 1} = \frac{2 \cos \pi T.(f_u - f_l) \cot \pi T.(f_u - f_l)}{\cot (f_u - f_l).(T + 1)},$$

and

$$L' = \frac{\omega_a - 1}{\omega_a + 1} = \frac{\cot \pi T.(f_u - f_l) - 1}{\cot \pi T.(f_u - f_l) + 1}.$$

The filter gain is $1/J'$.

2nd-order case. For the second-order case:

$$H(z) = \left[1 + \frac{2^{\frac{1}{2}}}{\omega_a}\left(\frac{z^2 - 1}{z^2 - 2z\alpha + 1}\right) + \frac{1}{\omega_a^2}.\left(\frac{z^2 - 1}{z^2 - 2z\alpha + 1}\right)^2\right]$$

Rearranging and dividing by z^4 gives:

$$H(z) = \frac{\omega_a^2(1 - 4z^{-1}.\alpha + z^{-2}(4\alpha^2 + 2) - 4z^{-3}.\alpha + 2z^{-4})}{(\omega_a^2 + 2^{\frac{1}{2}}\omega_a + 1) - z^{-1}(4\alpha\omega_a^2 + 2.2^{\frac{1}{2}}\omega_a\alpha) + z^{-2}(4\alpha^2\omega_a^2 + 2\omega_a^2 - 2)}$$

$$+ z^{-3}(-4\alpha\omega_a^2 + 2.2^{\frac{1}{2}}\alpha\omega_a) + z^{-4}(\omega_a^2 - 2^{\frac{1}{2}}\omega_a + 1)$$

from which the difference equation can be obtained as:

$$y_i = M'x_i - Sx_{i-1} + Tx_{i-2} - Sx_{i-3} + M'x_{i-4}$$

$$+ Oy_{i-1} - P'y_{i-2} - Q'y_{i-3} - R'y_{i-4} \qquad (7\text{-}108)$$

TABLE 7-2 s-plane to z-plane transformation from a **low-pass** continuous filter

	s is replaced by:	ω_a is replaced by:
Low-pass/Low-pass	$\dfrac{z - 1}{z + 1}$	$\tan (\pi f_c T)$
Low-pass/High-pass	$\dfrac{z + 1}{z - 1}$	$\cot (\pi f_c T)$
Low-pass/Band-pass	$\dfrac{z^2 - 2z\alpha + 1}{z^2 - 1}$	$\tan \pi T(f_u - f_l)$
Low-pass/Band-stop	$\dfrac{z^2 - 1}{z^2 - 2z\alpha + 1}$	$\cot \pi T(f_u - f_l).$

T = Sampling interval, f_c = cut-off frequency, f_l = lower cut-off frequency.
f_u = upper cut-off frequency, $\alpha = [\cos \pi T(f_u + f_l)]/[\cos \pi T(f_u - f_l)]$.

TABLE 7-3 Summary of filter difference equations derived from a continuous Butterworth low-pass filter design

Type	Order (n)	Equation	Coefficients
Low-pass	1	$y_i = Gx_i + Gx_{i-1} - Hy_{i-1}$	$G = [1 + \cot(\pi f_c T)]^{-1},\ H = [1 - \cot(\pi f_c T)]/[1 + \cot(\pi f_c T)]$
	2	$y_i = Cx_i + 2Cx_{i-1} + Cx_{i-2} - Dy_{i-1} - Ey_{i-2}$	$C = [1 + 2^{\frac{1}{2}}\cot(\pi f_c T) + [\cot(\pi f_c T)]^2]^{-1}$ $D = 2[1 - [\cot(\pi f_c T)]^2]$ $E = (1 - 2^{\frac{1}{2}}\cot(\pi f_c T) + [\cot(\pi f_c T)]^2)$
High-pass	1	$y_i = G'x_i - G'x_{i-1} - H'y_{i-1}$	$G' = [1 + \tan(\pi f_c T)]^{-1},\ H = [1 - \tan(\pi f_c T)]/[1 + \tan(\pi f_c T)]$
	2	$y_i = C'x_i - 2C'x_{i-1} + C'x_{i-2} + D'y_{i-1} - E'y_{i-2}$	$C' = [1 + 2^{\frac{1}{2}}\tan(\pi f_c T) + [\tan(\pi f_c T)]^2]^{-1}$ $D' = 2[1 - [\tan(\pi f_c T)]^2]$ $E' = 1 - 2^{\frac{1}{2}}\tan(\pi f_c T) + [\tan(\pi f_c T)]^2$
Band-pass	1	$y_i = Jx_i - Jx_{i-2} + Ky_{i-1} + Ly_{i-2}$	$J = [1 + \cot \pi T(f_u - f_l)]^{-1}$ $K = [2 \cos \pi T(f_u - f_l)]\cos \pi T(f_u - f_l).\cos \pi T(f_u - f_l)]$ $L = [\tan \pi T(f_u - f_l) - 1]/[\tan \pi T(f_u - f_l) + 1]$
	2	$y_i = Mx_i - 2Mx_{i-2} + Mx_{i-4}$ $\quad - Oy_{i-1} - Py_{i-2} - Qy_{i-3} - Ry_{i-4}$	$M = [\tan^2 \pi T(f_u - f_l)]/N$ $N = \tan^2 \pi T(f_u - f_l) + 2^{\frac{1}{2}}\tan \pi T(f_u - f_l) + 1$ $O = [-\cos \pi T(f_u + f_l)][2.2^{\frac{1}{2}}\tan T(f_u - f_l) + 4]/[\cos \pi T(f_u - f_l)]N$ $P = \left[-2 \tan^2 T(f_u - f_l) + \left(\dfrac{2\cos \pi T(f_u + f_l)}{\cos \pi T(f_u - f_l)}\right)^2 + 2\right]/N$ $Q = \cos \pi T(f_u + f_l)[2.2^{\frac{1}{2}}\tan \pi T(f_u - f_l) - 4]/[\cos \pi T(f_u - f_l)]N$ $R = [\tan^2 T(f_u - f_l) - 2^{\frac{1}{2}}\tan \pi T(f_u - f_l) + 1]/N$
Band-stop	1	$y_i = J'x_i - K'x_{i-1} + J'x_{i-2} + K'y_{i-1} - L'y_{i-2}$	$J' = [1 + \tan \pi T(f_u - f_l)]^{-1}$ $K' = 2 \cos \pi T(f_u - f_l) \cot \pi T(f_u - f_l)/[\cot \pi T(f_u - f_l) + 1]$ $L' = [\cot \pi T(f_u - f_l) - 1]/[\cot \pi T(f_u - f_l) + 1]$ $M' = [\cot^2 \pi T(f_u - f_l)]/N'$
	2	$y_i = M'x_i - Sx_{i-1} + Tx_{i-2} - Sx_{i-3}$ $\quad + M'x_{i-4} + O'y_{i-1} - P'y_{i-2}$ $\quad - Q'y_{i-3} - R'y_{i-4}$	$N' = \cot^2 \pi T(f_u - f_l) + 2^{\frac{1}{2}}\cot \pi T(f_u - f_l) + 1$ $O' = [\cos \pi T(f_u - f_l)\cot \pi T(f_u - f_l) + 4 \cot^2 \pi T(f_u - f_l)][\cos \pi T(f_u - f_l)]/[\cos \pi T(f_u - f_l)]N'$ $P' = \left[2 \cot \pi T(f_u - f_l)][2.2^{\frac{1}{2}}\cot \pi T(f_u - f_l) + \left(\dfrac{2 \cos \pi T(f_u + f_l)\cot \pi T(f_u - f_l)}{\cos \pi T(f_u - f_l)}\right)^2 - 2\right]N'$ $Q' = [\cos \pi T(f_u - f_l)][2.2^{\frac{1}{2}}\cot \pi T(f_u - f_l) - 4 \cot^2 \pi T(f_u - f_l)][\cos \pi T(f_u - f_l)]/[\cos \pi T(f_u - f_l)]N'$ $R' = [\cot^2 \pi T(f_u - f_l) - 2^{\frac{1}{2}}\cot \pi T(f_u - f_l) + 1]/N'$ $S = [4 \cos \pi T(f_u - f_l).\cot^2 \pi T(f_u - f_l)][\cos \pi T(f_u - f_l)]N'$ $T = \cot^2 \pi T(f_u - f_l)\left[\left(\dfrac{2 \cos \pi T(f_u + f_l)}{\cos \pi T(f_u - f_l)}\right)^2 + 2\right]N'$

T = sampling interval, f_c = cut-off frequency, f_u = upper cut-off frequency, f_l = lower cut-off frequency.

where: $M' = \omega_a^2/N'$, $N' = \omega_a^2 + 2^{\frac{1}{2}}\omega_a + 1$, $O' = (4\alpha^2 \cdot \omega_a^2 + 2 \cdot 2^{\frac{1}{2}}\omega_a\alpha)N'$, $P' = (4\alpha^2 \cdot \omega_a^2 + 2\omega_a^2 - 2)/N'$, $Q' = (-4\alpha \cdot \omega_a^2 + 2 \cdot 2^{\frac{1}{2}}\omega_a\alpha)/N'$, $R' = (\omega_a^2 - 2^{\frac{1}{2}}\omega_a + 1)/N'$, $S = 4\alpha\omega_a^2/N'$, $T = \omega_a^2(4\alpha^2 + 2)/N'$, and ω_a and α are defined by equations (7-101) and (7-105). The filter gain is $1/M'$.

The transformations and resulting difference equations for these filter designs are summarised in tables 7-2 and 7-3.

7-3-5   ERRORS IN DIGITAL FILTER REALISATION

A number of errors are associated with a given design of digital filter. Performance errors such as phase or amplitude error have already been considered. Errors peculiar to the use of sampled data are:

1. Quantisation of the input data.

2. Quantisation of the calculated values for the filter weights.

3. Quantisation of the results of iterative mathematic operations (i.e. round-off and truncation errors).

4. Aliasing errors.

5. Errors introduced due to the dynamic range of the input data (e.g. effects of overflow in the accumulator).

Quantisation of the input data has been considered in chapter 4, where it was shown to be equivalent to introducing an additional noise signal at the filter input. The average noise power varies with the filter realisation and becomes less as more zeros are introduced into the polynomial transfer function. A similar effect obtains from a consideration of the quantisation of the products and sums obtained at each iteration of the recursive equation. The cumulative effect of round-off error in digital filter calculation depends on the realisation of the filter transfer function. It will be shown in the next section that the minimal noise contribution is obtained with cascade representation.

Quantisation of the filter weights is necessary due to the finite word length of the computer. The effect of inaccuracy in the quantised values may be seen from the following example. If we consider a transfer function:

$$H(z) = \frac{1}{(1 - a_1 z^{-1})(1 - a_2 z^{-1})} \tag{7-109}$$

having two poles at distances $1/a_1$ and $1/a_2$ removed from the origin of the z-plane, then defining the desired values of the poles as A_1 and A_2 the partial derivatives against the coefficient values, a_1 and a_2 for each of the cascaded realisations:

$$H_1(z) = \frac{1}{1 - a_1 z^{-1}} \text{ and } H_2(z) = \frac{1}{1 - a_2 z^{-1}}$$

325

will be

$$\frac{\partial A_1}{\partial a_1} = 1, \quad \frac{\partial A_2}{\partial a_2} = 1$$

$$\frac{\partial A_1}{\partial a_2} = \frac{\partial A_2}{\partial a_1} = 0.$$

If equation (7-109) is now realised in direct form as:

$$H(z) = \frac{1}{1 - b_1 z^{-1} + b_2 z^{-1}} \tag{7-110}$$

where $b_1 = a_1 + a_2$ and $b_2 = a_1 \cdot a_2$, we can state the partial derivatives of the derived poles A_1 and A_2 against actual coefficients b_1 and b_2 as:

$$\frac{\partial A_1}{\partial b_1} = 1 \qquad \frac{\partial A_2}{\partial b_2} = \frac{1}{a_1}$$

$$\frac{\partial A_1}{\partial b_2} = \frac{1}{a_2} \qquad \frac{\partial A_2}{\partial b_1} = 1.$$

This shows that:

(a) change in the filter coefficients will produce a change in the pole positions,

(b) a change in the direct form coefficients will be productive of a larger pole position change than the cascaded form coefficients.

In some cases the inaccuracy in coefficient value can cause the pole positions to move outside the unit circle thus causing instability. As a general rule it is inadvisable to implement filters higher than the second order in direct form.

The digital computer will set a limit to the magnitude of numbers that can be represented. Overflow error can occur during the summing and product operations and will generally show itself as a repeated oscillation at the output. This effect will be more severe with fixed point arithmetic as compared with floating point operation and a limit may need to be imposed at the input of the filter to avoid these effects.

7-3-6 IMPLEMENTATION OF mth ORDER FILTER

Difficulties can be experienced when the coefficients of high-order recursive filters are determined. In particular with a very large change in gain at the pass band edges, then a wide range in filter weight coefficients will be realised. For computers having a small word length it will not be possible to represent this range adequately since the larger coefficients will be specified to a greater degree of accuracy than the smaller ones. In larger machines having floating-point hardware then double-precision working can be used to retain the dynamic range required.

The effects of this dynamic error, which is represented by a quantisation of the desired filter coefficients, has been described in the previous section in terms of pole position on the z-plane. It has been shown [21] that if the poles are located within 10^{-n} of the unit circle, the filter weights will need to be defined with an accuracy of n decimal places to achieve stability. Even within this stability margin a satisfactory performance may not be achieved, due to these coefficients inaccuracies.

A better solution is to reduce the filter order and achieve the desired performance by means of cascaded transfer functions (see chapter 3),

$$H(z) = H_1(z).H_2(z).H_3(z), \ldots, H_n(z). \tag{7-111}$$

An accurate stable filter can be designed for each of the fractional transforms so that the composite filter can be considered as comprising a number of smaller filters connected in series with the output of one filter forming the input data for the next.

Only two filter algorithms need be considered. A first-order (single real pole) and a second-order (complex pole pair) filter from which any combination can be obtained. The method has been generalised by Knowles and Edwards [22] and termed cascade programming. The z-transform of any complex filter can be expressed as:

$$H(z) = A \prod_{i=1}^{n} \frac{(1 - a_i z^{-1})(1 - a_i^* z^{-1})}{(1 - b_i z^{-1})(1 - b_i^* z^{-1})} \tag{7-112}$$

where A is the filter gain and * indicates a complex conjugate (since in a realisable filter any complex poles or zeros must exist in conjugate pairs).

Realisation of equation (7-111) consists of the cascade evaluation of simple polynomials having the form:

$$H_i(z) = \frac{1 - a_i z^{-1}}{1 - b_i z^{-1}} \tag{7-113}$$

or their conjugate equivalent, with a suitable gain factor interposed between each product term. In addition to the reduced value of quantisation noise resulting from the calculation of coefficient values in cascading operation the cumulative error due to round-off will also be reduced by the method of cascading.

7-3-7 PROGRAMMING A CASCADE FILTER

As indicated earlier only two forms of the elemental filter need be considered. Suitable first order and second order difference equations were developed in section 7-3-4. Specification of the four filter types can be defined by selection of two frequency values f_l and f_u, located at the band edges, thus for:

Low-pass: $f_l = 0, f_u = f_c$
High-pass: $f_l = f_c, f_u = 0$
Band-pass: $f_l = f_{c_1}, f_u = f_{c_u}$
Band-stop: $f_l = f_{c_u}, f_u = f_{c_1}$

where f_c = cut-off frequency, f_{c_1} = lower cut-off frequency, and f_{c_u} = higher cut-off frequency. A sign logic test can then act as a check on selected filter type.

Due to the close similarity between the low-pass and high-pass equations a single algorithm can be evolved to calculate either case. To obtain the desired cut-off slope, defined in decibels/octave, the order of the composite filter is calculated from equation (7-74) and hence the number of elemental filters required. Thus for a slope of S db/octave then

$$S = 10 \log_{10} (P_i/P_0)$$

where

$$P_i/P_0 = \frac{1}{|A|^2} = 1 + (\omega/\omega_c)^{2n}$$

so that for $(\omega/\omega_c) = 2$ (i.e. one octave)

$$S = 10 \log_{10} (1 + 2^{2n})$$

giving

$$n = 1 \cdot 662 \log_{10} (10^{0.1S} - 1). \tag{7-114}$$

The obtained value of n is rounded up to the next integer value and tested. If n is odd then a single first-order stage is required and $\frac{1}{2}(n - 1)$ second-order stages. If n is even then only $(n/2)$ second-order stages are required. A simplified flow diagram for a cascaded program is given in fig. 7-23. This consists of a main segment and a number of sub-routine segments to design the filter and carry out the filtering.

The master segment reads the data which specifies the filter type, band-edge frequencies, band-edge slope and sampling interval. Depending upon the filter type the appropriate overlaid subroutine HLPASS (which designs a high pass or a low pass filter), BANDPASS or BANDSTOP is brought into core. The design parameters are evaluated and the coefficients for the recursive filter model are computed. These are inserted into the difference equation, which operates on the input signal the required number of times, finally producing the filtered output. The output signal is copied onto the output tape and the OUTPUT subroutine is transferred into core.

328

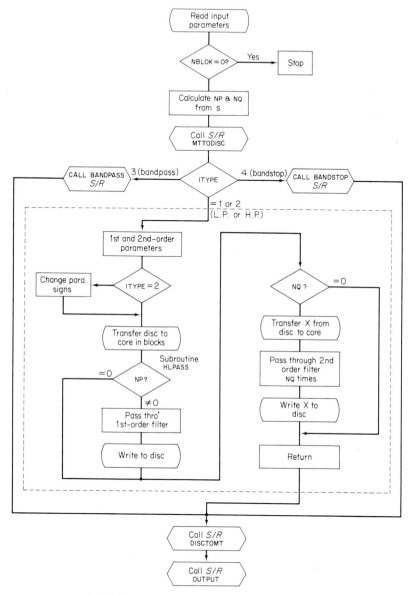

Fig. 7-23. Flow diagram for a cascade recursive filter.

In order to provide information on the realisation of the designed filter in the frequency domain the z-transform coefficients can be output. Substitution of these into the general equation (7-112) with $\exp(-j\omega T)$ substituted for z^{-1} can permit the frequency and phase characteristics of the realised filter to be

329

obtained and plotted. It may be noted that using the transformations from the low-pass filter given in table 7-3 only symmetrical attenuation slopes can be obtained for the bandpass or bandstop case. If asymmetrical response is desired then the data will need to be filtered twice using a combination of low-pass and high-pass filters, each having the desired slope for the appropriate band edge.

BIBLIOGRAPHY

1. JURY, E. I. *Theory and Application of the Z-Transform Method.* John Wiley, New York, 1964.
2. RADER, C. M. and GOLD, B. *Digital processing of signals.* McGraw-Hill, New York, 1969.
3. STOCKHAM, T. G. High-speed convolution and correlation. *A.F.I.P.S. Proc. Spring Joint Comp. Conf.* **28**, 229-33, 1966.
4. HELMS, H. D. Fast Fourier transform method of computing difference equations and simulating filters. *I.E.E.E. Trans.* (Audio and Electroacoust.) AU-15, 85-90, June 1967.
5. OPPENHEIM, A. V. Generalized superposition. *Inf. Control* **11**, 528-36, Nov. 1967.
6. DOLPH, C. L. A current distribution for broadside arrays which optimise the relationship between beam-width and side-lobe level. *Proc. I.R.E.* (Waves and Electrons) 335-48, June 1946.
7. HELMS, H. D. Designing digital filters with constrains. I.E.E. seminar on digital processing of analog signals. Zurich, Switzerland, March 1970.
8. KUO, F. F. and KAISER, J. F. *System Analysis by Digital Computer* (Chapter 7) John Wiley, New York, 1966.
9. ORMSBY, J. F. A. Design of numerical filters with applications to missile data processing. *J. A.C.M.* **8**, 3, 440-66, July 1961.
10. GOLD, B. and JORDAN, K. A note on digital filter synthesis. *Proc. I.E.E.E.* **56**, 1717-18, Oct. 1968.
11. HOLZ, H. and LEONDES, C. T. The synthesis of recursive digital filters. *J. A.C.M.* **13**, 2, 262-80, April 1966.
12. ENOCHSON, K. D. and OTNES, R. L. Programming and analysis for digital time series data. Shock and Vibration Information Center, U.S. Dept. Defence, 1968.
13. BOGNOR, R. E. Frequency sampling filters, Hilbert transformers and resonators. *Bell. Syst. Tech. J.,* 3, 501, March 1969.
14. LERNER, R. M. Band-pass filters with linear phase. *Proc. I.E.E.E.* **52**, 249-68, 1964.
15. LEWIS, M. Synthesis of sampled signal networks. *I.R.E. Trans.* (Circuit Theory) CT-7, March 1960.
16. OTNES, R. K. An elementary design procedure for digital filters. *I.E.E.E. Trans.* (Audio and Electroacoust.) AU-16, 3, 330-5, Sept. 1968.
17. KAISER, J. F. Design methods for sampled-data filters. *Proc. First Allerton Conf. on circuit and system theory.* Nov. 1963.
18. KAISER, J. F. Some practical considerations in the realisation of linear digital filters. *Proc. Third Allerton Conf. on circuit and system theory.* 621-33, Oct. 1965.
19. GUILLEMIN, E. A. *Synthesis of Passive Networks.* John Wiley, New York, 1957.

20. CONSTANTINIDES, A. G. Spectral transformations for digital filters. *Proc. I.E.E.* **117**, 8, 1585-90, Aug. 1970.
21. KNOWLES, J. B. and OLCAYNO, E. M. Coefficient accuracy and digital filter response *Elec. Letters* **1**, 6, 160-1, Aug. 1965.
22. KNOWLES, J. B. and EDWARDS, R. Effects of a finite word length computer in a sampled data feedback system. *Proc. I.E.E.E.* **12**, 6, June, 1965.

Additional References
23. WHITE, W. D. and RUVIN, A. E. Recent advances in the synthesis of comb filters. *I.R.E. Nat. Conv. Rec.* **5**, 186-99, 1957.
24. OPPENHEIM, A. V. Papers on digital signal processing. M.I.T. Press, Cambridge, Mass., 1969.
25. HUELSMAN, L. P. *Active Filters.* McGraw-Hill, New York, 1970.
26. GOLD, B. and RADAR, C. M. Effects of quantisation noise in digital filters. *A.F.I.P.S. Proc. Spring Joint Comp. Conf.* **28**, 213-9, 1966.
27. SANDBERG, I. W. Floating point round-off accumulation in digital filter realisations. *Bell. Syst. Tech, J.* **46**, 1775-91, Oct. 1967.
28. Special issue on digital filtering. *I.E.E.E. Trans.* (Audio and Electroacoust.) AU-18, June 1970.
29. RABINER, L. R. Techniques for designing finite-duration impulse-response digital filters. *I.E.E.E. Trans.* (Comm. Tech.) COM-19, 2, 188-95 April 1971.
30. RABINER, L. R. and SCHAFER, R. W. Recursive and non-recursive realisations of digital filters designed by frequency sampling techniques. *I.E.E.E. Trans.* (Audio and Electroacoust.) AU-19, **3**, 200-7, Sept. 1971.

Chapter 8

SPECTRAL ANALYSIS

8-1 INTRODUCTION

The statistical techniques developed in chapter 5 were broadly concerned with the characteristics of the signal expressed in the amplitude domain. The significance of time and frequency series was discussed in chapter 6 and their transform relationships developed. These will be considered further in this chapter as they apply to the derivation of the statistical characteristics of the signal in the frequency domain.

The results of a frequency analysis can be presented in various ways. Two of the most commonly used, express the instantaneous amplitude of the signal and the instantaneous power contained in the signal, in terms of the frequency spectrum. The term 'power' will be used throughout this chapter although it is recognised that data subject to analysis need not necessarily relate to energy considerations. The term 'auto-spectrum' has been widely used in this context and can be freely substituted for 'power-spectrum' without any change to the calculations involved.

Information concerning the power spectrum is required for a number of reasons associated with the physical phenomena under examination. We saw in chapter 5 that a difficulty is experienced in satisfactorily defining the characteristics of a random signal in terms of a finite time-history, since an infinitely long ensemble would be required. Such physical properties as mean square value and hence probability density function can be derived from a finite random signal (including noise) via the power spectral density function. Consideration of the signal in terms of its power or energy content distributed over the frequency spectrum permits integration of the power spectral density to yield valuable information about the total power involved in the process. This concept is particularly valuable in the shock and vibration field, where excitation by periodic (sinusoidal) wave-forms can no longer be assumed. Finally, the relationship between spectral density and correlation, which forms a further important statistical property of random data, provides a close link with earlier work in statistical physics. Two advantages should be noted concerning this relationship.

It is possible to obtain mathematically the transform of a correlation function whereas the transform of a time history may not be possible. This is because the correlation will decay to zero after a finite length of time so that a finite transform will be obtained. This is not true of the transform of a truly random signal which will persist indefinitely. Secondly the route to the correlation function via the power spectral density function, although indirect, is a very powerful and economic method when associated with the fast Fourier transform algorithm implemented on the digital machine [1].

At least three basic techniques exist for the derivation of power spectral density.

1. Direct Fourier transform
2. Indirect method via the Fourier transform of the autocorrelation function.
3. Band-pass filtering.

The direct method involves computing the Fourier transform of the time-history series and taking the mean absolute value of the function. Until fairly recently the direct method was not applied generally to experimental data because of the large number of calculations required, although it was the earliest method to be used by pure mathematicians and theoretical physicists.

Calculation of correlation functions has been carried out extensively on random data since its introduction by G. I. Taylor in the 1920s. The later work of Wiener [2] and Khintchine [3] showed that the power spectral density and the autocorrelation form a Fourier transform pair and this has provided a method of derivation of the power spectral density by first calculating the autocorrelation function and then taking its Fourier transform. A practical basis for this method was developed by Blackman and Tukey [4] and has been widely used in all fields of engineering and physics. This technique proved to be considerably faster to carry out using digital computers than the direct Fourier transform method, although until the mid-1950s very large amounts of computing time were required to obtain the complete power spectral density function.

Traditionally filtering methods have been carried out on the analog computer and generally take the form of a set of parallel band-pass filters extending over the frequency spectrum of interest. The signal to be analysed is applied simultaneously to all the filter inputs and the output signals are then squared and averaged to provide discrete points on the power spectral density frequency spectrum. In the case of random signals it is necessary to include division by the filter bandwidth in this averaging process. This is not necessary for periodic signals, since all the energy is contained in a single Dirac peak. Other analog methods, such as heterodyning the signal with a variable frequency oscillator using a single narrow-band filter are also in use and can be considered as equivalent to the filter-bank method.

The improvement in cost and speed of digital computing hardware and software developments, such as the fast Fourier transform algorithm described in chapter 6, have radically changed the situation with regard to the calculation of power spectral density. Whilst analog filtering methods continue to be used because of their cheapness and ease of simulation on a general-purpose computer, current methods of power spectral density evaluation using the direct method of calculation from the F.F.T. of the time-history series, are now dominant.

The foremost methods at present in use and described later in this chapter are:

1. Direct F.F.T. method.
2. Digital filtering.
3. Analog filtering.
4. Hybrid methods.

Recent development in digital hardware [5] indicate that it is now economic to assemble fast parallel/multiplying arithmetic units in sufficient quantities to permit very rapid calculation. Under these conditions the indirect method may become prominent yet again. Although the direct methods are better from the point of view of efficiency, their study is of less value as a means of understanding the fundamental problems of this type of analysis. For this reason the theory and error analysis presented in the following pages will constantly refer to spectrum analysis using filtering techniques.

8-2 POWER SPECTRUM ESTIMATION

The power spectrum of a random signal describes the energy characteristics of the signal in the frequency domain. A direct way of obtaining a working relationship for this is via an electrical analogy in which the available time-history series $V = x(t)$ of the signal is considered as representing a potential applied across a unit resistance for the time duration of the record, T sec. This is shown in fig. 8-1. The average power in R is:

$$\bar{P} = \frac{1}{T} \int_0^T x^2(t) \,.\, dt. \tag{8-1}$$

Substitution of the Fourier series for $x(t)$ (equation (6-13)) gives the expansion:

$$\bar{P} = \frac{a_0^2}{4} + \sum_{k=1}^{n} (a_k \cos k\,\omega_0 t + b_k \sin k\omega_0 t)^2$$

which reduces to

$$\bar{P} = C_0^2 + \tfrac{1}{2} \sum_{k=1}^{n} C_k^2 \tag{8-2}$$

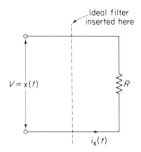

Fig. 8-1. Derivation of average power.

due to the orthogonal nature of the $\sin k \, \omega_0 t$ and $\cos k \, \omega_0 t$ terms (equations (6-14) to (6-19)). In terms of the Fourier constituents of $x(t)$ we can see that the average power:

(a) is independent of the phase relationships between the $\cos \omega t$ and $\sin \omega t$ terms,

(b) is always positive,

(c) increases with k, the number of frequency terms taken.

If we now define the frequency spectrum by its integral Fourier terms, where the frequency interval, δf is made equal to $1/2T$ (fig. 8-2), then the average power contained in any frequency interval is derived from equation (8-2) as:

$$\delta \bar{P}_k = (C_0^2 + \tfrac{1}{2} \sum_{k=1}^{n} C_k^2) - (C_0^2 + \tfrac{1}{2} \sum_{k=1}^{n-1} C_{k+1}^2) = \tfrac{1}{2} C_k^2. \quad (k \neq 0) \quad (8\text{-}3)$$

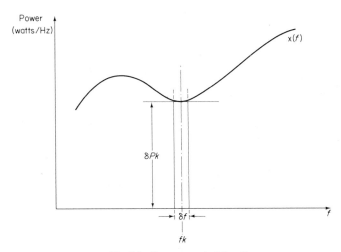

Fig. 8-2. Power spectral density.

335

This is equivalent to inserting an ideal filter in our electrical analogy, as shown in fig. 8-1. This filter is assumed to pass only the frequencies contained in the interval defined by k so that the average power developed in R will be proportional to the power contained in the signal over the frequency interval. The density of the average power, defined at a given frequency, f_k will be:

$$G(f_k) = \lim_{\delta f \to 0} \frac{\overline{\delta p_k}}{\delta f} \text{ W/Hz} \tag{8-4}$$

where $G(f_k)$ represents the Power Spectral Density function of $x(t)$ at the frequency f_k. Hence the power spectral density function represents the total power contained in a narrow 'slice' of the power spectrum, one frequency unit in width and centred at a given frequency value f_k. Units for $G(f_k)$ are usually volts^2/Hertz which is directly applicable to our electrical analog. Other units are admissible, dependent upon the nature of the function $x(t)$. For example where $x(t)$ represents an acceleration the units will be g^2/Hertz.

From fig. 8-2 we can see that a practical value for power spectral density can be obtained by filtering the signal $x(t)$ through an ideal band-pass filter and obtaining the average mean-squared value of the output of the filter, i.e.

$$G(f) = \lim_{\substack{T \to \infty \\ \delta f \to 0}} \frac{1}{T \delta f} \int_0^T x^2(t, f, \delta f) \, dt \tag{8-5}$$

where $x(t, f, \delta f,)$ is that portion of $x(t)$ lying in the frequency band δf, centred frequency f. Since we have neither infinite time in which to obtain an estimate for $G(f)$, nor records of infinite length, nor infinitely narrow band-pass filters, equation (8-5) reduces to:

$$G(f) \simeq \frac{\overline{x_B^2(t)}}{B} \tag{8-6}$$

where B is the effective band-width of the filter and $\overline{x_B^2(t)}$ is the average of the square of the amplitudes within the analysis band-width. This is the fundamental filter method of estimating spectral density and is commonly implemented on the analog computer. As we shall see later the definition given by equation (8-6) raises a number of practical points concerning averaging time, filter bandwidth, shape and record duration which will need to be considered in some detail. The practical form of equation (8-5) is also determined by the characteristics of the signal $x(t)$. If this is periodic then it may be represented by the Fourier series for a set of harmonically related sine and cosine wave-forms thus:

$$x(t) = \int_{k=-\infty}^{\infty} X(f) \exp\left(j2\pi \frac{kt}{T}\right). \, df \tag{8-7}$$

where T is also the period of the fundamental frequency. A plot of $G(f)$ will show the power concentrated as a series of discrete lines spaced over the frequency spectrum at intervals of $2\pi/T$ and in this form is referred to as a Line Power Spectrum. If $x(t)$ is non-periodic it can often be represented by the Fourier integral:

$$x(t) = \int_{-\infty}^{\infty} X(f) \exp\ (j\omega t) \,. \, df \qquad (8\text{-}8)$$

which represents the sum of sine and cosine wave-forms extending over all frequencies. The resulting plot of $G(f)$ will show the power over a continuous range of frequencies thus forming a continuous power spectrum.

The relationship between the power contained in a signal in both time and frequency domains can be obtained by an application of Parsevals' theorem. This is derived below for two real functions $x(t)$ and $y(t)$. Their product can be expressed in Fourier transform terms as:

$$x(t)y(t) = x(t) \int_{-\infty}^{\infty} Y(f) \exp\ (j\omega t) \,. \, df$$

so that

$$\int_{-\infty}^{\infty} x(t)y(t) \,. \, dt = \int_{-\infty}^{\infty} \left[\int_{-\infty}^{\infty} x(t)\, Y(f) \exp\ (j\omega t) \,. \, df \right] dt \qquad (8\text{-}9)$$

which can be rearranged as

$$\int_{-\infty}^{\infty} x(t)y(t) \,. \, dt = \int_{-\infty}^{\infty} Y(f) \left[\int_{-\infty}^{\infty} x(t) \exp\ (j\omega t) \,. \, dt \right] df.$$

We recognise the bracketed quantity as the complex conjugate, $X^*(f)$. Hence:

$$\int_{-\infty}^{\infty} x(t)y(t) \,. \, dt = \int_{-\infty}^{\infty} X^*(f)\, Y(f) \,. \, df \qquad (8\text{-}10)$$

and if $x(t) = y(t)$ and $X(f) = Y(f)$ then

$$X^*(f)Y(f) = |X(f)|^2$$

so that we can write

$$\int_{-\infty}^{\infty} x^2(t) \,. \, dt = \int_{-\infty}^{\infty} |X(f)|^2\, df \qquad (8\text{-}11)$$

which is known as Parseval's Theorem and expresses the equality of the power contained in the two domains.

We can make use of this relationship by substitution in equation (8-1) to obtain:

$$\bar{P} = \frac{1}{T} \int_{0}^{T} |X(f)|^2 \,. \, df. \qquad (8\text{-}12)$$

337

Since

$$\bar{P} = \delta P_0 + \sum_{k=1}^{n} \delta P_k$$

then from equation (8-4):

$$\bar{P} = Gf_{(0)} \cdot \delta f + \sum_{k=1}^{n} Gf_{(k)} \cdot \delta f \qquad (8\text{-}13)$$

As $\delta f \to 0$ then:

$$\bar{P} = \int_0^T G(f) \cdot df \qquad (8\text{-}14)$$

since, in the limit, the sum becomes an integral. Comparing equations (8-12) and (8-14) we can express the power spectral density, $G(f)$ in terms of the Fourier transform, $X(f)$ of the signal, $x(t)$ as:

$$G(f) = \frac{|X(f)|^2}{T} \qquad (8\text{-}15)$$

This gives a second method of frequency decomposition of the power contained in $x(t)$ and is known as the Periodogram. The application of the periodogram requires more care than the direct filter method of spectral density evaluation (equation (8-6)) or the indirect method via the autocorrelogram (equation (8-23)). It gives a very 'raw' estimation of spectral density which must be smoothed in the frequency domain to obtain a reasonable estimate. The difficulty is that its variance does not become zero as large numbers of values are taken and, in particular, it can be shown that if $x(t)$ is a Gaussian process then, in the limit (as T becomes very large), the estimation is proportional to a chi-squared random variable having two degrees of freedom. However the method is important and, as we shall see later, is in wide use as a method of spectral analysis using the fast Fourier transform. Sometimes adequate smoothing of such estimates is difficult and the choice of suitable frequency windows will be discussed later in this chapter.

Fig. 8-3 shows a comparison of the power spectral density of a random noise signal with its power spectrum. The finite value of the power spectral density at zero frequency does not necessarily imply a finite d.c. component since the height of the curve at this point represents the average power contributed by frequencies on either side of zero. On the other hand, the integrated power spectrum represents the summation of power in increasingly larger samples of frequency as ω become greater and will be zero before the first sample is reached (no matter how small δf becomes). Fig. 8-3 represents the spectrum of a particular type of narrow band random noise. Truly wide-band noise (white noise) will have a power density spectrum that is uniform over all frequencies, but since this implies infinite power a practical spectrum will show a reduction

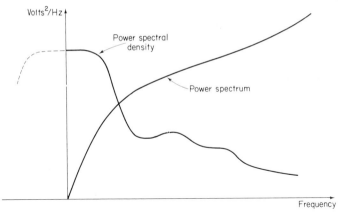

Fig. 8-3. Power-spectra.

in value at the extremes of frequency. By contrast the power spectral density of a sinusoidal function is defined by:

$$G(f) = \tfrac{1}{2}x^2\,\delta(f_n) \qquad\qquad (8\text{-}16)$$

where $\tfrac{1}{2}x^2$ is the mean square value of the sine wave and $\delta(f_n)$ is a Dirac function at $f = f_n$. This presents an infinitely large value density function at the frequency of the sine wave and zero at all other frequencies. The power spectrum, on the other hand, represents the integral of equation (8-16), which is equal to the mean square value of the sine wave. From these two representations we can see that a combination of random noise plus periodic components will result in a power density spectrum consisting of a continuous function upon which are superimposed infinitely large and infinitely narrow peaks, located at the fundamental and harmonic frequencies of the periodic component whereas the power spectrum will give finite pulses for the periodic components but will inadequately represent the random components (fig. 8-4). This difference in the

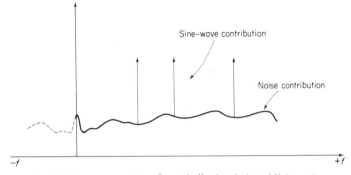

Fig. 8-4. Power-spectrum of a periodic signal plus additive noise.

339

power distribution characteristics for periodic and non-periodic signals is fundamental to the ideal representation of $x(t)$. Lee [6] has suggested a way of overcoming this difficulty by the use of an integrated power spectrum:

$$S(f) = \int_{-\infty}^{f} G(f) \, . \, \mathrm{d}f. \tag{8-17}$$

which is simply the integral of the power spectral density extending to an infinite negative frequency. The result is a positive increasing monotonic function which becomes identical to the autocorrelation function $R_x(0)$ of the random variable at an infinite positive frequency. Periodic components of the random signal are represented by steps in the spectrum at the fundamental and harmonics of the periodic function, where the height of the steps indicates their relative power content. These representations are, however, ideal ones and only fully relevant to very long data ensembles, where the period of the repeated wave-form, $t_n \ll T$. In many practical conditions this will not apply and, as will be shown later, the power spectral density will be found to exhibit a continuous distribution of frequencies about the fundamental period frequency f_n, and a similar cluster about the harmonics, thus tending to put a finite height and width on the Dirac peaks which occur with infinite record lengths.

It will have been noticed that the description of power spectrum representations has included both positive and negative frequencies. The spectrum in this case is called 'double-sided'. A spectrum containing positive frequencies only is referred to as a 'single-sided' spectrum. The inclusion of negative frequencies is a mathematical notation which makes possible the expression of Fourier series and integrals in exponential form and, as shown in chapter 6 and below, leads to a more compact form of expression. Correct interpretation of expressions involving negative frequencies is necessary for real physical problems and will involve the application of power scaling factors. The average power is distributed symmetrically over the complete frequency range such that:

$$G(f_k) = G(-f_k). \tag{8-18}$$

Therefore the terms containing $G(f_k)$ in equation (8-13) may be replaced by:

$$G(f_k) \, \delta f = \tfrac{1}{2} [G(f_k) + G(-f_k)] \, \delta f. \tag{8-19}$$

Single and double-sided spectral density functions are shown in fig. 8-5. Their relationship is given by:

$$G(f) = 2S(f). \quad \text{(for } 0 \leqslant f < \infty) \tag{8-20}$$

Practical density evaluations derive $G(f)$ and it is necessary to include the scaling factor when referring to mathematical calculations in which integration over negative frequency takes place.

A third derivation of the power spectral density function can be made via the autocorrelation function. Since the autocorrelation function and the power

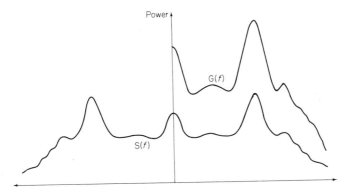

Fig. 8-5. Single and double-sided spectra.

spectral density function are respectively time domain and frequency domain descriptions of the same random signal, we should expect to find them related by the Fourier transform. A fuller discussion and derivation of these relationships will be deferred until the next chapter. For the present they will be stated as:

$$S(f) = \int_{-\infty}^{\infty} R(\tau) \cdot \exp(-j2\pi f\tau)\, d\tau \qquad (8\text{-}21)$$

and

$$R(\tau) = \int_{-\infty}^{\infty} S(f) \cdot \exp(j2\pi f\tau)\, df \qquad (8\text{-}22)$$

where $S(f)$ is the double-sided spectral density and $R(\tau)$ is the autocorrelation function. These are known as the Wiener–Khintchine relationships and by their use the power spectral density can be obtained from the autocorrelation of the single variable $x(t)$. A necessary condition is that $R(\tau)$ be integrable over the range $-\infty$ to $+\infty$. Physically realisable one-sided power spectral density functions, where f varies over the range 0 to $+\infty$ are defined from the autocorrelation cosine Fourier transform, i.e.

$$G(f) = 2 \int_{-\infty}^{\infty} R(\tau) \cos(2\pi f\tau)\cdot d\tau = 4 \int_{0}^{\infty} R(\tau) \cos(2\pi f\tau)\cdot d\tau. \qquad (8\text{-}23)$$

The second relationship is valid due to the symmetry of the autocorrelation function.

The derivation of power spectral density via the autocorrelogram is almost always carried out digitally. The autocorrelation function is first estimated from sampled data and then used as input to a Fourier transform program as described in section 8-6. It should be noted that this indirect method of obtaining the power spectral density is a matter of convenience. No more information regarding the analysed signal can be gained by the use of correlation function

341

techniques than may be obtained from the power spectral density derived by direct methods as long as the process is stationary. However, it should be noted that since the autocorrelation function will always reduce to a zero value after a finite time, it is essentially realisable and stable. On the other hand, a direct Fourier transform may not be theoretically realisable and can lead to an instability which does not reduce, even after a long period.

8-3   CROSS-SPECTRA

A common requirement is to discover degrees of similarity between two signals obtained, for example, from separated points in a physical system subject to some form of random vibration. This information may lead to the determination of the degree of transmissability between the common source and point of measurement, and locate sources of resonance causing fatigue damage. Comparison between the two signals can take place in the time domain or the frequency domain. Time domain estimates are obtained by finding the mean of the cross-products of the two signals, point-by-point along the time axis. If the mean products are large the signals contain a degree of similarity, the actual value being dependent on the size of the mean product. By shifting the time axis of one of the signals relative to the other, the calculated mean product can be repeated at various time delays to discover whether any similarity exists between time-separated sections of the signals. This is roughly the method of cross-correlation discussed in the next chapter. A disadvantage of this method is that it tells us nothing about the relative phase characteristics of the two signals. This is illustrated in fig. 8-6. Two composite signals A and B are shown, each derived

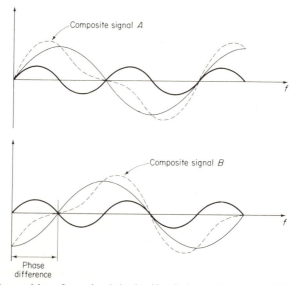

Fig. 8-6. Composition of two signals having identical constituents but differing in phase.

342

from identical constituents in frequency and amplitude, and differing only in the relative phase of the two constituents. A cross-correlation of the type described above will clearly not identify their similarity. We need to include phase information in our measurement and this implies analysis in the frequency domain.

A frequency domain method, analogous to the time-domain mean product evaluation is shown in fig. 8-7. Here a narrow 'slice' of both signals, A and B, is selected at a centre frequency, f. The width of the slice is given as δf. The cross-product of the mean values of the signals over the interval δf is obtained and shown as Cab in the third curve. The process is repeated for different values

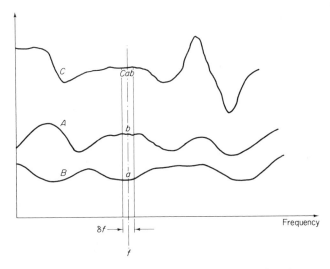

Fig. 8-7. Product of two signals in the frequency domain.

of centre frequency, f, to give the frequency-product curve C. This is one form of the cross-spectrum of the two signals, obtained in the frequency domain. It will be effective in indicating pronounced similarity in the frequency domain and only to this extent will the technique be successful. It will still fail to identify the degree of similarity between a large number of composite signals, all having identical constituent elements and differing only in the relative phase of these constituents. However, if we now make a second measurement, similar to the first, but with a fixed phase difference included in one of the signals, a comparison between these two sets of measurements will tell us something about the phase similarity between the two signals, since the effects of this added phase shift will differ for the different signals. It is convenient for mathematical analysis if we make the fixed phase shift $90°$ as in the simple example that follows. We will assume that each of the two signals consists of a single

343

frequency, $f = \omega/2\pi$ existing during the interval δf, and that one of these signals has a phase shift component θ_1 and the other a phase shift θ_2 viz.

$$x(f) = X \sin (\omega t - \theta_1) \tag{8-24}$$

$$y(f) = Y \sin (\omega t - \theta_2). \tag{8-25}$$

The product is:

$$\begin{aligned}
C_{xy} &= XY \sin (\omega t - \theta_1) \sin (\omega t - \theta_2) \\
&= XY(\sin \omega t \cos \theta_1 - \cos \omega t \sin \theta_1)(\sin \omega t \cos \theta_2 - \cos \omega t \sin \theta_2) \\
&= XY(\sin^2 \omega t \cdot \cos \theta_1 \cos \theta_2 + \cos^2 \omega t \cdot \sin \theta_1 \sin \theta_2 \\
&\quad - \sin \omega t \cdot \cos \omega t \cdot \cos \theta_1 \sin \theta_2 - \cos \omega t \cdot \sin \omega t \cdot \sin \theta_1 \cos \theta_2
\end{aligned}$$

which can be averaged over an integral number of periods to give:

$$\begin{aligned}
C_{xy} &= XY(\cos \theta_1 \cos \theta_2 + \sin \theta_1 \sin \theta_2) \\
&= XY \cos (\theta_1 - \theta_2).
\end{aligned} \tag{8-26}$$

If we assume that a second average product value is taken with a phase shift of $90°$ added to $y(f)$ the mean value of this will be:

$$Q_{xy} = XY \sin (\theta_1 - \theta_2). \tag{8-27}$$

The complete cross-power spectrum can now be expressed as:

$$G_{xy}(f) = C_{xy}(f) - jQ_{xy}(f) \tag{8-28}$$

where $j = \sqrt{-1}$ indicates the $90°$ phase shift relationship between the two mean products. This can be expressed as a function of a modulus and a phase shift:

$$|G_{xy}| = \sqrt{[C_{xy}^2(f) + Q_{xy}^2(f)]} \tag{8-29}$$

and,

$$\theta_{xy} = \arctan \frac{Q_{xy}(f)}{C_{xy}(f)} \tag{8-30}$$

which completely defines the degree of similarity between the two sinusoidal signals in terms of two product terms.

A more general derivation can be obtained using the complex Fourier series for the two signals which leads to the definition of the Cross Power Spectral Density. From equation (6-38) we can write:

$$x(t) = \sum_{k=-n}^{n} X_k(t) \exp (jk \, \omega_0 t) \tag{8-31}$$

$$y(t) = \sum_{k=-n}^{n} Y_k(t) \exp (jk \, \omega_0 t) \tag{8-32}$$

where $X_k(t)$ and $Y_k(t)$ are complex numbers representing the Fourier coefficients of the time-history series. Since for any given value of k the values of $X_k(t)$ and $Y_k(t)$ will be unique, we can regard these as the result of passing the total time-history series, $x(t)$ and $y(t)$ through a narrow filter whose centre frequency is varied as coefficient k changes. In order to take into account the desired $90°$ phase shift in the product we can multiply $X_k(t)$ by the complex conjugate of $Y_k(t)$. Thus if:

$$X_k = |X_k| \exp(j\theta_1)$$
$$Y_k = |Y_k| \exp(j\theta_2)$$
$$Y_k^* = |Y_k| \exp(-j\theta_2)$$

then the product of the filtered $x(t)$ signal and the filtered and phase-shifted $y(t)$ signal will be:

$$X_k Y_k^* = |X_k||Y_k| \exp[j(\theta_1 - \theta_2)]$$
$$= |X_k||Y_k| \cos(\theta_1 - \theta_2) + j|X_k||Y_k| \sin(\theta_1 - \theta_2) \qquad (8\text{-}33)$$

which is the cross-spectrum of the two signals. The real part involves the cosine of the relative phase, and the imaginary part the sine as given in equations (8-26) and (8-27). Referring to fig. 8-7 we can now define the cross-power spectral density in a similar way to power spectral density as:

$$G_{xy}(f) = \lim_{\substack{T \to \infty \\ \delta f \to 0}} \cdot \frac{1}{T\delta f} \sum_{k=-n}^{n} X_k Y_k^*. \qquad (8\text{-}34)$$

That is, we divide the cross-power spectrum, defined by the average value of the sample area shown in fig. 8-7, by the sample width, δf. The result is the density of the cross-power obtained at frequency f, and will be a complex number.

Alternative expressions, suitable for analog filter representation, can be derived from the above as:

$$C_{xy}(f) = \lim_{\substack{T \to \infty \\ \delta f \to 0}} \frac{1}{T\delta f} \int_0^T x(t, f, \delta f) y(t, f, \delta f) \, dt \qquad (8\text{-}35)$$

which is the Co-spectral Density Function and:

$$Q_{xy}(f) = \lim_{\substack{T \to \infty \\ \delta f \to 0}} \frac{1}{T\delta f} \int_0^T x(t, f, \delta f) y^0(t, f, \delta f) \, dt \qquad (8\text{-}36)$$

which is the Quadrature Spectral Density Function, where $x(t,f,\delta f)$ and $y(t,f,\delta f)$ represent those portions of $x(t)$ and $y(t)$ that lie within the frequency band, δf centred at frequency f, i.e. the filter outputs; and $y^0(t,f,\delta f)$ represents the

345

phase-shifted filter output. The complex cross-spectral density function is represented graphically by two curves as shown in fig. 8-8. One curve expressing the magnitude $|G_{xy}|$, and the other the phase angle θ_{xy}, of the function given by equations (8-29) and (8-30).

The cross-spectral power density may also be derived from the cross-correlation. In this case a double-sided spectrum is obtained, viz.

$$S_{xy}(f) = \int_{-\infty}^{\infty} R_{xy}(\tau) \exp(-j2\pi f \tau) \, . \, d\tau. \qquad (8\text{-}37)$$

The physically realisable single-sided spectral density function corresponding to equation (8-37) is defined for $0 \leqslant f < \infty$ as:

$$G_{xy}(f) = 2S_{xy}(f) \qquad (8\text{-}38)$$

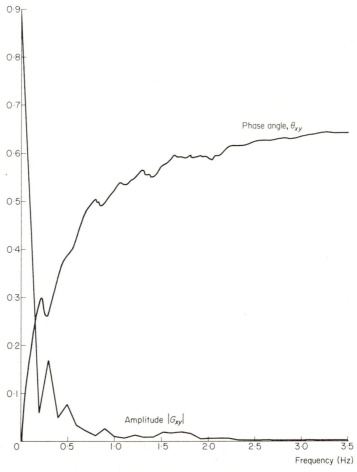

Fig. 8-8. Cross-spectral density.

so that

$$G_{xy}(f) = 2 \int_{-\infty}^{\infty} R_{xy}(\tau) \exp(-j2\pi f\tau) . d\tau. \qquad (8\text{-}39)$$

Whilst the similarity between two signals can be expressed in the frequency domain using these results it is convenient to be able to give a quantitative figure in normalised form. We can do this by defining a Coherence Function as:

$$Y_{xy}^2(f) = \frac{|G_{xy}(f)|^2}{G_x(f) . G_y(f)} \qquad (8\text{-}40)$$

where $G_x(f)$ and $G_y(f)$ are smoothed spectral density values for the individual signals, $x(t)$ and $y(t)$. The value of the coherence function cannot be greater than 1. If $Y_{xy}^2(f) = 0$ the two signals are said to be completely incoherent at this particular frequency. If $Y_{xy}^2(f) = 1$ the two signals are fully coherent. It is important to realise that the estimates for $G_x(f)$, $G_y(f)$, and $G_{xy}(f)$ must be obtained under identical conditions (i.e. filter bandwidth, record length etc.), otherwise gross errors in coherence estimation will be obtained.

An important use for the cross-spectrum is in system measurement. The calculation of the transfer function of a system simply by the ratio of the input and output signals either directly, or as a ratio of their Fourier transforms, can include a major source of error if the output signal is not totally due to the input signal. This would occur, for example, if the system contributes its own source of noise to the resultant output. If we take the frequency response function, $H(\omega) = Y(\omega)/X(\omega)$ and multiply the numerator and denominator by the complex conjugate of the input response $X^*(\omega)$, we can define the frequency response function as the ratio of the cross-spectrum to the input power spectrum, viz.

$$H(\omega) = \frac{Y(\omega)X^*(\omega)}{X(\omega)X^*(\omega)} = \frac{G_{xy}(\omega)}{G_x(\omega)}. \qquad (8\text{-}41)$$

(This result is obtained more rigorously in chapter 9.)

Two important advantages are obtained by deriving the system transfer function through this calculation for the system response function. The first is that the phase relationship between input and output is preserved. Secondly the measurement is not limited to any particular input, such as a sinusoidal signal. The input signal may, in fact, be random noise or whatever is the normal operational input for the system. These advantages lead to the ability to make the measurement accurately despite the presence of internal noise generated by the system and to carry this out without interfering with the system operating normally. Similar advantages can be obtained using cross-correlation and the subject will be considered further in the next chapter.

8-4   ERRORS IN SPECTRAL DENSITY ESTIMATES

Any practical measurement of power-spectra and spectral density from finite data samples may only be considered as an estimate having an error relationship to the true value. An assumption is generally made that the process is stationary and that the finite data samples are representative of the random function over a large interval of time (theoretically an infinite time). As indicated earlier it is extremely difficult to be certain of the stationary nature of the process when only one random variable is considered. However, the assumption that the random variable itself is stationary will often have to be made on purely practical grounds, since an estimate (albeit, an approximate result) must be derived from the finite samples available. Even if this assumption is valid there are a number of other factors which can contribute to the total error of estimations made. To arrive at a confidence limit for these estimates it is necessary to study the nature and extent of the errors. Errors can arise due to the inability to apply the complete features demanded by the mathematical method (such as inability to integrate over an infinite time), inaccuracies of the method of data representation (such as the need to estimate with sampled data), and the imperfect nature of the acquisition itself. This latter factor relates to the equipment performance, e.g. amplifier non-linearity, frequency distortion etc., and will not be considered here. Some of the factors upon which the reliability of the estimate will depend are listed below:

(*a*) record length,
(*b*) filter bandwidth,
(*c*) filter shape,
(*d*) averaging time,
(*e*) scan rate,
(*f*) extraneous noise,
and (in the case of digital data)
(*g*) sampling rate and quantising level.

The most important of these are record length and filter bandwidth. Both are finite in value and when considered together give rise to problems in achieving satisfactory resolution and good statistical accuracy. The situation is formally expressed by stating that the product of record length and bandwidth must be equal to or greater than a given constant, i.e.

$$BT \geqslant C. \tag{8-42}$$

The finite length of record also has the effect of introducing frequencies into the spectral estimate which interact to produce a distorted version of the 'true spectral estimate'. Other errors due to averaging, aliasing and noise are considered below, together with the difficulties encountered when digitised information is subject to analysis. For reasons given earlier, the discussion is limited to stationary ergodic random processes and, in order to avoid difficulties and errors caused by the presence of large amplitude pulses at zero frequency, a zero mean value for the signal is assumed. This can normally be arranged during the pre-processing stage. Errors arising out of the correlation calculations used in the indirect method are only briefly mentioned and will be fully considered in the next chapter.

8-4-1 RESOLUTION

From equation (8-5) we see that the resolution of the power-spectral density measurement increases as δf becomes smaller. If the effective analysis filter bandwidth is wider than a given signal peak then the measured power contained within the peak will be distributed over the analysis bandwidth. A reduction in measured power will be apparent at the centre frequency of the peak roughly proportional to the ratio of the areas of the filter response and signal peak in the frequency domain. Hence in order to achieve high resolution we must use a filter bandwidth of the same order as the narrowest peak in the signal to be analysed. Unfortunately we will then find that the statistical reliability of the result will be impaired unless the record length, T, is proportionately increased. This can be illustrated quantitatively if we consider a finite sample of a sinusoidal waveform extending over period T, and assumed to have zero value at all times outside this period (fig. 8-9).

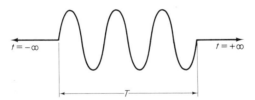

Fig. 8-9. A finite sinusoidal wave-form.

From equation (8-15) the power spectral density can be written in terms of the complex spectrum (equation 6-36)) as:

$$G(f) = \frac{\left| \int_0^T x(t) \exp(-j\omega t).\, dt \right|^2}{T} \quad \text{(where } x(t) = \sin t)$$

$$= \frac{1}{T} \left| \left[\frac{\exp(-j\omega t)}{\omega^2 - 1} (\cos t + j\omega \sin t) \right]_0^T \right|^2$$

$$= \frac{1}{T} \left[\frac{\exp(-j\omega T)}{\omega^2 - 1} (\cos T + j\omega \sin T) - \frac{1}{\omega^2 - 1} \right]^2$$

349

replacing exp $(-j\omega T)$ by $(\cos \omega T - j \sin \omega T)$

$$G(f) = \frac{\cos^2 T + \omega^2 \sin^2 T - 2 \cos T \cos \omega T - 2\omega \sin T \sin \omega T + 1}{T(\omega^2 - 1)^2}. \quad (8\text{-}43)$$

A plot of this relationship for increasing values of Tf (ratio of length of record to fundamental period of the sinusoidal waveform) is given in fig. 8-10 and illustrates the frequency spreading that can occur where the record length is finite. From this we see that for a fixed record length T, a decrease in filter

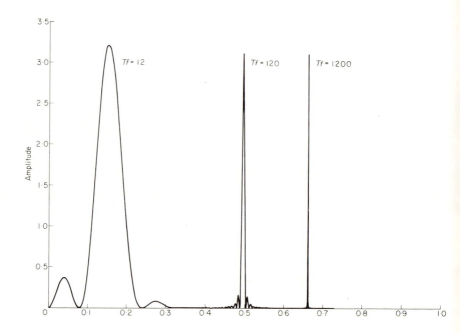

Fig. 8-10. Effect of frequency spreading due to a finite record length.

bandwidth, δf, will increase the resolution, but only at the expense of statistical reliability, so that a compromise will become necessary. A similar result obtains if the number of digital samples, N, taken from a finite record is too small. Again a frequency spreading occurs for narrow bandwidth peaks contained within the signal.

The size of filter bandwidth required for a known signal characteristic can easily be obtained. If we let a be the width of the narrowest expected peak in the spectra and consider the elementary frequency band δf, resulting from a

350

sampling interval, h, we shall require $n = a/\delta f = ah$ of these to define the spectral density of this peak. Thus the filter bandwidth required will be $n\delta f$ Hz. However, a wider bandwidth may be dictated by the need to reduce the effects of uncertainty. A practical criterion suggested by Forlifer [7] indicates that an acceptable resolution is obtained when the narrowest peak in the spectra being measured contains no greater than four filter bandwidths, thus:

$$\delta f \leqslant \tfrac{1}{4}B. \tag{8-44}$$

8-4-2 UNCERTAINTY

It has been stated earlier that a fundamental difficulty preventing the precise determination of power spectral density is the conflicting requirements of infinitely long record lengths and a zero bandwidth filter. Let us consider the situation relating to a specific time, t. We assume that the direct filter method is in use and that a measurement has been taken at a small, but finite time, δt before t. Then at time t we expect to make a further measurement statistically independent of the measurement taken at $(t - \delta t)$. To achieve this the filter must not contain any information previously stored in it, i.e. its output must fall to zero during time δt. However, as δt tend to zero, the Q-factor of the filter must also tend to zero if this condition is to be met. Under these conditions it would possess infinite bandwidth and hence fail to discriminate or resolve a particular frequency f. Similarly if we desire to resolve power spectral density at a particular frequency f, and at no other frequency, an infinitely narrow filter and hence infinite Q-factor is required which provides an infinitely long storage of information at this frequency. Clearly the two requirements are mutually incompatible. The conflict has been referred to as being due to an Uncertainty Principle [8], which states the impossibility of obtaining simultaneously an arbitrarily high degree of resolution in both time and frequency domains.

The choice of resolution and accuracy is something of an arbitrary one and a compromise decision will have to be made subject to the needs of the analysis and such criteria as that given by equation (8-44).

8-4-3 STATISTICAL ERRORS

We can recognise three different values of the power spectral density function from which to define the measurement error:

(a) The true power spectral density $G(f)$, obtained over an infinite time.

(b) The measured value of power spectral density $\hat{G}(f)$, obtained over a finite time for a single ensemble.

(c) The expected value of power spectral density $E(\hat{G}(f))$, resulting from the averaging of a large number of ensemble measurements.

351

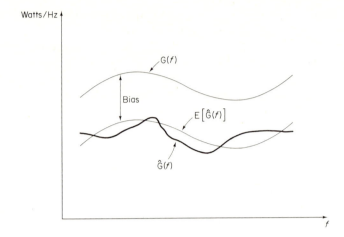

Fig. 8-11. Estimates of power spectral density function.

Thus we have the situation where it is not possible to measure (*a*) and where we may not have enough data to measure (*c*) effectively. In a practical case we will only have measurement (*b*) available and wish to estimate its error relationship to the unattainable value of (*a*). We have in fact a bias level and a variance to consider in our estimates, as indicated in fig. 8-11. The mean square error, $\overline{e^2}$, for the measured value of $G(f)$ can be written:

$$\overline{e^2} = E[(\hat{G}(f) - G(f))^2] \qquad (8\text{-}45)$$

where, as already defined,

$$G(f) = \lim_{\substack{T \to \infty \\ B \to 0}} \hat{G}(f) \qquad (8\text{-}46)$$

and $\hat{G}(f)$ is the estimated value for $G(f)$.
This can be expanded by adding and subtracting the expected value of $E(\hat{G}(f))$ to equation (8-45) to result in an expression for the mean square error in terms of the variance and bias of the estimate:

$$\overline{e^2} = \text{var}[\hat{G}(f)] + \text{bias}^2[\hat{G}(f)]. \qquad (8\text{-}47)$$

The variance term represents the dynamic error between the measured and expected value and is given as the mean square value of the error about the mean value, i.e.

$$\text{var} = E[(\hat{G}(f) - E[\hat{G}(f)])^2] \qquad (8\text{-}48)$$

and the bias error represents the expected value of the difference between the true and estimated value of the measurement:

$$\text{bias} = E[\hat{G}(f) - G(f)]. \tag{8-49}$$

It can be shown [9] that these errors may be related by the filter bandwidth, B, and record time, T, of random signals to give:

$$\text{variance} \simeq \frac{G^2(f)}{BT} \tag{8-50}$$

and

$$\text{bias} \simeq \frac{B^2}{24} \cdot \ddot{G}(f) \tag{8-51}$$

where $\ddot{G}(f)$ is the second derivative of $G(f)$.

Hence the total mean-square error derived from equation (8-47) is:

$$\overline{e^2} = \frac{G^2(f)}{BT} + \frac{B^4 \ddot{G}(f)^2}{576} \tag{8-52}$$

which, as we would expect, approaches zero for very long record times, T, and very small filter bandwidths, B.

For a properly resolved power spectral density (equation 8-44), then the bias error will be negligible and we can write a normalised version of the mean square error from (8-52) as:

$$\epsilon^2 = \frac{E[(\hat{G}(f) - G(f))^2]}{\hat{G}^2(f)} \simeq \frac{1}{BT}. \tag{8-53}$$

The square root of equation (8-53) is known as the Normalised Standard Error, ϵ, of the power spectral density estimate, $\hat{G}(f)$, obtained from a single record:

$$\epsilon = \frac{1}{\sqrt{(BT)}} \tag{8-54}$$

and is always less than unity. It is applicable only where the signal can be assumed to have a Gaussian distribution.

It can be shown [4] that the error (ϵ) between the true power spectral density $G(f)$ and the measurement from a single rêcord, $\hat{G}(f)$, will follow a Gaussian distribution and will assume a value of $\pm \epsilon$ with a confidence factor of 67%, assuming ϵ is no greater than 0·2. This means that if we make a series of measurements we can expect that about $\frac{2}{3}$ of these will have an estimation error of no more than $\pm \epsilon$. Larger error values are distributed according to a chi-squared probability and are not particularly meaningful. Equation (8-54) not only enables the accuracy of the measurements to be stated explicitly, but also permits comparison with measurements taken under different known conditions.

This method of specifying statistical accuracy relates particularly to analog methods, where B is stated explicitly. When sampled data is involved the filter bandwidth is stated implicitly and the accuracy given by the 'degrees of freedom' of the measurement. For reasons which were considered in chapter 4 sampling of continuous data occurs at time intervals of: $h \leqslant 1/2f_N$ where f_N is the Nyquist frequency. Assuming that the probability distribution of the random signal is similar to band-limited white noise and has a normal distribution, then these sample points can be shown to be statistically independent of each other. Thus we can regard f_N as representing the bandwidth of the sampling process in the frequency domain, i.e. $f_N = B$.

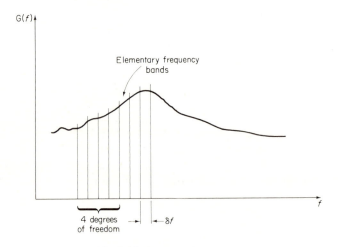

Fig. 8-12. Degrees of freedom.

The number of statistically independent and discrete samples required, spaced at intervals, $1/(2B)$ apart in the record of length, T sec, is thus:

$$n = \frac{T}{1/(2B)} = 2BT \qquad (8\text{-}55)$$

where n is known as the degrees of freedom for the data.

Now if $n = 1$, then $B = 1/(2T) = \delta f$ and we can consider the frequency spectrum of the power spectral density as divided into elementary bands, each of width δf, and each having one degree of freedom (fig. 8-12). Thus we can use 'degrees of freedom' as a way of expressing the sharpness of the filter used to examine the frequency spectrum. This is illustrated in fig. 8-13 which gives the frequency spectrum for the same random data analysed for different degrees of freedom.

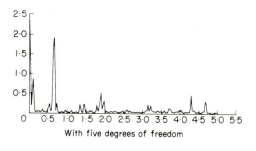

With five degrees of freedom

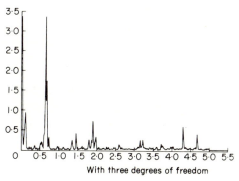

With three degrees of freedom

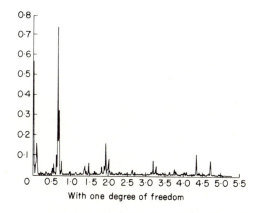

With one degree of freedom

Fig. 8-13. Showing effect of different degrees of freedom.

Where a correlation procedure is used to estimate the frequency spectrum the number of degrees of freedom is equal to the number of observations appearing within the estimation bandwidth, so that:

$$n = \frac{2N}{m} \qquad (8\text{-}56)$$

355

where m is the number of autocorrelation lags. The use of a spectral or lag window will modify equations (8-55) and (8-56) so that the number of degrees of freedom should be multiplied by a coefficient B_s given in table 8-2. This is equivalent to the inverse of the noise bandwidth of a filter as referred to in chapter 3. The statistical accuracy of estimation will decrease as n becomes smaller. The relationship between the normalised standard error ϵ, and n, is obtained from equations (8-54) and (8-55) and (8-56) as:

$$\epsilon^2 = \frac{2}{n} = \frac{m}{N} \tag{8-57}$$

In a practical case we may know little about the signal before analysis and will need to develop some empirical way of determining the effective analysis bandwidth to use. One such technique is to carry out a preliminary analysis using a wide bandwidth (large number of degrees of freedom), and to repeat the analysis with progressively narrower bandwidths. If little change of shape occurs over a wide range of bandwidths then a satisfactory resolution and accuracy will have been obtained. If the result of narrowing the filter bandwidth gives an increasingly fluctuating spectrum this could be due to insufficient record length, that is, number of discrete samples available. A fairly general experience is to find that the spectra will converge to a constant pattern as the filter bandwidth narrows until a given value is reached, beyond which a divergence of results will be obtained. The 'true' spectra can be taken at the filter bandwidth obtained just prior to this value. A suitable starting point for this procedure is found by examination of the autocorrelogram for the signal to be analysed. The delay is measured at the point where little significant change in the correlogram is seen and the filter bandwidth is taken as the reciprocal of this period.

8-4-4 CONFIDENCE LIMITS

An alternative way of defining the error for spectral estimates is through the chi-squared probability distribution. It will be shown that the power spectrum can be derived from a covariance correlation;

$$C(k) = E[X_j . X_{j+k}] \tag{8-58}$$

where the expected value is the product of two independent variables, X_j and X_{j+k}, separated by a time lag, k. Since for random data we can generally assume that these variables will have a Gaussian distribution in amplitude, the probability distribution for the power spectrum will be chi-squared having two degrees of freedom, i.e.

$$\chi_2^2 = X_j^2 + X_{j+k}^2. \tag{8-59}$$

This is the probability distribution for the periodogram estimation.

More generally, since spectral power estimates are mean-square value estimates obtained within a narrow-band, $n\delta f$ (fig. 8-12), and since each elementary band estimate, $(n=1)$, can be considered as independent and Gaussian the probability distribution of the estimate will be chi-squared having n degrees of freedom. From this a confidence limit can be placed on the value of the spectral estimate obtained, based on the degrees of freedom used. If $\hat{G}(f)$ is an estimate of the spectral density obtained at a centre frequency, f, with an effective analysis bandwidth given by n degrees of freedom, then the estimate can be expected to lie between two limits L_1 and L_2 with a probability of 100 P%, i.e.

$$\text{prob} \ [L_1 < \hat{G}(f) < L_2] = P \tag{8-60}$$

where

$$L_1 = \frac{n}{\chi^2_{n;\frac{1}{2}(1+P)}} \ \hat{G}(f) \tag{8-61}$$

and

$$L_2 = \frac{n}{\chi^2_{n;\frac{1}{2}(1-P)}} \ \hat{G}(f). \tag{8-62}$$

The modifying chi-squared parameters,

$$A = \frac{n}{\chi^2_{n;\frac{1}{2}(1+P)}}, \ B = \frac{n}{\chi^2_{n;\frac{1}{2}(1-P)}}$$

are obtained from standard tables for χ^2 distribution for $\alpha = \frac{1}{2}(P+1)$ and $\alpha = \frac{1}{2}(P-1)$ (e.g. reference [10]). The statistical tables available generally will only permit values up to about $n = 100$ to be obtained directly. Higher values can be calculated from:

$$\chi^2_{n;\alpha} \simeq n + Z_\alpha \sqrt{(2n)} + \tfrac{2}{3}(Z^2_\alpha - 1) + \frac{1}{9\sqrt{(2n)}} (Z^3_\alpha - 7Z_\alpha) \tag{8-63}$$

where Z_α is the ordinate obtained from tables of the normal probability integral. Selected values of A and B for several different probability levels are given in table 8-1.

A procedure for using confidence limits is as follows:

1. The number of degrees of freedom required for an estimate is determined from record length, bandwidth and spectral window (equation (8-55) and table 8-2).

2. The confidence level is decided, e.g. 90%, 80% etc.

3. Confidence parameters, A and B, determined from table 8-1.

4. Each value of $\hat{G}(f)$ is multiplied by A and B to determine accuracy limits at a given frequency.

TABLE 8-1 Table of confidence limits for auto-spectrum evaluation

N	P = 99% A	B	P = 90% A	B	P = 80% A	B	P = 50% A	B
1	25641·0256	0·1269	256·4103	0·2604	63·2911	0·3690	9·4340	0·7364
2	200·0000	0·1887	19·4175	0·3339	9·4787	0·4338	3·4542	0·7994
3	41·8410	0·2336	8·5227	0·3841	5·1370	0·4800	2·4712	0·7184
4	19·3237	0·2692	5·6259	0·4215	3·7736	0·5141	2·0812	0·7362
5	12·1359	0·2985	4·3478	0·4517	3·1056	0·5411	1·8720	0·7490
6	8·8757	0·3235	3·6585	0·4766	2·7273	0·5639	1·7396	0·7649
7	7·0779	0·3452	3·2258	0·4975	2·4735	0·5842	1·6486	0·7698
8	5·9701	0·3643	2·9304	0·5158	2·2923	0·5988	1·5810	0·7784
9	5·2023	0·3815	2·7027	0·5319	2·1583	0·6131	1·5291	0·7860
10	4·6296	0·3970	2·5381	0·5461	2·0534	0·6254	1·4874	0·7930
20	2·6918	0·5000	1·8433	0·6367	1·6077	0·7040	1·2968	0·8366
30	2·1755	0·5590	1·6225	0·6854	1·4563	0·7452	1·2277	0·8598
40	1·9314	0·5991	1·5089	0·7174	1·3769	0·7721	1·1902	0·8749
50	1·7848	0·6290	1·4382	0·7406	1·3274	0·7912	1·1665	0·8861
60	1·6875	0·6525	1·3892	0·7587	1·2921	0·8061	1·1493	0·8944
70	1·6168	0·6717	1·3529	0·7732	1·2657	0·8181	1·1363	0·9011
80	1·5628	0·6878	1·3247	0·7852	1·2451	0·8280	1·1261	0·9066
90	1·5199	0·7015	1·3020	0·7954	1·2285	0·8364	1·1178	0·9112
100	1·4849	0·7135	1·2832	0·8042	1·2147	0·8436	1·1109	0·9152
200	1·3136	0·7836	1·1885	0·8547	1·1442	0·8847	1·0752	0·9377
300	1·2464	0·8178	1·1500	0·8787	1·1152	0·9040	1·0603	0·9483
400	1·2087	0·8393	1·1279	0·8936	1·0985	0·9159	1·0517	0·9548
500	1·1839	0·8544	1·1132	0·9039	1·0873	0·9242	1·0459	0·9593
600	1·1661	0·8659	1·1026	0·9117	1·0792	0·9304	1·0417	0·9626
700	1·1524	0·8749	1·0944	0·9178	1·0729	0·9353	1·0384	0·9653
800	1·1416	0·8823	1·0879	0·9228	1·0680	0·9392	1·0358	0·9674
900	1·1328	0·8885	1·0826	0·9270	1·0639	0·9425	1·0337	0·9692
1000	1·1254	0·8937	1·0781	0·9305	1·0604	0·9453	1·0319	0·9707

A = coefficient of upper limit, B = coefficient of lower limit, P = percentage accuracy of auto spectrum value between these limits, N = degrees of freedom for spectral evaluation.

Example

It is required to estimate the power in a spectrum for a record of length $T = 1$ sec, and which is band-limited to contain no frequencies higher than 2 kHz. It is assumed that the width of the narrowest peak in the spectrum will be no greater than 40 Hz. The data is to be smoothed before spectral estimation by means of a Hanning window. The sampling rate must be $\geqslant$ 4 kHz and the sampling interval $\leqslant 0·250$ msec. Hence the number of degrees of freedom will be obtained from equation (8-55) and modified by the bandwidth of the equivalent smoothing filter using table (8-2) to obtain $n = 2 \times 40 \times 1·3 = 100$. For a confidence level of 90% the modifying values are: $A = 1·2821$ and $B = 0·7813$, so that we need to multiply our estimates by these values to determine the error band. Note that in

TABLE 8.1 Characteristics of smoothing windows

Type	Lag window ($W(\tau)$)	Spectral window ($W(f)$)	Equivalent bandwidth (B_s)
Rectangular	$= 1$ for $|\tau| \leqslant T_m$ $= 0$ for $|\tau| > T_m$	$= 2T_m\left[\dfrac{\sin 2\pi f T_m}{2\pi f T_m}\right] = W_0$	0·5
Bartlett	$= 1 - \dfrac{|\tau|}{T_m}$ for $|\tau| \leqslant T_m$ $= 0$ for $|\tau| > T_m$	$= T_m\left[\dfrac{\sin \pi f T_m}{\pi f T_m}\right]^2$	1·5
Parzen	$= 1 - 6\left(\dfrac{|\tau|}{T_m}\right)^2 + 6\left(\dfrac{|\tau|}{T_m}\right)^3$ for $|\tau| \leqslant \dfrac{T_m}{2}$ $= 2\left(1 - \dfrac{|\tau|}{T_m}\right)^3$ for $\dfrac{T_m}{2} < |\tau| \leqslant T_m$ $= 0$ for $|\tau| > T_m$	$= 0.75 T_m\left[\dfrac{\sin \pi f \frac{1}{2} T_m}{\pi f \frac{1}{2} T_m}\right]^4$	1·9
Hanning	$= 0.5 + 0.5\cos\left(\dfrac{\Pi\tau}{T_m}\right)$ for $|\tau| \leqslant T_m$ $= 0$ for $|\tau| > T_m$	$= 0.5 W_0(f)$ $+ 0.25 W_0\left[f + \dfrac{1}{2T_m}\right] + 0.25 W_0\left[f - \dfrac{1}{2T_m}\right]$	1·3
Hamming	$= 0.54 + 0.46\cos\left(\dfrac{\pi\tau}{T_m}\right)$ for $|\tau| \leqslant T_m$ $= 0$ for $|\tau| > T_m$	$= 0.54 W_0(f)$ $+ 0.23 W_0\left[f + \dfrac{1}{2T_m}\right] + 0.23 W_0\left[f - \dfrac{1}{2T_m}\right]$	1·3
Blackman	$= 0.42 + 0.5\cos\left(\dfrac{\pi\tau}{T_m}\right)$ $+ 0.08\cos\left(\dfrac{2\pi\tau}{T_m}\right)$ for $|\tau| \leqslant T_m$ $= 0$ for $|\tau| > T_m$	$= 0.42 W_0(f) + 0.25 W_0\left[f + \dfrac{1}{2T_m}\right]$ $+ 0.25 W_0\left[f - \dfrac{1}{2T_m}\right] + 0.04 W_0\left[f + \dfrac{1}{T_m}\right]$ $+ 0.04 W_0\left[f - \dfrac{1}{T_m}\right]$	1·4

where: $T_m = h_m$ = total time lag for window, h = sampling interval, and m = number of samples in lag window.

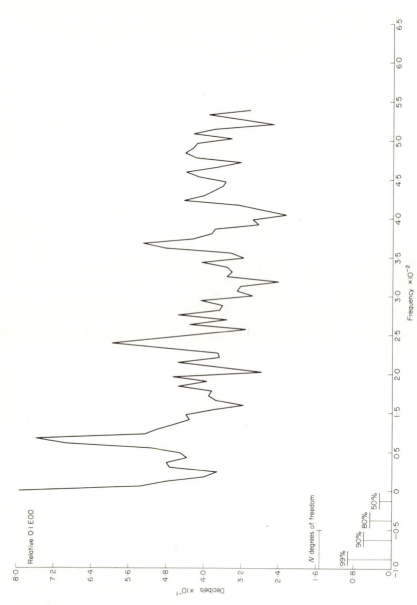

Fig. 8-14. Calibrated plot of P.S.D. on a logarithmic scale.

a practical determination of an error band, for a given percentage probability, the width of the band will be dependent on the instantaneous value of $\hat{G}(f)$ at any frequency. If the spectral estimates are plotted on a logarithmic scale the confidence interval for the spectral estimate can be indicated as a constant vertical ordinate applicable to any part of the spectrum. Thus from (8-61) and (8-62) the limits L_1 and L_2 will become:

$$\log \hat{G}(f) + \log \frac{n}{\chi^2_{n;\frac{1}{2}(1+P)}} \tag{8-64}$$

and

$$\log \hat{G}(f) + \log \frac{n}{\chi^2_{n;\frac{1}{2}(1-P)}} \tag{8-65}$$

permitting the confidence interval simply to be added to the estimate obtained. An example is given in fig. 8-14 which indicates the analysis bandwidth by means of a short horizontal line and the accuracy by means of a series of vertical ordinates for different confidence levels. The presentation of power spectrum in logarithmic form is also valuable in the engineering sense, since it is proportional changes of power that are important.

The use of confidence levels is only one of the many attempts that have been made to indicate the acceptable limits of spectral estimates. Another technique that has been proposed by Wonnacott [11] is to construct two extreme limit band-pass spectral windows, known as the 'inner window' and the 'outer window', which fall on either side of the ideal band-pass window. The two spectra that are obtained from the use of these windows can be considered as representing the limits of the spectral estimates.

8-4-5 SCANNING RATE AND ANALYSIS TIME

The rate at which analysis proceeds in the frequency domain is critical and must be slow enough to permit all the information relating to a single analysis point to be collected before the frequency is changed. This is particularly relevant to filter methods of analysis, either analog or digital, although similar considerations apply for the indirect or autocorrelation methods. Scanning rate depends primarily on the characteristic of the averaging method used and, to a lesser extent, on the filter performance. The averaging time can be varied during the analysis to maintain a constant value of the normalised standard error and this procedure would be carried out, for example, in a proportional bandwidth system (see later section 8-7). This does however, raise practical problems of implementation, and a more general approach is to use an integration time, T_a, as long as the analysis record to give the minimum error for all frequencies. An averaging time longer than this will have very little effect in reducing averaging error. This is only applicable to true integration averaging or discrete value summation. For RC averaging (see section 8-7) the minimum error is obtained as the time constant K, of the averaging circuit tends to infinity. The relationship is an inverse exponential one and the error due to averaging becomes less than 5% when RC = K = T_a so that there is also little advantage in using an RC time constant greater than the record length.

In order that the scanning rate will be sufficiently long to include information contained in the entire filter bandwidth B, the rate should be

$$\text{Rate} \leqslant B/T_a \text{ Hz.} \qquad (8\text{-}66)$$

Where RC averaging is used the signal requires a period of at least four time constants in order to arrive at a steady state averaging value (fig. 8-15). Hence for RC averaging the scanning rate is:

$$\text{Rate} \leqslant B/4K. \qquad (8\text{-}67)$$

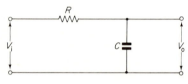

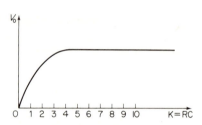

Fig. 8-15. R–C averaging.

A similar information storage occurs in the band-pass filter used in the analysis method. The build-up of the potential at the output of a filter having a shape factor of four has been determined empirically as:

$$T_f \simeq 4/B \text{ sec} \qquad (8\text{-}68)$$

where T_f is the time required to obtain 63% of the final output. Thus from equation (8-66) the scanning rate for time filter averaging must be less than:

$$B/T_f = \tfrac{1}{4}B^2 \text{ Hz.} \qquad (8\text{-}69)$$

Since $BT \geqslant 1$ for acceptable error (equation 8-54), this restriction can usually be ignored.

The time taken for analysis of a record of length T can be obtained from the scanning rate and bandwidth as a function of the frequency range selected.

Assuming a linear scanning rate, R, this may be derived from equation (8-66) as:

$$\frac{dB}{dT_a} = R \text{ Hz/sec.} \tag{8-70}$$

Integrating over the analysis frequency range and time for analysis, T_t, gives:

$$\int_{F_1}^{F_2} df = R \int_0^{T_t} dt$$

so that solving for T_t gives:

$$(F_2 - F_1) = R(T_t - 0)$$

and

$$T_t = \frac{1}{R}(F_2 - F_1) \tag{8-71}$$

where F_2 is the highest analysis frequency (Hz), and F_1 is the lowest analysis frequency (Hz). This can be restated for true averaging in terms of record length and filter bandwidth from equation (8-66):

$$T_t = \frac{T_a F}{B} \tag{8-72}$$

and for RC averaging

$$T_t = 4KF/B \tag{8-73}$$

where $F = F_2 - F_1$.

These fixed bandwidth methods of analysis result in lengthy computation times, with a resolution often unnecessarily narrow for most of the frequency range. A substantial improvement is obtained if a continuous change in filter bandwidth can be made during the process of analysis. The method is to vary the bandwidth as a percentage of the filter centre frequency, $B = Cf_0$ where C is the percentage bandwidth/100, and f_0 is the filter centre frequency in Hz.

From equation (8-70) the scanning rate can be obtained as:

$$R \leqslant \frac{Cf_0}{T_a} = df/dt. \tag{8-74}$$

The analysis time can be derived, as before, by integrating with the relative variable over time T_t and frequency F:

$$\int_{F_1}^{F_2} \frac{df}{f_0} = \int_0^{T_t} C/T_a \,.\, dt$$

therefore $\log_e (F_2/F_1) = (C/T_a)T_t$
and

$$T_t = T_a/C \,.\, \log_e \frac{F_2}{F_1}. \tag{8-75}$$

A disadvantage of the percentage bandwidth analyser is that where the signal contains additive white noise the effect of this will increase with frequency, since the power contribution will be proportional to the filter bandwidth. This can be compensated for in the scaling of the final averaged values but will limit the dynamic range possible by the method.

8-5 SMOOTHING FUNCTIONS

Smoothing forms an indispensable part of power spectral density evaluation so that it is pertinent to consider general methods for smoothing in both time and frequency domains in this chapter.

Two fundamental problems are associated with any of the methods of power spectral density evaluation discussed so far. These are:

(*a*) Leakage. That is, a true power-spectral density estimate for a given frequency always contains in addition some element of power derived from other adjacent frequencies. This is referred to as a 'Leakage' of power from outside the desired area of measurement and reduces the accuracy and value of the estimate. No matter how the power spectrum is calculated we find that, although a value of power spectral density is desired at a given frequency value, power at other frequencies will be involved in the calculation.

(*b*) Smearing. This represents the conflict between desired resolution and resulting statistical accuracy. We saw earlier that this presents us with a choice of good statistical reliability using a wide frequency window, or fine resolution using a narrow frequency window, or some compromise decision involving an attainable resolution and accuracy. If a wide analysis bandwidth is employed to obtain a good statistical reliability, an indeterminate spectral discrimination is obtained, and we say that the result is 'smeared'. If the opposite extreme is taken, then, in the limit, the variance of the result equals the variability of the data and it is difficult to derive useful information from the measurement.

The nature of this conflict has been discussed in some detail earlier. Here we are concerned with the implementation of windows and their effect on these two problems. It will be seen that we will need to use both lag (time) and spectral (frequency), windows to obtain acceptable results.

Smoothing can be carried out in the time or frequency domain and be effective on continuous or discrete data. Continuous smoothing in the frequency domain is achieved by analog low-pass filtering and has been considered in chapter 3. Continuous smoothing in the time domain is equivalent to a convolution operation and for reasons which will be made clear later is sometimes referred to as Tapering of the signal. Here the time-history series is modified by multiplication by a weighting function defining the smoothing characteristics required. Smoothing of discrete data is obtained by a process of digital filtering, which in its simplest form, is represented by the summation of adjacent sample values. There is no real distinction here between discrete series smoothing in the time or frequency domains since they both represent identical operations on a series of discrete sample values. The meaning ascribed to these values is known by reference to their source but is otherwise irrelevant to the smoothing process. Continuous operations, on the other hand, involve domain-transform hardware such as analog filters, and are applied in the time domain due to the fundamental characteristics of these elements.

This section will consider first the process of tapering and continuous convolution. Smoothing of discrete data by simple summation routines and its relation to low-pass filtering will then be considered. Finally the effect of a choice of spectral window shape on the validity of spectral estimates will be discussed.

8-5-1   LAG AND SPECTRAL WINDOWS

A finite record available for analysis, $x(t)$, can be considered as equivalent to the result of multiplying an infinite extent of random data, $y(t)$, with a finite rectangular window (fig. 8-16). The finite record invariably begins and ends

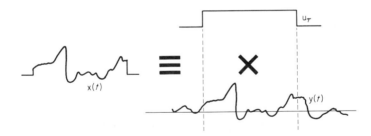

Fig. 8-16. Equivalent finite length record.

abruptly and will introduce additional frequencies (side-lobes), not present in the original data. The Fourier transform of the modified time series $x(t)$ may be expressed as:

$$X(f) = \int_0^T x(t) \exp(-j\omega t)\, dt = \int_{-\infty}^{\infty} y(t) u_\tau . \exp(-j\omega t)\, dt \quad (8\text{-}76)$$

where $u_\tau = 1\cdot0$ for $0 < t < T$, and $0\cdot0$ for $0 > t > T$.

This will be seen to represent the product of the Fourier transform of $y(t)$ with that of u_τ (since convolution in the time domain transforms to a product in the frequency domain (see equation (7-14)). Hence the transformation of u_T into

365

the frequency domain $U(f)$, will give a measure of the distortion experienced by truncation. The Fourier transform of u_T is:

$$U(f) = \int_{-\infty}^{\infty} u_\tau \exp(-j\omega\tau) . d\tau$$

$$= 2T\frac{\sin \omega\tau}{\omega\tau} \qquad (8\text{-}77)$$

This has the form, $\sin x/x$, shown in fig. 8-17. The large side-lobes will be reduced if a gradual change in amplitude of the signal, known as tapering, is imposed at

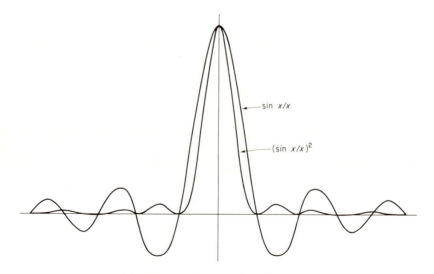

Fig. 8-17. $\operatorname{Sin} x/x$ and $\sin^2 x/x^2$ windows.

the beginning and end of the record. Some such process is necessary whenever a finite Fourier transformation takes place (e.g. the use of the fast Fourier transform in digital calculations), since this implies integration over a finite time period. An example is given in fig. 8-18 which shows clearly the reduction of side lobes for $U(f)$ when tapering is carried out. Truncation of the autocorrelation function, where this is used as a route to the power spectral density produces a similar spectral distortion [12]. This is because the transformation process, indicated in equation (8-21), involves an infinite integral, which demands a knowledge of R_τ for all values of τ up to infinity. In practice we must truncate τ at some point which may not necessarily correspond to zero value for

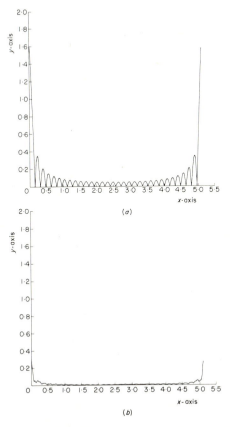

Fig. 8-18. Effect of tapering applied to the Fourier transform of a rectangular wave-form, (a) without tapering; (b) with tapering.

R_τ. We have again to introduce a rectangular window weighting function and the practical equation for a real value of power-spectral density using this method will be:

$$G(f) = 4 \int_0^\infty R(\tau) u_\tau \cos(2\pi f\tau) \, d\tau \qquad (8\text{-}78)$$

where $u_\tau = 1\cdot0$ for $0 < \tau < T$, and $0\cdot0$ for $0 > \tau > T$, which may be compared with the one-sided spectral density estimate given by equation (8-23).

The method of compensating for finite record length is by modifying the rectangular characteristics of the weighting function u_τ, by introducing a second multiplying function of time under the integral, known as a lag window, $W(\tau)$. The effect of the introduction of the lag window is to reduce the amplitude of the frequency side-lobes (particularly the negative excursions), and to broaden the width of the main lobe.

367

The lag window will first be considered in relation to power spectral density from the correlation function (equation 8-78). The most widely used window is that due to Hann which takes the form:

$$W'(\tau) = \frac{1}{2}\left[1 + \cos\frac{\pi\tau}{T_m}\right] \text{ for } |\tau| \leqslant T_m$$

$$= 0 \text{ for } |\tau| > T_m \qquad (8\text{-}79)$$

where T_m is the maximum value of the autocorrelation delay and not necessarily the length of the record. The Fourier transform of equation (8-79) can be shown to consist of the summation of three impulse response functions each having the $\sin x/x$ form of a rectangular function, i.e.

$$W'(f) = \frac{1}{2}W_0(f) + \frac{1}{4}W_0\left(f + \frac{1}{2T_m}\right) + \frac{1}{4}W_0\left(f - \frac{1}{2T_m}\right) \qquad (8\text{-}80)$$

where W_0 is the Fourier transform of the rectangular window derived earlier as:

$$W_0 = 2T_m\left[\frac{\sin 2\pi f T_m}{2\pi f T_m}\right].$$

The averaging effect of this summation has been described earlier in chapter 7 (fig. 7.20), as an illustration of a frequency sampling filter. A second form of lag window is attributable to Bartlett [13] and defined as:

$$W''(\tau) = 1 - \frac{|\tau|}{T_m} \text{ (for } |\tau| \leqslant T_m)$$

$$= 0 \text{ (for } |\tau| > T_m) \qquad (8\text{-}81)$$

for which the Fourier transform is:

$$W''(f) = T_m\left[\frac{\sin \pi f T_m}{\pi f T_m}\right]^2 \qquad (8\text{-}82)$$

This window produces only positive value side lobes and has a lower variance than the Hanning Window.

A third window known as the Parzen [14] window takes the form:

$$W'''(\tau) = 1 - 6\left(\frac{|\tau|}{T_m}\right)^2 + 6\left(\frac{|\tau|}{T_m}\right)^3 \text{ (for } |\tau| \leqslant \tfrac{1}{2}T_m)$$

$$= 2\left(1 - \frac{|\tau|}{T_m}\right)^3 \text{ (for } \tfrac{1}{2}T_m < |\tau| \leqslant T_m) \qquad (8\text{-}83)$$

$$= 0 \text{ (for } |\tau| > T_m)$$

for which the Fourier transform is

$$W'''(f) = 0.75\, T_m \left[\frac{\sin \pi f(T_m/2)}{\pi f(T_m/2)} \right]^4 \qquad (8\text{-}84)$$

This window is also productive of positive value estimates and has a smaller variance than either the Hann or Bartlett window. The three windows are compared with that of the rectangular window for the time and frequency domain in fig. 8-19 and their characteristics given in table 8-2. The choice of lag window for the derivation of spectral estimates via the autocorrelation function largely depends on the signal characteristics and accuracy of estimation required. For example the Parzen window would be chosen on the basis of minimum

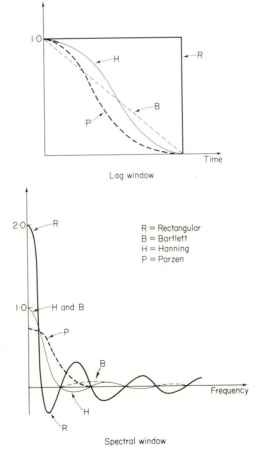

Lag window

R = Rectangular
B = Bartlett
H = Hanning
P = Parzen

Spectral window

Fig. 8-19. Smoothing functions.

369

variance but would produce a larger bias than any of the others. The oscillatory nature of the Hann window about zero value could achieve a limited amount of power cancellation for equal numbers of positive and negative side lobes. The Bartlett and Parzen windows on the other hand will guarantee a positive estimate value. For most purposes the Hann window will be found to be adequate and can be very simply programmed for the digital machine.

8-5-2 TAPERING THE SIGNAL

A similar smoothing operation must be carried out on the time history series when using the periodogram approach to the power spectral density. An effective form of tapering of the series is to multiply values at each end by a Hanning function. It has been found sufficient to perform this operation over 10% of the record at each end (fig. 8-20). If we let M be the fraction of the total signal to be

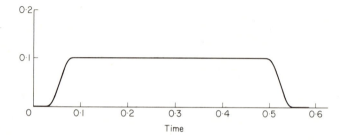

Fig. 8-20. Cosine tapering.

tapered ($M = 0\cdot1\ N$), then the amplitude of the modified series, x_i', will be:

$$\left.\begin{aligned}
x_i' &= \frac{x_i}{2}\left(1 - \cos \pi\,\frac{(i-1)}{(M-1)}\right) &&\text{(for } 1 < i < M)\\[2mm]
x_i' &= x_i &&\text{(for } N-M > i > M+1)\\[2mm]
x_i' &= \frac{x_i}{2}\left(1 - \cos \pi\frac{(N-i)}{(M-1)}\right) &&\text{(for } N-M+1 < i < N)
\end{aligned}\right\}(i = 1, 2, \ldots, N)$$

$$(8\text{-}85)$$

A similar result can be obtained by convolution of the impulse response function for the Hanning window with the Fourier transform of the signal before squaring is carried out to obtain a power-spectral density estimation. Whether tapering is carried out before transformation or convolution afterwards is a matter of computational convenience. It is important to scale the signal correctly after tapering in order to compensate for the reduction in the total power contained in the spectral estimates. The problem is essentially the same as that found in digital filtering where a zero insertion loss is required. This involves normalising the result of the convolution by the area under the impulse response function of the lag window. Using the tapering routine given by equation (8-85) then this involves multiplying each modified value by $1/0{\cdot}875$.

An exponential window is particularly easy to implement using analog methods. Otterman [15] has suggested the use of a double exponential window, implemented by calculating the difference between the E.M.P. Fourier estimates, taken for time constants α_1 and α_2.

$$X_{(\alpha_1 - \alpha_2)}(f) = X_{\alpha_1}(f) - X_{\alpha_2}(f)$$

$$= (\alpha_1 - \alpha_2) \int_{-\infty}^{\infty} x(\tau) \left[\exp(\alpha_1 t) - \exp(\alpha_2 t) \right]$$

$$\times \exp(-\tau) . u(t - \tau) . \exp(-j\omega\tau) . d\tau \qquad (8\text{-}86)$$

where $[\exp(\alpha_1 t) - \exp(\alpha_2 t)]$, is the exponential window function.

Where analog methods of spectral analysis are used, the preparation of tape loops provides an area of application for the generation of E.M.P. time window techniques. An example is shown in fig. 8-21 where exponential tapering takes

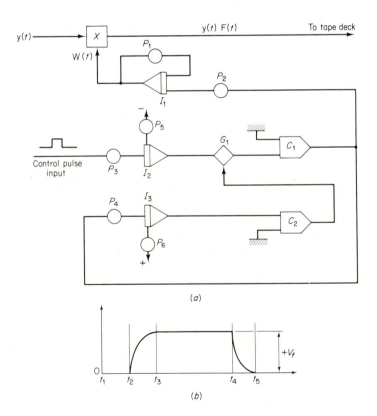

(a)

(b)

Fig. 8-21. Generation of E.M.P. time window.

371

place prior to recording on the tape loop. This uses a simple integrator/multiplier combination controlled by two ramp generators. The signal to be processed, $y(t)$, is multiplied by a window function, $W(t)$, which is generated in the following way. A control pulse of duration τ is provided from the input tape transport at a time t_1, prior to the commencement of the record. This initiates a run-down of integrator I_2, which is applied via a closed gate G_1 to the comparator C_1. When the ramp generated at I_2 reaches zero level, C_1 is initiated to provide a positive level (logic 1) to both of the integrators I_1 and I_3. The delay $(t_2 - t_1)$ in providing this level is set by potentiometer P_3 at a value large enough for the tape splice to pass through the tape reading head. The exponential function, $I(t_2 - t_3)$, is then generated by I_1 and multiplied by the signal function, $y(t)$. At the same time the integrator I_3 is initiated by the logic input, and, after a delay period determined by the setting of potentiometer P_4 and the length of the record, the comparator, C_2, is operated to open gate G_1, thus permitting the E.M.P. integrator, I_1, to fall exponentially to its original state to complete the generation of the window function, $W(t)$. The control pulse, τ, should be of such a duration that it will have reverted to zero before time t_4 has been reached.

8-5-3   SMOOTHING THE SPECTRAL ESTIMATE

Smoothing, required to overcome the leakage effects of the finite length record, has been discussed above. A somewhat similar situation exists with the leakage caused by the window shape of the analysis filter. Even with well-designed narrow-band filters, we could have the situation where the signal exhibits a large dynamic range of energy over the frequency spectrum, due to a large amount of power located at the low frequency end of the spectrum. The filter used in the spectral density estimation may well have insufficient attenuation to deal with this situation, i.e. we have a 'leakage' problem. In such cases the signal is frequency smoothed before processing (see chapter 3), to limit the dynamic range of the signal. Under these circumstances it will be necessary to re-scale the data after processing using the known characteristics of the pre-whitening filter.

Where the spectral estimate is derived from the periodogram considerable leakage is experienced due to the side lobes of the effective analysis filter. From equation (8-15) we can write:

$$\hat{G}_x(k) = \frac{1}{N} \left| \sum_{i=0}^{N-1} x_i . \exp\left(-j2\pi in/N\right) \right|^2 \qquad (8\text{-}87)$$

for a discrete time series.

We recognise this as an estimation of the power spectral density derived from the continuous Fourier transform of the time series $x(t)$. Used in this form, the periodogram gives a very 'raw' estimation of spectral density [16] since the expected value will contain an additional term representing a spectral window:

$$W(f) = N \left[\frac{\sin (\pi f N)}{(\pi f N)} \right]^2 . \tag{8-88}$$

This term is also shown in fig. 8-17 and represents the factor by which the true spectrum is modified. In statistical terms the periodogram is a function in which the variance does not approach zero as the record length increases and a large number of samples are taken. As applied to real Gaussian processes, in the limit the estimate obtained will be proportional to a chi-squared random variable, having only two degrees of freedom. To obtain a better estimate, having a greater number of degrees of freedom and hence reduction in frequency side-lobes it is necessary to smooth the periodogram by carrying out a summation of weighted adjacent values before squaring takes place. To obtain $2v$ degrees of freedom for the estimate it is necessary to add together v neighbouring values, thus:

$$\hat{G}_{x(i)} = \frac{1}{m} \sum_{k=1}^{v} G_{x(i+k)} \quad (i = 0, v, 2v, 3v, \ldots, mv) \tag{8-89}$$

where m is an integer of value N/v. This represents a simple form of moving average filter where the weighting attached to the values being summed determines the frequency characteristic of the spectral window. It will now be shown that this smoothing routine is equivalent to a digital low-pass filter.

A simple three-point smoothing routine for x_n values can take the form:

$$x_n' = \frac{A x_{n-1} + x_n + B x_{n+1}}{1 + A + B} \tag{8-90}$$

where A and B are constants and $n = 1, 2, \ldots, N$.

In order to determine the effect of this smoothing routine in the frequency domain we can let $x_n = X \sin \omega t$ and substitute this in equation (8-90).

$$x_n' = \frac{A X \sin \omega (t - T) + X \sin \omega t + B X \sin \omega (t + T)}{1 + A + B}$$

where T is the sampling interval, $\omega = 2\pi f$ and $f = 1/(2T)$.
This can be expanded to give:

$$x_n' = \frac{X}{1 + A + B} \cdot \sin \omega(t + \phi) \tag{8-91}$$

where

$$\phi = \arctan \frac{\sin \omega t (A - B)}{1 + (A + B) \cos \omega t} \tag{8-92}$$

is the phase shift introduced by the filter. If ϕ is required to be zero (the usual case) then A must be made equal to B. The assumption is made here that the

373

signal frequency $f \leqslant 1/(2T)$ in accordance with the sampling theorem. Consequently any phase shift at frequencies beyond $f = 1/(2T)$ can be neglected. Using this as a criterion it can easily be shown that a maximum value of 0·5 is required for A. Hence:

$$x'_n = \tfrac{1}{2}[0{\cdot}5 \, X \sin \omega(t - T) + X \sin \omega t + 0{\cdot}5 \, X \sin \omega(t + T)]. \qquad (8\text{-}93)$$

Expanding equation (8-93) gives a simplified form: $x'_n = x_n \tfrac{1}{2}(1 + \cos \omega t)$ so that the filter transfer function may be expressed as:

$$H(\omega) = \frac{x'_n}{x_n} = \tfrac{1}{2}(1 + \cos \omega t) \qquad (8\text{-}94)$$

which describes a 'cosine bell' function shape given earlier in equation (8-85). The cut-off frequency, f_c, is obtained by making $H(\omega) = 0{\cdot}707$ (-3 db amplitude response), when $f_c = 0{\cdot}183/T$.

Equation (8-94) can be seen to express a low-pass frequency filter whose cut-off frequency is a function of the sampling interval, T. For a given sampling interval, the cut-off frequency can be reduced (and the slope proportionately increased) by n repetitions of equation (8-94), so that if we now equate $H(\omega)_n = [\tfrac{1}{2}(1 + \cos \omega t)]^n = 0{\cdot}707$, as before, the new cut-off frequency obtained will be:

$$f_c(n) = \frac{0{\cdot}159}{T} \text{ arcos } [2^{(1 - \tfrac{1}{2}n)} - 1]. \qquad (8\text{-}95)$$

For large values of n this results in a Gaussian frequency characteristic so that a sharp cut-off cannot be expected. Equation (8-93) is generally written as a three sample smoothing routine,

$$x'_n = 0{\cdot}25x_{n-1} + 0{\cdot}5x_n + 0{\cdot}25x_{n+1} \qquad (8\text{-}96)$$

which has been referred to earlier as Hanning.

An alternative weighting, due to Hamming [17], achieves a lower amplitude for the immediate side-lobes but a slower falling off for subsequent lobes. The weighting is:

$$x''_n = 0{\cdot}23x_{n-1} + 0{\cdot}54x_n + 0{\cdot}23x_{n+1}. \qquad (8\text{-}97)$$

Other forms of moving average smoothing can be evolved to give a higher order of smoothing. For example the Blackman window [4] will give a very low side-lobe amplitude, equivalent to a filter having ten degrees of freedom. The weighting function is:

$$x'''_n = 0{\cdot}42x_n + 0{\cdot}25x_{n+1} + 0{\cdot}25x_{n-1} + 0{\cdot}04x_{n+2} + 0{\cdot}04x_{n-2}. \qquad (8\text{-}98)$$

These are, of course, special examples of non-recursive digital filters and equivalent to a convolution operation.

8-6 DIGITAL METHODS

Spectrum analysis using the digital computer is currently dominated by the influence of the fast Fourier transform algorithm. Using this technique a considerable increase in computational speed is obtained over earlier methods. In the case of hardware F.F.T. units for example, it is possible to obtain a real-time spectrum of a radar signal with a resolution unequalled by other methods [18]. When implemented on a general-purpose digital computer the use of the algorithm leads to considerable reductions in computing costs.

Three methods of spectral density estimation will be considered in this section:

(*a*) Direct estimation via the periodogram. This method is not without its difficulties which arise as a consequence of large side-lobes present with the periodogram estimate. To use it effectively we must taper the signal and carry out smoothing in the frequency domain before squaring and averaging the data. Much controversy still exists about the exact nature of the tapering and smoothing routines to employ.

(*b*) Indirect estimation from the autocorrelogram. Much more information is available, and consequently a more general acceptance found for the indirect estimation via the autocorrelogram of the signal. For reasons noted earlier, the autocorrelogram may be derived from the Fourier transform.

(*c*) The fundamental parallel filter method of spectral estimation, which finds its digital equivalent in the use of multiple digital filters. Use can be made of Fourier transform techniques in simulating these filters [19], although for reasons of computational economy the filter design can use a recursive filter bank with advantage.

A particularly interesting modular method, due to Enochson and Otnes [20], achieves a third octave analysis by repeated decimation of the signal and application of the succesive filtered outputs to a limited group of weighted filters. For large numbers of data samples this can be considerably faster than the F.F.T. method via the periodogram.

8-6-1 THE PERIODOGRAM APPROACH

Referring to equation (8-87) direct estimation of the power spectrum can be obtained from the relationship:

$$\hat{G}_x(k) = \frac{1}{N} | X_k |^2 \tag{8-99}$$

where X_k is the Fourier transform of the time series x_i and $\hat{G}_x(k)$ is a single-sided spectrum of the power contained in the signal. This is implemented

375

from the summation of the squares of the real and imaginary coefficients of the Fourier transform:

$$\hat{G}_x(k) = \frac{1}{N} ([\text{Re } X_k]^2 + [\text{Im } X_k]^2) \tag{8-100}$$

where N is the total number of data points. Cross spectra are obtained from the relationship:

$$\hat{G}_{xy}(k) = \frac{1}{N} |X_k^* Y_k| \tag{8-101}$$

where * indicates the complex conjugate. If used directly, these estimations will lead to inaccurate results due to the presence of large side-lobes in the equivalent filter response. Tapering the signal before transformation will reduce the amplitude of these side-lobes. In an earlier section, (8-5), this was shown to be equivalent to convolution of the 'raw' spectrum with a smoothing spectral window. Excessive tapering must be avoided as it will result in loss of data and effectively reduce the available data length. It is better to achieve a moderate reduction in side-lobe response by tapering 10% of the signal wave-form at either end (fig. 8-20) and to carry out further smoothing in the frequency domain.

Since the estimate for $G(f)$ from the periodogram will have a probability distribution that is chi-squared having two degrees of freedom (section 8-4-4), the standard error will be found from equation (8-57), to be:

$$\epsilon = (2/n)^{\frac{1}{2}} = 1 \tag{8-102}$$

or 100%. Clearly this amount of error is unacceptable and before these raw estimates can be used some smoothing in the frequency domain will be essential. A number of different smoothing routines have been used for this purpose some of which were described earlier in this chapter. There is a considerable lack of agreement as to an acceptable criterion of optimum performance for spectral density analysis in respect of the choice of smoothing window to use. The three-point Hanning routine (equation (8-96)) is used often due to its simplicity and ease of computer implementation. This smoothing operation is carried out on the raw Fourier coefficients before the values are squared, added and normalised to produce the power-spectra. Hanning of the modulus values given by equation (8-99) is not nearly so effective in smoothing the Fourier transform.

The F.F.T. algorithm requires that the number of data samples be a power of two and this imposes a constraint either at the data acquisition level (which is undesirable), or in the pre-processing of the data before analysis. This can often be arranged by suitable choice of digitisation rate and record length or by the addition of samples of zero value to make up the number of coefficients. To summarise the procedure for spectral density estimates via the periodogram the essential steps can be stated as:

1. Remove any mean value from the signal to be analysed.
2. Taper the ends of the time series using equation (8-94).

3. Augment the data samples with zeros to obtain a length $N=2^P$ where P is an integer.

4. Carry out discrete Fourier transformation of the augmented series.

5. Smooth the values, obtained from the raw estimates, using a Hanning routine.

6. Using equation (8-100), compute the spectral estimate $\hat{G}(k)$ from the square of the modulus of the Fourier transform.

7. Scale the result to allow for tapering and plot on a linear/log scale.

8. Compute the confidence factor and bandwidth (degree of freedom) and add calibration ordinates to the graph.

Steps 1, 2 and 3 may form part of a pre-processing operation and 7 and 8 may be included in a calibration and plotting routine so that computational speed comparisons between this method and any other should be based on steps 4, 5 and 6 only.

A typical spectral density based on this method was shown in fig. 8-14. The estimate for power spectral density will be limited to 2 degrees of freedom and to obtain better statistical properties a measure of frequency smoothing is carried out. This involves summing and averaging the results of $n/2$ such spectral estimates to give:

$$\hat{G} = \frac{2}{n} [\hat{G}(k) + \hat{G}(k + 1), \ldots, + \hat{G}_{(k+\frac{1}{2}n-1)}] \qquad (8\text{-}103)$$

where n = degrees of freedom required. The effective resolution bandwidth is now $n/2T$ instead of $\frac{1}{2T}$.

It was shown in chapter 6 that the in-core requirements of the F.F.T. precludes the analysis of very long records using the above procedure. Instead an analysis by parts may be carried out.

The signal is divided into m equal segments, each segment considered as two equal sections k_1 and k_2. The first segment m_1 is tapered and its transform obtained. The second half of m_1 is taken together with the first section of m_2 to form a second segment for analysis $[k_2(m_1) + k_1(m_2)]$.

After tapering its transform is found and summed with the first transform to obtain an average value.

This procedure is continued for the following segment, m_2, and the next overlapping segment $[k_2(m_2) + k_1(m_3)]$ and the results averaged with the summated transform obtained previously. Continuing in this way an averaged transform can be generated for the entire signal using in-core storage little larger than would be required for one segment of the signal. The final summated transform is divided by the number of segment operations ($=2m-1$) and the raw spectral estimate obtained from the square of the modulus of the averaged transform. Overlapping of the segments for transformation is included to reduce the effects of amplitude weighting of the data, caused by tapering at the ends of each segment.

8-6-2 THE MEAN-LAGGED PRODUCT APPROACH

Spectral estimation via the correlation coefficient removes the uncertainty related to the large side-lobe characteristic of the periodogram and enables a statistically satisfactory estimate to be obtained. The procedure is longer but less smoothing of the final result is required. The raw periodogram is first obtained as described above and then, instead of using this directly, it is inversely transformed to obtain the autocorrelation function. This is then multiplied by a smoothing or lag window to obtain a weighted autocorrelation function. The amount of the correlation function which will need to be retained is given by the smallest power of 2 which is greater than the number of lags, *m,* used. The truncated autocorrelation function is then directly transformed to obtain the smoothed power spectrum.

To satisfactorily implement this method on the digital computer the procedure requires some modification, namely addition of zeros to the raw spectrum before multiplication by the lag values and scaling of the results by a ramp function of *N.* This is a consequence of circular correlation that results from the operation of the transform algorithm and will be fully considered in the next chapter.

8-6-3 USE OF A DIGITAL FILTER

A fundamental method of spectral analysis consists of passing the signal through a band-pass filter to determine the amplitude coefficient and squaring and averaging the output to provide a frequency point for a power spectral density analysis. Either the filter centre frequency can be changed after each measurement to obtain a complete spectral estimation or a contiguous approach carried out. The method can be expressed as:

$$\hat{G}_i(f) = \frac{1}{N} \sum_{i=0}^{N} \left[\sum_{k=0}^{M} h_k x_{i+k} \right]^2 \qquad (8\text{-}104)$$

where h_k is the impulse response function of a band-pass filter comprising M points. This expression represents the time average of the squared convolution integral for the filtered data. The bracketed expression will be recognised as a non-recursive digital filter. In a practical case a recursive filter would be chosen as this enables a reduction in analysis time to be made.

The Lerner filter, referred to in chapter 7, has an advantage in contiguous operation since some of its poles can be shared by adjacent filters [21]. It has also an equiripple amplitude characteristic and linear phase characteristics. This filter may be considered [22] in terms of the summation of a number of simpler sub-filters consisting of a complex pole and single zero (fig. 8-22a). Each of these sub-filters is characterised by a complex pole pair at the same radius but a different angle in the z-plane, so that they can be considered as individual narrow-band filters, each covering a section of the pass-band. The composite band-pass filter characteristic is obtained by adding and subtracting alternative sub-filter outputs (fig. 8-22b). Cut-off rates of some 60 db/octave are obtained with as few as eight sub-filters of this type.

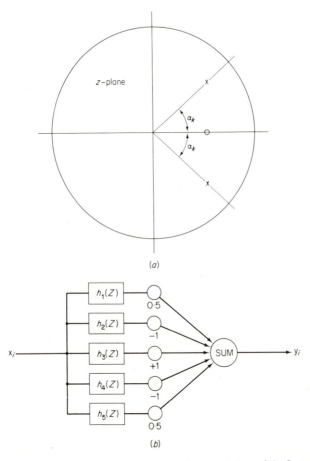

Fig. 8-22. (a) Pole-zeros of the Lerner filter. (b) Representation of the Lerner filter.

The use of constant bandwidth digital filters suffers from the same disadvantages as the use of constant bandwidth analog filters, namely unequal ratios of centre frequency/resolution bandwidth over the total analysis band, and a filter settling time which limits attainable scanning rate. These effects render constant bandwidth filter methods inferior to F.F.T. methods for values of N greater than about 50.

The position is improved considerably if a proportional bandwidth filter is used as will be shown for the analog case. A simple method is available, using decimation as a technique for controlling the filter bandwidth. Referring to section 7-2-2, if the sampling interval remains constant the effective filter frequency (cut-off or centre) can be halved by simply discarding every other point. Enochson and Otnes [20] have used this technique effectively to produce a one-third octave analysis algorithm which is faster than the F.F.T. for values of N less than 40 000. Four recursive filters need to be implemented. Three of these are arranged each to cover one octave of the signal to be analysed and a

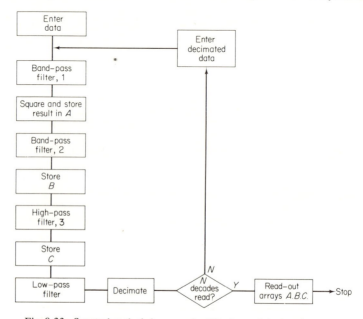

Fig. 8-23. Spectral analysis by recursive filtering and decimation.

fourth to carry out low-pass filtering before decimation (fig. 8-23). Following each decimation by two the filter will react on the data at frequencies one octave lower to produce the next set of $\frac{1}{3}$ octave values and so on. The time required for analysis is the summation of a geometric progression since each successive operation on the data will be carried out on half the preceding number of points.

8-7 ANALOG METHODS

The fundamental analog method of power spectral density analysis is shown in fig. 8-24. The time-history function, $x(t)$, is passed through a band-pass filter tuned to a given frequency and the result squared and averaged to give the power spectral density value at the one frequency at which the filter is tuned. Practical elaborations of this method are almost entirely concerned with filter design and control to cover the required analysis band.

Broadly, we can distinguish two types of analysis systems:

(a) Contiguous band systems,
(b) swept filter systems.

The first of these requires that the signal be applied through a number of parallel band-pass filters, each having a different centre frequency. Swept filter systems can be implemented by direct incrementation of filter centre frequency over the band of interest, or alternatively, the input function can be heterodyned with a variable frequency oscillator and the sum and difference terms passed through a fixed frequency filter.

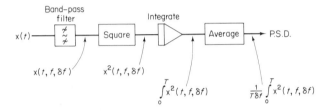

Fig. 8-24. Analog method of power spectral density estimation.

Input functions can also be subject to modification in either the time or the frequency domain as part of the power spectral density evaluation procedure. There are advantages in filtering the data through a broad band-pass filter before carrying out a power spectral density analysis. This will reduce the various forms of distortion which can arise when a very wide band signal is processed. It will also give a reduced analysis time by removal of the very lowest frequencies in the neighbourhood of zero frequency. The subject has been considered earlier in chapter 3, in connection with pre-whitening techniques. Modification in the time domain can form an integral part of spectral evaluation for certain types of signal. The method is known as E.M.P. power spectral density evaluation and is of value where the signal to be analysed is weighted such that the most recent samples are more significant than earlier ones, or where non-stationary processes are being studied.

As indicated in fig. 8-24 the output at the filter is subject to a mean square calculation to extract the power spectral density function. Methods of calculation of the mean square value do not vary widely and are generally limited to a squaring circuit, followed by true integration using an operational amplifier, or the use of passive resistance/capacitance (R.C.), integrator networks.

8-7-1 CONTIGUOUS BAND ANALYSIS

A schematic diagram showing the contiguous band method is given in fig. 8-25. The signal is applied to all the filters in parallel. Each filter is centred on a separate frequency and the frequencies distributed over the analysis band. The distribution can be linear, in the case of a fixed bandwidth filter, or logarithmic in the case of a constant percentage bandwidth filter. Here the filter bandwidth B is made a constant percentage, C, of the filter centre frequency f_0, so that $B = Cf_0/100$. One arrangement is to extend each filter bandwidth over an octave or $\frac{1}{3}$ octave. The reason for this stems from early work in acoustics when it was found that the loudness response of the human ear follows a $\frac{1}{3}$ octave type of logarithmic response. Consequently it proved more convenient to express the

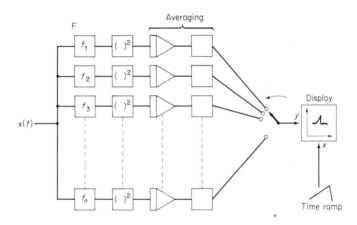

Fig. 8-25. The contiguous filter analyser.

data in the form of $\frac{1}{3}$-octave band pressure levels than as a continuous power spectral density function. The characteristics of a contiguous bandwidth system, using a group of such filters, are shown in fig. 8-26. The cross-over points for the filters are at -6 db.

The main advantage of this method is in speed of analysis. Power spectral density estimates at each frequency are obtained simultaneously and read out sequentially to the display or storage device (e.g. X/Y plotter, display oscilloscope, or digital storage), so that the time for analysis is the time necessary to filter at the lowest analysis frequency band. This presents a limitation in the speed of analysis determined by the settling time of the filter (equation 8-67). Where the minimum analysis frequency is very low, and particularly if a small number of lag values are used, the advantage of the contiguous over the swept filter method may not be very great. Additionally, where the record length is short this may be inadequate to allow for filter settling time and derivation through the autocorrelogram may be necessary. The facility of individual record-ing for each channel output can provide a means of analysis for non-stationary processes such as geophysical phenomena, speech patterns etc., by division into short-term spectra over which stationarity can be assumed. An example of this is given in chapter 11.

A difficulty with the proportional bandwidth method of analysis is that the scaling by B of the squared filter outputs (equation (8-6)), means that a different scaling factor is required for each contiguous band. This adds to the amount of analog equipment required to obtain a relatively coarse resolution over the frequency band. If the limited number of filters which can be used are distri-buted too widely over the band, then the effect is that of sampling below the

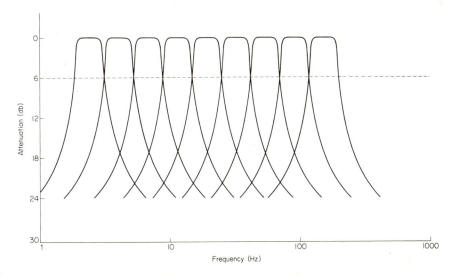

Fig. 8-26. Contiguous filter bank frequency characteristics.

Nyquist limit in the frequency domain, resulting in a form of aliasing error, although the method is an analog one. As a consequence the method, in its analog form, finds its widest use as a 'quick-look' facility prior to detailed spectral analysis using other methods.

8-7-2   USE OF A SWEPT FILTER

The simplest technique is to alter the centre frequency of the filter either continuously, or in steps to cover the analysis band of interest. This is shown in fig. 8-27. The principle difficulty of the method is the filter complexity neces-sary in order to carry this out without altering its shape and bandwidth. Various forms of motor drive have been used to control the filter elements but are limited to the analysis of very low frequencies. Alternative methods using analog

383

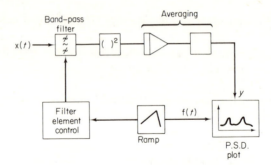

Fig. 8-27. Swept frequency filter.

multipliers or digital control of integrators have been discussed in chapter 3 and permit analysis to be carried out at frequencies up to about 1 kHz. The resolution with this method is always poor, due to the conflicting design requirements of narrow bandwidth filters having an easily adjustable centre frequency. In addition it is difficult to use this method when analysis is to be carried out at extremely low frequencies since narrow-band filters having a low centre frequency are not feasible using analog methods. In some cases it may be possible to record the data on magnetic tape and then to play this back at a higher speed in order to translate the frequency band to a higher region and thus ease this problem of filter design.

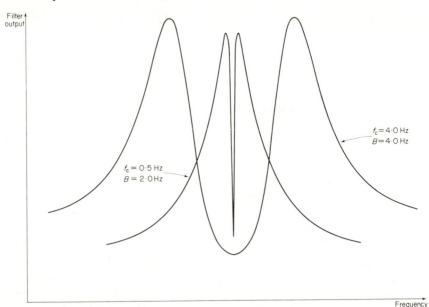

Fig. 8-28. Effect of narrow bandwidth filter on coherence.

A heterodyne method of Fourier analysis was mentioned briefly in chapter 6. The output of a local oscillator, of frequency, f_0, is multiplied with that of the signal to be analysed, f_a, such that sum and difference terms of the two signals and their harmonics are produced. These are passed through a low-pass filter which will pass those components of the signal which lie between $(f_0 - B)$ and $(f_0 + B)$, where B is the bandwidth of the filter. Thus by altering the value of f_0 a range of analysis frequencies can be covered with a fixed frequency filter of

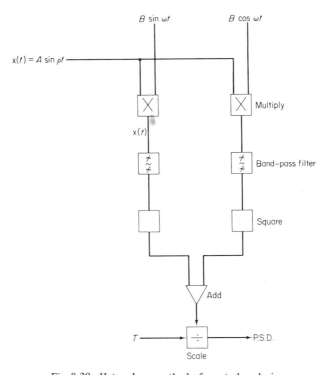

Fig. 8-29. Heterodyne method of spectral analysis.

effective bandwidth, $2B$. Unfortunately this method results in a reduction in filter output when $f_0 = f_a$, known as the Coherence Frequency, since the output amplitude will be dependant on the relative phase angles of the two components of the difference terms near to f_0. This effect is particularly severe when the oscillator frequency is close to the filter bandwidth, as shown in fig. 8-28, and only a single frequency is applied to the analyser.

It is necessary, therefore, to use two multiply/filter units arranged to measure the in-phase and out-of-phase components of the signal and to add these two components vectorially. The method is shown in fig. 8-29 and is similar to the

385

direct Fourier integral method described in section 6-8-2. In effect this calculates the power spectral density as:

$$\text{P.S.D.} = \frac{1}{TB} \int_{-\infty}^{\infty} (F[(x(t) \sin \omega t)]^2 + F[(x(t) \cos \omega t)]^2) \, dt \quad (8\text{-}105)$$

where $F[\,]$ represents a filtering operation over a bandwidth, B. The terms $(x(t) \sin \omega t$ and $x(t) \cos \omega t)$ are calculated separately and subject to a narrow-band filtering operation. Following squaring and summation the power spectral density is obtained by normalising and averaging the result.

As noted earlier, the analysis bandwidth is determined by the heterodyne frequency, $\omega/2\pi$, whilst the bandwidth is determined quite separately from the filter characteristics. To show that this method is free from phase ambiguity let the signal input, $x(t)$, consist of a series of harmonic terms:

$$x_i(t) = \sum_{i=1}^{M} A_i \sin (\omega_i t + \phi_i) \quad (8\text{-}106)$$

and the heterodyne frequency be given as:

$$y(t) = C \sin \omega_0 t \quad (8\text{-}107)$$

for the in-phase component and,

$$y^0(t) = C \cos \omega_0 t \quad (8\text{-}108)$$

for the out-of-phase component. The products will be:

$$x_i(t)y(t) = \tfrac{1}{2}A_i C \left[\cos (\omega_i - \omega_0 t + \phi_i) - \cos (\omega_i t + \omega_0 t + \phi_i) \right] \quad (8\text{-}109)$$

and $\quad x_i(t)y^0(t) = \tfrac{1}{2}A_i C \left[\sin (\omega_i - \omega_0 t + \phi_i) + \sin (\omega_i t + \omega_0 t + \phi_i) \right].$

The difference terms are retained by passing the products through a narrow-band filter (which can be low-pass) giving for $G_x(f)$:

$$G_x(f) = \frac{2}{BT} \int_0^T [X_{i(F)}{}^2 + X_i^0{}_{(F)}{}^2] \quad (8\text{-}110)$$

where

$$X_{i(F)} = \tfrac{1}{2}A_i C . \cos (\omega_i t - \omega_0 t + \phi_i)$$
$$X_{i(F)}^0 = \tfrac{1}{2}A_i C . \sin (\omega_i t - \omega_0 t + \phi_i)$$

hence

$$G_x(f) = \frac{2}{BM} \sum_{i=1}^{M} (\tfrac{1}{2}A_i C)^2. \quad (8\text{-}111)$$

Thus the output is given as a mean-square value for random complex signals as well as periodic signals. A similar derivation can be made for cross-spectrum calculation.

386

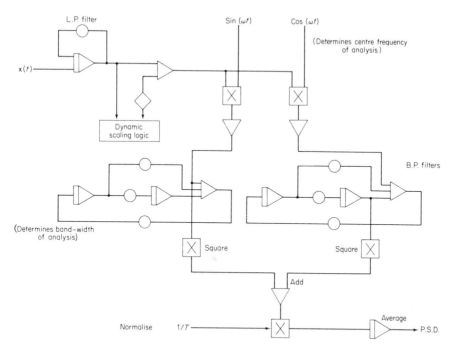

Fig. 8-30. Analog spectral analysis by heterodyne method.

This method is shown mechanised for the analog computer in fig. 8-30. The sine/cosine generator is controlled by means of a ramp voltage generator operating between two voltages, proportional to the signal limit frequencies f_1 and f_2. The X-coordinate of the plotting device may also be derived from this ramp voltage. It is sometimes desirable for this frequency control to be proportional to $\log(f)$ and this case will be considered later. A second-order Butterworth low-pass filter is included for each product term and allows independent control of analysis bandwidth. Should a constant resolution be required, it is possible to replace the potentiometers by multipliers, and so permit a proportional bandwidth analysis by suitable control of the attenuation value during the analysis period. The characteristics of analog multipliers are such, however, that their inclusion will inevitably degrade the performance of the analyser and limit the rate of analysis possible.

Spectral density analysis, for periodic signals only, can be carried out using the forcing function oscillator described in section 6-8. The Fourier transform of the signal is derived from equation (6-41) and substituted in equation (8-15) to give:

$$G(f) = \lim_{T \to \infty} \left| \frac{(\text{cospec})^2 + (\text{sinspec})^2}{T} \right|. \tag{8-112}$$

387

The basic circuit given in fig. 6-9 would be used with the inclusion of squaring and scaling by T to obtain a power spectral density estimate. As pointed out previously the filter bandwidth used in this method is, theoretically, infinitely narrow, so that the power spectral density is taken at discrete frequency points only. Thus if we consider a 10-second record ($= T$), then the lowest frequency of analysis possible will be $1/T = 0.1$ Hz, so that the complete spectrum will be defined by integral multiples of this frequency, e.g. $0.1, 0.2, 0.3$, up to say, 100 Hz, giving 1000 points. With long records, for example, $T = 300$ seconds, then the lowest frequency of analysis for this record will be 0.003 Hz, requiring 30 000 samples to fully define the spectrum. The analysis time will be very long and will, in all probability, be unacceptable. Under these conditions we will need to average over a frequency interval as well as a time interval and the previous methods described are better for this.

8-7-3   ANALYSIS CONTROL

Variation of effective filter centre frequency in the case of the swept filter methods, requires a voltage/frequency converter for plotter X-coordinate control. In both cases the relationship can be linear and derived from a simple integrator ramp circuit. A frequency scale change can be obtained at pre-determined levels as indicated in fig. 6-11. A stair-step control waveform would be appropriate where filter frequency is controlled in discrete steps and may be derived from a sample/hold pair shown in fig. 8-31. Logarithmic change of filter frequency and plotter control will be necessary for constant percentage bandwidth methods and a scaling routine required in order to cover several decades. One such method will be described for the mode-controlled filter, described in chapter 3 and used in a simple serial filter method of spectral analysis. The control arrangements are shown in fig. 8-32. The signal for analysis is assumed to be obtained repetitively from a previously recorded magnetic tape by using the shuttle control system.

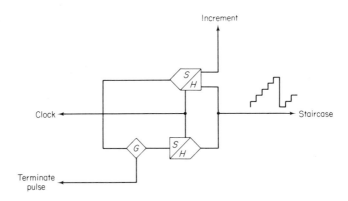

Fig. 8-31. Staircase generator.

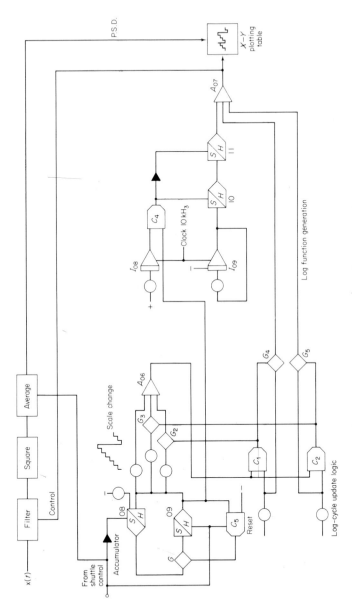

Fig. 8-32. Logarithmic control of spectral analysers.

389

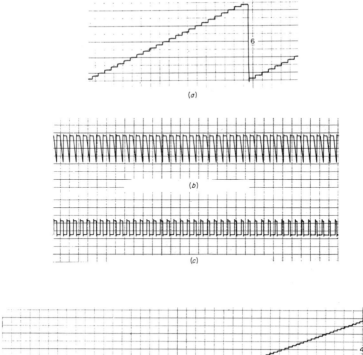

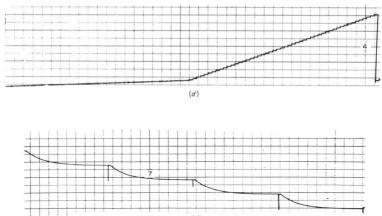

Fig. 8-33. Control wave-forms. (*a*) Stair-step wave-form. (*b*) High-speed saw-tooth. (*c*) Variable pulse width control. (*d*) Two decade linear stair-step. (*e*) Multi-cycle exponential generator output.

A pulse is obtained from the tape transport system at the end of each run and used to initiate the control process. With each data shuttle a set voltage is applied to an accumulator circuit S/H08/9, and results in the generation of a stair-step control wave-form, shown in fig. 8-33*a*. This is compared with a high speed linear saw-tooth wave-form (*b*) in order to produce the repetitive control

390

wave-form (c) for filter mode control. A scale change occurs with each decade of control voltage. A logarithmic control voltage is required for the plotter and is provided by an iterative exponential generator comprising I09, C4, S/H10/11. In this arrangement the equation $dx/dt = 2\cdot3y$ is implemented in integrator circuit I09 and the solution obtained is: $x = \exp(-2\cdot3T)$ which equals $x = \exp(-2\cdot3y)$, if $y = T$.

This comparison is repeatedly carried out at a frequency very much higher than the stair-step change. The output of the stair-step, y, is compared with the ramp T from integrator I08 and the output of the exponential generator I09 is followed when equality is reached. S/H10 and 11 form a sample-hold pair so that the stepped output of S/H11 represents a logarithmic transform of the linear step input y. The final output is retained and added to the logarithmic stepped wave-form at the beginning of each cycle (e).

8-7-4   CROSS SPECTRAL DENSITY

Practical measurements imply the calculation of the physically realisable one-sided cross spectrum. Equations (8-26) and (8-27) would generally be implemented on an analog computer to give the cospectral density and the quadrature spectral density in polar form as a magnitude factor and a phase factor (equations (8-29) and (8-30)), which can simply be derived from the first two calculations.

One method of implementation is to use two similar power spectral density configurations in the manner shown in fig. 8-34. Given the inputs $x(t)$ and $y(t)$ then the cospectral density is obtained. If $y(t)$ is subject to a 90° phase shift (designated $y^0(t)$) before being applied to F_2 then the quadrature spectral density is obtained. It is important that the characteristics of the two filters are identical. A small amount of phase distortion present in each of the two filters will not affect the results providing that the distortion characteristic is the same in both cases.

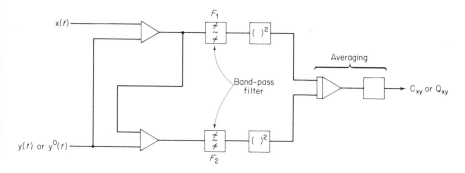

Fig. 8-34. Cross-spectral analysis using two identical analysers.

From fig. 8-34 we see that the main operations necessary in cross-spectral density analysis are filtering, phase shift of one of the signals, multiplication and averaging. The major variations in the methods are concerned with ways of realising the 90° phase shift and change in analysis frequency. Fig. 8-35 shows a common method of obtaining values for C_{xy} and Q_{xy} simultaneously. The 90° phase shift can be obtained by the use of an RC network, fig. 8-36. It can be shown that when $\omega RC = 1$, the network attenuates the input to one third of its

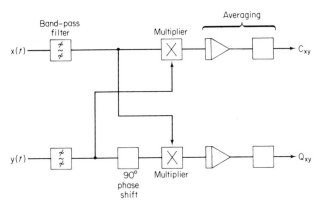

Fig. 8-35. Cross-spectral analyser incorporating 90° phase shift.

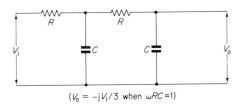

$(V_o = -jV_i/3$ when $\omega RC = 1)$

Fig. 8-36. Method of obtaining 90°phase shift.

value and introduces a phase lag of minus 90°. This has the disadvantage of requiring a change in the filter time constants corresponding to change in the analysis frequency.

As with the power spectral density methods described earlier, a change in analysis frequency can be obtained by the use of heterodyne techniques. By using an oscillator providing the sine and cosine outputs simultaneously, the problem of obtaining the necessary 90° phase shift between signals without the addition of passive elements, can be solved. A schematic diagram for the method is given in fig. 8-37. With this method, the filters can be low-pass only. The output of the multipliers contains the oscillator sum and difference frequencies and the low-pass filter is arranged to pass only the difference terms. Thus the apparent bandwidth of the filter extends from $f_o - f_c$ to $f_o + f_c$, where f_o is the oscillator frequency and f_c is the cut-off frequency of the low-pass filter. The $90°$ phase shift is obtained by multiplying the y term by both the sine and cosine terms. This can be seen if we carry out the cross-multiplication of $x(t)$, $y(t)$, cos ωt and sin ωt (ignoring the modifying effect of the filters), which gives:

$$[y(t) \cos \omega t][x(t) \cos \omega t] \text{ for the cospectral term, and}$$

$$[x(t) \cos \omega t][y(t) \sin \omega t] \text{ for the quadrature term,}$$

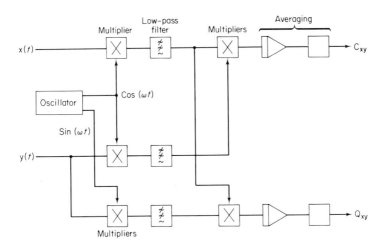

Fig. 8-37. Heterodyne method of cross-spectral analysis.

which differ only in that the last terms represent a phase shift of $90°$. As with the equivalent power spectral density method some coherence distortion is experienced at and close to the effective filter centre frequency. This does not however represent a serious limitation of the method.

The errors found with cross spectral density estimation are similar in form to those discussed in connection with power spectral density, with the exception of statistical accuracy which is less well defined. Analysis time is longer since a complete joint spectrum makes it necessary to calculate the cross spectral density for each lag value of the $x(t)$ channel at each lag value of the $y(t)$ channel. Thus, in the case of true averaging, we can write for the complete analysis time from equation (8-72)

$$T_t \geqslant T_a \left(\frac{F}{B}\right)^2 \tag{8-113}$$

and, for *RC* averaging:

$$T_t \geqslant 4\text{K} \left(\frac{F}{B}\right)^2 \tag{8-114}$$

where F is the spectral range of interest, and B is the filter bandwidth (identical values of F and B are assumed for both x and y channels).

8-8 HYBRID METHODS

In this context hybrid methods are considered to include the combination of an analog computer linked to a digital computer, or at least an analog computer which includes some form of digital memory. Analog-to-digital and digital-to-analog conversion form an important part of the linkage system and give rise to a number of problems which have been discussed in chapter 4. The application of logic control to analog methods, such as that described in section 8-7-3, does not in itself constitute a hybrid method and will not be referred to further.

The main reasons for considering hybrid methods as opposed to analog methods or digital processes are:

1. Need for operation in real-time.
2. Economy in hardware or computing time.
3. Reduction in the engineering problems associated with filter design, particularly for very low frequencies.

The choice can also be regarded as an application of optimisation in which the appropriate hardware is selected to match the particular operation and signal characteristics. As with other techniques the fundamental methods used for power spectral density evaluation are by direct calculation or via the autocorrelation.

A direct method in wide use is known as the Time Compression Method. This uses the storage capabilities of the digital computer, not only to retain the signal being analysed, but also to permit an upwards translation of frequency content and reduction in read-out time to achieve the same effects as the increase in replay speed obtained using an analog tape unit.

This is illustrated in fig. 8-38. The signal is sampled at a rate determined by the highest frequency of interest and converted into digital form. Successive samples are stored in sequential locations in a memory store. This 'read-in' operation need not be carried out in real-time. In the case of information recorded on analog magnetic tape, it may be advantageous to adjust the speed of replay to be faster or slower than real-time. Read-in, however, need only be carried out once to transfer a digitised version of the signal to memory storage. Read-out of the data is made to a digital-to-analog converter in the same order as the samples were originally read in, but can be carried out at any speed up to the limit of the memory cycle time. Thus under favourable conditions data can be time-compressed, since the time interval between consecutive samples taken from the input signal is considerably longer than the interval between read-out samples. This situation results in a contraction of the time-base for the signal being analysed, which is equivalent to a reduction in the spectral bandwidth. Thus repeated read-out of the digitised signal can be made at a very high rate to achieve the same effect as accelerated replay of analog tape without the practical difficulties attendant on the production and use of magnetic tape loops.

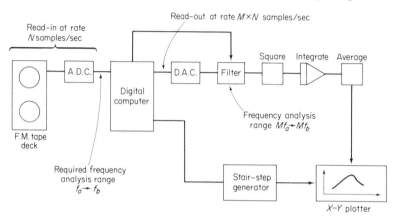

Fig. 8-38. Hybrid spectral analysis.

This can be seen if we compare the Fourier transforms of the original signal and its accelerated version. From equation (6-36):

$$X(f) = \int_{-\infty}^{\infty} x(t) \exp\left(-j2\pi f_o t\right) dt \qquad (8\text{-}115)$$

and if the increase in read-out speed is given by a factor, k, then the Fourier transform for $x(kt)$ is:

$$X_k(f) = \int_{-\infty}^{\infty} x(kt) \exp\left(-j2\pi f_o t\right) dt$$

$$= \frac{1}{k} \int_{-\infty}^{\infty} x(kt) \exp\left(-j2\pi \frac{f_o}{k} kt\right) d(kt)$$

$$= \frac{1}{k} X(f/k) \qquad (8\text{-}116)$$

i.e., the new spectrum $X(f/k)$, is identical in form to the original spectrum $X(f)$, but with wider frequency separation for the spectral components. Storage can be in conventional core memory or use made of special-purpose re-circulating delay lines. The former is used in hybrid computing systems and the latter is often incorporated in portable power spectral density analysers. The core storage method offers greater flexibility and we will consider this first.

One of the difficulties of the analog methods described earlier is the need to vary the filter frequency during the production of a complete power spectrum. If the speed of access from the stored data is adjusted for each complete set of data read-out a fixed frequency filter can be used. As an example of the analysis of a record, originally recorded on F.M. magnetic tape, the following analysis conditions will be taken.

Let the recorded signal have a duration of $T = 10$ sec, assume the frequency band of interest extends to $f_c = 100$ Hz, and let the number of samples for the signal be $N = 3000$. Then the sampling interval $h = 1/(2f_c) = 0.005$ sec. If we take the statistical accuracy for fifty degrees of freedom i.e., $n = 50$, then the number of lag values will be $m = 2N/n = 120$, and the filter bandwidth will be $B = 1/mh$ $= 1.65$ Hz (in real time).

The data samples can be read-out at a maximum speed of R microseconds, corresponding to the cycle time of the computer core store (we will take $R = 4$ μsec in this example), thus the minimum time to read-out the stored record will be $RN = 12$ msec.

The corresponding time compression will be $T/RN = 833 = T_c$. If we now let the filter centre frequency $= 300$ Hz, then the minimum analysis frequency $f_c/T_c = 0.36$ Hz. It will be necessary to read-out the record 120 times, each time increasing the read-out loop time by 8 microseconds so that the last read-out time should take $T_1 \simeq 3$ seconds. This corresponds to the real time conditions present when we are analysing the highest frequency recorded in the signal (since the filter centre frequency is increased by a factor of 3). The total time of read-out is an arithmetic progression and is given by:

$$\sum_{r=1}^{r=n} [a + (r-1)d] = \tfrac{1}{2}n[2a + (n-1)d]. \tag{8-117}$$

Using the symbols given above, the total analysis time:

$$T_s = m/2 \, (2RN/1000 + T_1) \ \sec = (60/1000)(24 + 3000) = 3 \ \text{min.}$$

The actual filter bandwidth used will be $B = 3 \times 1.65 = 4.95$ Hz, since a reduction of 3 times over real-time operations will have been achieved. The bandwidth will remain constant during the analysis period and the method therefore behaves as a fixed band-width system. Some improvement in analysis time can be obtained if the analysis range is divided into subgroups, as stated previously.

The fixed frequency technique, although simple from the point of view of hardware design, does not make best use of the accelerated read-out capabilities of the method. A heterodyne method using a fixed frequency filter will give an improvement in speed. Here repeated read-out of the stored digital signal is carried out at the fastest rate and the swept frequency oscillator controlled in step with each read-out of the signal. The new frequency from the controlled oscillator is heterodyned with the converted analog signal at read-out operation and passed through a fixed frequency filter. Apart from the digital read-out, the method is similar to the heterodyne method described previously.

The advantage obtained by this hybrid method is reduced time of analysis by the factor T/RN, so that the total analysis time will be:

$$T_s = \frac{Tm}{(T/RN)} = RNm \tag{8-118}$$

which, for the example quoted previously, will be 1·44 secs. In this case the filter bandwidth can be increased by a factor T/RN, since a constant speed-up factor is maintained. One advantage of the large frequency translation obtained with this method is that very favourable filter characteristics can be obtained

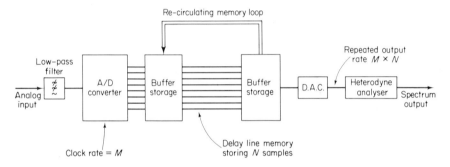

Fig. 8-39. Hybrid analysis using re-circulating delay lines.

using a crystal lattice filter. In particular, a fast settling time is obtained, so that the filter output is reduced to a very small or zero value at the end of each sampling period, giving minimum residual error for the next calculation point.

Where storage is accomplished using delay lines a re-circulating mechanism is used (fig. 8-39). The digital samples are fed into the lines as a parallel word, one digit per line, and the output is obtained fed into a parallel digital-to-analog converter. Feedback into the memory input is arranged such that the lines contain the data in the same order as originally entered. Each line stores N samples of the original signal which is continually re-circulated to permit continuous read-out at a rate very much faster than the read-in rate.

Use of the re-circulating memory is exactly the same as that of the core store version. Both systems suffer from limited storage which results in low BT products. As a consequence the normalised standard error of the results will be high (see equation (8-53)). Some improvement can be obtained if a large number of sample records are analysed and the results averaged using one of the techniques to be described in the following chapter.

397

BIBLIOGRAPHY

1. BINGHAM, C., GODFREY, M. D. and TUKEY, J. W. Modern techniques of power spectrum estimation. *I.E.E.E. Trans.* (Audio and Electroacoust.) AU-15, **2**, 56-66, June 1967.
2. WIENER, N. Extrapolation, interpolation and smoothing of stationary time series. M.I.T. Press, Cambridge, Mass., 1949.
3. KHINTCHINE, A. Y. Korrelationstheorie der stationaren stockastischen prozesse. *Math. Ann.* **109**, 604, 1934.
4. BLACKMAN, R. B. and TUKEY, J. M. *The Measurement of Power Spectra.* Dover, New York, 1959.
5. BERGLAND, G. D. Fast Fourier transform hardware implementation – a survey. *I.E.E.E. Trans.* (Audio and Electroacoust.) AU-17, **2**, 109, 1969.
6. LEE, Y. W. *Statistical Theory of Communication.* John Wiley, New York, 1960.
7. FORLIFER, W. R. The effects of filter bandwidth in spectrum analysis of random vibration. *Shock Vib.* and assoc. *Env. Bull.* **33**, 2, 1964.
8. DANIELS, H. E. Contribution to a discussion on evolutionary spectra and non-stationary processes. *J. Roy. Stat. Soc.,* B, **27**, 234, 1965.
9. BENDAT, J. S. and PIERSOL, A. G. *Random Data: Analysis and Measurement Procedures.* John Wiley, New York, 1971.
10. FISHER, A. R. and YATES, F. *Statistical Tables.* Oliver and Boyd, London, 1967.
11. WONNACOTT, T. H. Spectral analysis combining a Bartlett window with an associated inner window. *Technometr.* **3**, 235-43, 1961.
12. WEINREB, S. A digital spectral analysis technique and its applications to radio astronomy. *M.I.T. Tech. Rep.* **412**, Aug. 1963.
13. BARTLETT, M. S. Periodogram analysis and continuous spectral. *Biometrika* **37**, 1-16, 1950.
14. PARZEN, E. Mathematical considerations in the estimation of spectra. *Technometr.* **3**, 67-190, 1961.
15. OTTERMAN, J. The properties and methods of computation of exponentially mapped past statistical variables. *I.R.E. Trans.* (Auto. Control) AG5, **7**, *11-17, 1960.*
16. COOLEY, J. W., LEWIS, P. A. W. and WELCH, P. D. The application of the fast Fourier transform algorithm to the estimation of spectra and cross spectra. Lecture series, Applications and methods of random data analysis. Southampton University, July 1969.
17. HAMMING, R. W. Error detecting and error correcting codes. *Bell. Syst. Tech. J.* **26**, 147-60, April 1969.
18. GROGINSKY, H. L. and WORKS, G. A. A pipe-line fast Fourier transform. *I.E.E.E. Trans.* (Comp.) C-19, 11, Nov. 1970.
19. ENOCHSON, L. D. and PIERSOL, A. G. *Application of Fast Fourier Transform. Procedures to Shock and Vibration Data Analysis.* Soc. Automot. Eng., U.S.A., 16 pp, 1968.
20. ENOCHSON, L. D. and OTNES, R. E. An algorithm for digital one-third octave analysis. *J. Sound Vib.* **2**, 4, April 1968.
21. DROUILHET, P. R. and GOODMAN, L. M., Pole-shared linear-phase band-pass filter bank. *Proc. I.E.E.E.* **54**, 4, April 1966.
22. LERNER, R. M., Band-pass filters with linear phase. *Proc. I.E.E.E.* **52**, 249-68, 1964.

Additional References

23. JONES, R. H. A reappraisal of the periodogram in spectral analysis. *Technometr.* 7, 4, 531-42, 1965.
24. HARRIS, B. *Advanced Seminar on Spectral Analysis of Time Series.* John Wiley, New York, 1967.
25. KORN, G. A. *Random Process Simulation and Measurements.* McGraw-Hill, New York, 1966.
26. MARTIN, M. A. Frequency domain applications to data processing. *I.R.E. Trans.* (Space, Elec. Telem.) March 1959.
27. PIERSOL, A. G. The measurement and interpretation of ordinary power spectra for vibration problems. N.A.S.A. CR-90 Washington, D.C., and M.A.C. – 305 –01, Los Angeles, 1964.
28. TUKEY, J. W. Discussion emphasising the connection between analysis of variance and spectrum analysis. *Technometr.* 3, 2, 191-219, May, 1961.
29. KELLEY, R. D., ENOCHSON, L. D. and RONDIVELLI, L. A. Techniques and errors in measuring cross-correlation and cross spectral density function. N.A.S.A. CR-74505 Washington, D.C., 1967.
30. INOUYE, T. *et al.* Applications of Fourier transforms to the analysis of spectral data. *Nuclear Instr. Methods* 67, 125-32, 1969.
31. MORROW, C. T. and MUCHMORE, R. B. Shortcomings of present methods of measuring and simulating vibration environments. *J. Appl. Mech.* 22, 367-71, Sept. 1955.
32. WELCH, P. D. A direct digital method of power spectrum estimation. *I.B.M. J. Res. Dev.* 5, 2, 1961.
33. WATTS, D. G. *Optimal Windows for Power Spectra Estimation.* N.R.C. Res. Center, University of Wisconsin, 1964.
34. INSTON, H. H. Analog computer program for the frequency analysis of transient waveforms. *Proc. Third A.I.C.A. Congress.* Opatija 456-63, Sept. 1961.
35. LARROWE, V. L. and CRABTREE, R. E. Analog computation of time varying power spectra of seismic waves. Rep. 3708-15-T/5178-8-T University of Michigan, 1963.
36. BROCK, J. T. Analog cross-spectral density analysis. Lecture series, Applications and methods of random data analysis. Southampton University, July 1969.
37. NORSWORTHY, K. H. An improved random signal analyser. Lecture series, Applications and methods of random data analysis, Southampton University, July 1969.
38. CAPON, J., High-speed Fourier analysis with recirculating delay-line feed-back loops. (Inf. Theory) 1-10, 32-7, *I.R.E. Trans.* June 1961.
39. BEKEY, G. A. and KARPLUS, W. J. *Hybrid Computation.* John Wiley, New York, 1968.
40. LIM, R. S. and CAMERON, W. D. Power and cross-power spectrum analysis for hybrid computers. N.A.S.A. Rep. TMX-1324, Springfield, Virginia, 1966.

Chapter 9

CORRELATION ANALYSIS

9-1 INTRODUCTION

This chapter is concerned with those mathematical methods and computational techniques used to determine the similarity between events and to distinguish between signal and noise in a composite wave-form. Statisticians have long been using measurement techniques which establish the dependence of one set of numbers upon another set. Historically the first ideas leading to a quantitative measure of similarity stem from this work, which was originally confined to real data. The development of electrical communications and the need to improve signal/noise ratios have led engineers and scientists to develop correlation methods applicable to complex data. These have been derived from their work on Fourier series and transforms and, in particular, methods of using these to transform from a real (time) domain into a complex (frequency) domain. The realisation that, under given conditions, power spectral density functions and correlation functions are Fourier transforms of one another gave considerable impetus to this work, with the result that correlation techniques have been accepted during the last two decades as a major scientific tool, applicable to many areas of scientific research. It is interesting to note that within the last few years developments in electronic equipment and hybrid computation have stimulated considerable interest in correlation using fundamental (non-transform) methods giving fast alternative methods of correlation estimation.

Following a consideration of the mathematical basis for correlation a number of analysis techniques will be discussed and will include both transform and direct evaluation methods. Finally several important applications of correlation will be described including that of signal/noise enhancement.

9-2 STATISTICAL BASIS FOR CORRELATION

In statistical terms correlation is a measure of the linear relationship between two variables. In this context we generally mean random variables, where the samples are statistically independent of one another although correlation is also applicable to deterministic functions such as sinusoids.

A basic definition for the correlation between two variables x and y could be the sum of their cross-products, $\sum xy$. Consider the set of independant data samples given in table 9-1a. This same set is repeated in 9-1b but with the x and y values arranged in descending order of magnitude—large x values with large y values.

TABLE 9-1 Relation between statistically independent samples

(a) x	y	xy	(b) x	y	xy
1	4	4	0	1	0
2	0	0	1	0	1
0	6	0	2	2	4
7	2	14	3	4	12
5	4	20	4	4	16
3	7	21	5	5	25
6	5	30	6	6	36
4	1	4	7	7	49
		93			143

It will be seen (fig. 9-1) that a linear relationship between x and y can be established with the second set of data which is also distinguished by a high xy product sum. This is characteristic of correlated data and expresses a practical basis for a technique to separate a random relationship, (table a) from a linear relationship, (table b) which may co-exist in the same set of data.

A definition of correlation can be arrived at by way of the Covariance Function which has a precise meaning for statisticians and is often loosely

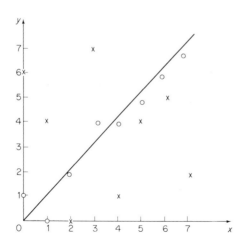

Fig. 9-1. Correlated and uncorrelated data sets.

referred to as correlation by engineers for reasons which will become apparent later. Let us first define two series of numbers as:

$$X_0, X_1, X_2, X_3, \ldots, X_n$$
$$Y_0, Y_1, Y_2, Y_3, \ldots, Y_n$$

If

$$\bar{x} = \frac{1}{N} \sum_{n=0}^{N} X_n = \text{average value for } X,$$

and

$$\bar{y} = \frac{1}{N} \sum_{n=0}^{N} Y_n = \text{average value for } Y$$

then

$$x_n = X_n - \bar{x}$$

$$(9\text{-}1)$$

and

$$y_n = Y_n - \bar{y}$$

are the deviations from the average values. Based on these deviation values the covariance of the cross-product may be defined as:

$$C_{xy}(t) = E[x_n(t).y_n(t)]. \qquad (9\text{-}2)$$

That is, an expected value for the covariance at time t is the product of the zero mean values available at that time. An average value is taken over the series length N, to give:

$$C_{xy} = \frac{1}{N} \sum_{n=0}^{N} x_n.y_n. \qquad (9\text{-}3)$$

The term covariance refers to the variance operation on the cross-product $x_n y_n$ (compare equation (9-3) with (5-6)) and is strictly restricted to the average cross-product of two quantities, x_n and y_n which have zero mean properties (equation (9-1)). This is often the case in practice since many transducers have no d.c. response or are a.c.-coupled so that a zero mean value time history is automatically obtained. As a consequence of this the covariance function, as defined above, is often referred to as the correlation function. The true correlation coefficient function is, however, a normalised version of the covariance function and is defined as the covariance divided by the product of the standard deviations for the series x and y, i.e.

$$K_{xy} = \frac{C_{xy}}{\sigma_x \sigma_y} \qquad (9\text{-}4)$$

where

$$\sigma_x = \sqrt{(C_{xx})} \text{ and } \sigma_y = \sqrt{(C_{yy})}.$$

The correlation function coefficients defined in this way are actually covariances normalised to lie within the range +1 to −1. When K_{xy} is near unity (either positive or negative), the correlation between the x series and the y series is said to be high. If K_{xy} tends to zero then the two series are independent giving little correlation.

9-3 APPLICABILITY TO SIGNAL PROCESSING

In the application of correlation methods to scientific or engineering problems, the features whose similarity are to be assessed are generally expressed as wave-forms rather than a discrete time series. These wave-forms usually represent a function of time, although they may relate to some other variable such as distance, pressure etc. The following treatment is based on the consideration of wave-forms as functions of time. In the case where $x(t)$ and $y(t)$ are continuous and stationary functions of time, then the average values are:

$$\bar{x} = \lim_{T \to \infty} \frac{1}{T} \int_0^T x(t)\, dt$$

$$\bar{y} = \lim_{T \to \infty} \frac{1}{T} \int_0^T y(t)\, dt$$

(9-5)

and the deviations,

$$X(t) = x(t) - \bar{x}$$

$$Y(t) = y(t) - \bar{y}.$$

(9-6)

In a practical covariance (or correlation) estimation, values will be required over a range of time delays existing between the two wave-forms, since the similarity may become apparent only after a given delay value. For example a similarity may be sought between a driving wave-form applied as the input to a system and an output value delayed in time by the lag of the system. A high correlation will only become possible if the input wave-form is delayed by an amount equivalent to the system lag before being correlated with the output signal. Since the system delay will, in general, be unknown the convariance is expressed as a function of delay, τ namely:

$$C_{xy}(\tau) = \lim_{T \to \infty} \frac{1}{T} \int_0^T [x(t) - \bar{x}]\,[y(t + \tau) - \bar{y}]\, dt.$$

(9-7)

The correlation coefficient function corresponding to equation (9-4) is the normalised form of (9-7) given as:

$$K_{xy}(\tau) = \frac{C_{xy}(\tau)}{\sigma_x \sigma_y}.$$

(9-8)

If the mean and variance for $x(t)$ and $y(t)$ are both zero, (9-7) reduces to:

$$R_{xy}(\tau) = \lim_{T \to \infty} \frac{1}{T} \int_0^T x(t)y(t + \tau)\, dt \qquad (9\text{-}9)$$

which is called the Cross-Correlation Function of the two variables. A normalised version of equation (9-9) is sometimes used in which the cross-correlation function is referred to zero time delay $(\tau = 0)$ for the individual variables, i.e.

$$R_{xy}^N(\tau) = \frac{R_{xy}(\tau)}{\sqrt{[R_x(0)R_y(0)]}}. \qquad (9\text{-}10)$$

This is not generally carried out for engineering calculations since the magnitude of the function proves to be more informative than its normalised coefficient value.

9-4 AUTOCORRELATION

The average values discussed above relating to a specific time t do not give information on typical periods or frequencies present in the time history records for x or y. To determine these it is necessary to correlate one wave-form with a time-shifted version of the other. Let us take the case of a wave-form, $x(t)$, correlated with a time-shifted version of itself.

Referring to equation (9-2), we can write:

$$C_x(\tau) = E[x(t).x(t + \tau)] \qquad (9\text{-}11)$$

where τ is the time-shift, lag, or delay.

If the record is stationary and ergodic we can replace this ensemble average by the time average over a real-time period T as:

$$R_x(\tau) = \lim_{T \to \infty} \frac{1}{T} \int_0^T x(t).x(t + \tau)\, dt \qquad (9\text{-}12)$$

which corresponds to the form of equation (9-7) if $x(t)$ has zero mean value and unit variance. A zero mean value can often be assumed or obtained by subtraction of the mean from the signal before processing. Correction after correlation may be desirable (e.g. to reduce computational time), when the square of the mean value should be subtracted from the mean value autocorrelation function thus:

$$R_x'(\tau) = R_x(\tau) - (\bar{x})^2. \qquad (9\text{-}13)$$

For convenience the condition of zero mean and unit variance will be assumed in

404

the following treatment. The practical form of equation (9-12) replaces T by $(T - \tau)$ giving:

$$\hat{R}_x(\tau) = \frac{1}{T - \tau} \int_0^{T-\tau} x(t)x(t + \tau) \, dt \qquad (9\text{-}14)$$

since an infinitely long record does not exist.

Equation (9-14) is an estimate for the autocorrelation function having an accuracy which decreases as τ becomes closer to T (the record length) since the average is taken over a continuously decreasing time period. We can deal with this estimation in two ways. The range of values for τ can be limited by truncation. This is generally carried out for digital computation since it will incidentally reduce the computational time needed. A practical value is to limit the maximum excursion of τ to $\pm T/2$. Alternatively the signal can be tapered by means of a window $W(\tau)$ designed to reduce the contribution to the average as (τ) gets closer to T. This window technique has been discussed previously in connection with power spectral density estimations.

One or other assumption will be made in the following discussions so that we can write for the autocorrelation function:

$$\hat{R}_x(\tau) = \frac{1}{T} \int_0^T x(t)x(t + \tau) \, dt \qquad (9\text{-}15)$$

with very little error. A practical meaning is given to equation (9-15) if we consider the derivation of $R_x(\tau)$ as a time history function, fig. 9-2. Each point on this functional curve results from the summation and averaging of the product of $x(t)$ and the delayed waveform $x(t + \tau)$ viewed over a limited 'window' of T seconds duration for a given value of τ. If we consider a finite number of delay values $m\tau = T$ then the first point on the function, $R(\tau)$, is obtained from the summation of the products of all the pairs of $x(t)$ and $x(t + \tau)$, the second point from the summation of the $x(t)$ and $x(t + 2\tau)$ product pairs and so on. The process continues until the delayed wave-form has been subjected to a maximum delay period $m\tau$ equal to the correlation 'window' period. For the continuous case expressed by equation (9-12) $m \to \infty$ and $\tau \to 0$ and the summation is replaced by integration. Equation (9-12) is known as the autocorrelation function of $x(t)$. It may be considered as a form of 'moving average' operation on the signal which emphasises the periodicities present in a given time series, and restricts the contribution of random added noise to a region close to $\tau = 0$. It tells us how much future (or past) events depend on present events and the dependence of present events on past (or future) events. Thus if our waveform, $x(t)$, contains a periodic signal as part of its data ensemble certain instantaneous values of $x(t)$ are capable of being determined by their relationship to other values in the time history sequence. We can determine this relationship by exploring the product, $x(t) \cdot x(t + \tau)$, over a range of values of τ.

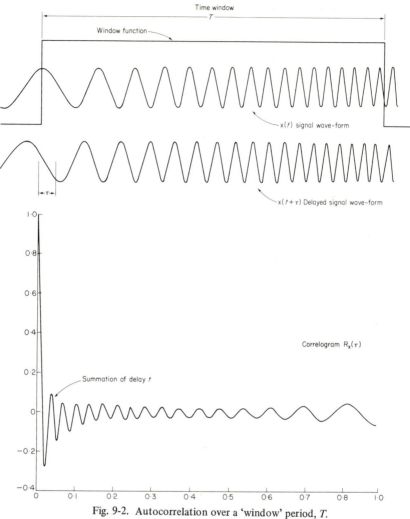

Fig. 9-2. Autocorrelation over a 'window' period, T.

This is illustrated in fig. 9-3 which shows a plot of the autocorrelation function of a sinusoidal wave-form and of a square wave-form. These plots are called autocorrelograms. It will be observed that these functions are periodic, having the same frequency as the signal, since a time-delay shift of a whole number of periods of the wave-form will not effect the values of the integral. The autocorrelation of a sine wave, $x = A \sin(\omega t + \phi)$ shown in fig. 9-3a, is of special interest. The function is given as:

$$R_x(\tau) = \lim_{T \to \infty} \frac{A^2}{2T} \int_{-T}^{T} \sin(\omega t + \phi) \sin[\omega(t + \tau) + \phi]\, dt.$$

406

If this is expanded using trigonometrical identities we find that all terms excluding the delay term will vanish, so that:

$$R_x(\tau) = \frac{A^2}{2}.\cos \omega\tau. \qquad (9\text{-}16)$$

Thus the autocorrelation of a sinusoidal wave-form is a cosinusoidal wave-form of peak amplitude equal to the mean-squared value of the original signal.

An important feature of this correlogram is that is persists periodically

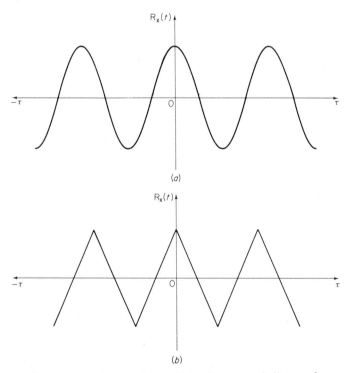

Fig. 9-3. Autocorrelogram of sinusoidal and square periodic wave-forms.

over all time displacements, with the same period as the original sine wave, but without its phase angle information. From fig. 9-3b we can also see that an autocorrelation function is an even function and is symmetrical about $\tau = 0$ so that,

$$R_x(\tau) = R_x(-\tau). \qquad (9\text{-}17)$$

Hence we can write an equivalent version of the autocorrelation function as:

$$R_x(\tau) = \lim_{T\to\infty} \frac{1}{T}\int_0^T x(t).x(t-\tau).\mathrm{d}t \qquad (9\text{-}18)$$

407

which has practical implications in that a real time signal can only be delayed relative to itself. The peak value of the function will be proportional to the mean square amplitude of the signal, since:

$$R_x(0) = \frac{1}{T} \int_0^T x^2(t)\, dt = \overline{x^2 t}. \tag{9-19}$$

We can state that $R_x(0)$ must be greater than or equal to $R_x(\tau)$. This is derived from the following considerations. If we consider the integral:

$$\frac{1}{T} \int_0^T [x(t) \pm x(t + \tau)]^2\, dt \tag{9-20}$$

we know that $x(t)$ cannot be equal to $x(t + \tau)$, so that if this condition is excluded this integral must be positive and greater than zero. Expansion of (9-20) gives:

$$\frac{1}{T} \int_0^T x^2 t.dt + \frac{1}{T} \int_0^T x^2(t + \tau).dt \pm \frac{2}{T} \int_0^T x(t).x(t + \tau).dt > 0$$

which from equation (9-12) and equation (9-17) $R_x(0) + R_x(0) \pm 2R_x(\tau) > 0$ therefore $R_x(0) > R_x(\tau)$ for all values of τ.

Using equations (9-15) and (9-19) we can restate the normalised version of the autocorrelation function as:

$$R_x^N(\tau) = \frac{\hat{R}_x(\tau)}{R_x(0)} = \frac{\dfrac{1}{T} \displaystyle\int_0^T x(t).x(t + \tau).dt}{\dfrac{1}{T} \displaystyle\int_0^T x^2(t).dt}$$

$$= \frac{\displaystyle\int_0^T x(t).x(t + \tau).dt}{\displaystyle\int_0^T x^2(t).dt}. \tag{9-21}$$

The particular value of the autocorrelogram is that it is a time domain wave-form which gives information concerning the behaviour of the correlated function in the frequency domain. This will be formalised later in terms of a mathematical relationship. For the present we will consider a further important example of autocorrelation which will illustrate this behaviour. The autocorrelogram of a wide-band noise wave-form is shown in fig. 9-4a. When compared with a time-shifted version of itself, only a small time-shift is required to destroy the similarity, which does not recur. In the limiting case of hypothetical white noise, the autocorrelogram is a Dirac delta function at zero phase shift ($\tau = 0$). It is important to note that the autocorrelogram of a random noise signal will reduce with time displacement and become zero for large time displacements (assuming a normalised signal having zero mean value). In contrast a sinusoidal signal has a correlation function which persists indefinitely over an infinite range of time lags. The decay of the autocorrelogram in this example gives a measure of the bandwidth of the system producing the random signal. Thus the length of time over which all the constituent frequency components of the signal can preserve the phase relationships which exist between them when $\tau = 0$ is inversely

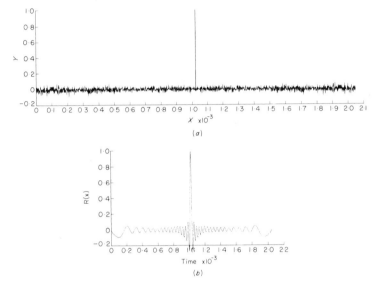

Fig. 9-4. (a) Autocorrelogram of a wide band noise wave-form. (b) Autocorrelogram of a narrow band signal with added noise.

proportional to the maximum frequency difference between them. This is expressed by the relationship derived by Wainstein and Zubakov [1]:

$$\delta\tau . \delta\omega \simeq \Pi/2. \tag{9-22}$$

where $\delta\tau$ = length of time over which the correlation is significant, and $\delta\omega$ = bandwidth of the process.

If the wave-form has a random plus a periodic component, the autocorrelation function will take on a form similar to that given in fig. 9-4b. The correlation function exhibits the periodicity of the repetitive signal whilst reducing the contribution of the additive noise except in the region of minimum delay.

It will be seen from these examples that the autocorrelation function provides a powerful means of identifying hidden periodicities in random noise using the fundamental difference in the shape of the autocorrelogram for periodic and random data.

409

The definition given in equation (9-12) is applicable to periodic wave-forms where the integration interval T is chosen to be equal to an integral number of whole periods for the wave-form. When applied to transient wave-forms the mean value of the integral tends to zero over a period much greater than the transient period so that the limits of integration must be considered as extending over all time (in the limiting case) so that we can write:

$$R_x(\tau) = \int_{-\infty}^{\infty} x(t).x(t + \tau).dt. \tag{9-23}$$

The peak value of this expression,

$$R_x(0) = \int_{-\infty}^{\infty} x^2(t).dt. \tag{9-24}$$

can be considered as expressing the total energy contained in the signal. The necessary limitation of this integral in the practical case gives rise to some problems in correlation estimates which will be considered later.

9-4-1 WIENER'S THEOREM

It is useful to consider the relation between this correlation process, expressed in the time domain, and an equivalent process in the frequency domain. We have seen earlier that the application of Fourier transform operations to random wave-forms will result in related transform pairs. This technique can also be applied to a correlation series, to allow us to express correlation in frequency terms. Applying a discrete Fourier transform to equation (9-15) gives:

$$F(\tau) = \frac{1}{N} \sum_{k=0}^{N-1} \frac{1}{T} \left[\int_0^T x(t).x(t + \tau).dt \right] \exp(-jk\omega_0\tau). \tag{9-25}$$

By inverting the order of integration we obtain:

$$F(\tau) = \frac{1}{TN} \sum_{k=0}^{N-1} x(t) \left[\int_0^T x(t + \tau).\exp(-jk\omega_0\tau).dt \right].$$

A simplification is possible if we define a new variable $\theta = t + \tau$ so that dt becomes $d\theta$ inside the bracket and, we can write:

$$F(\tau) = \frac{1}{TN} \sum_{k=0}^{N-1} x(t) \left[\int_0^T x(\theta).\exp(-jk\omega_0\theta).\exp(jk\omega_0 t).d\theta \right]$$

$$= \frac{1}{TN} \sum_{k=0}^{N-1} x(t).\exp(jk\omega_0 t) \int_0^T x(\theta).\exp(-jk\omega_0\theta).d\theta \tag{9-26}$$

and since T is related to N by $T = Nh$, where h is the sampling interval, then we can recognise the summation and integral terms as complex conjugates of one

410

another. Also, since both terms are equal to the amplitude spectrum, equation (9-26) becomes:

$$F(\tau) = \frac{1}{T} \{X_k X_k^*\} = \frac{|X_k|^2}{T} \qquad (9\text{-}27)$$

where * = complex conjugate.

From equation (8-15) we see that the term, $|X_k|^2/T$ is equal to the power-spectrum $G_x(f)$ of the original time series, so that we may write:

$$G_x(f) = \frac{1}{N} \sum_{k=0}^{N-1} R_x(\tau).\exp(-jk\omega_0\tau) \qquad (9\text{-}28)$$

for which the inverse transform is:

$$R_x(\tau) = \sum_{k=0}^{N-1} G_x(f).\exp(jk\omega_0\tau). \qquad (9\text{-}29)$$

This transform pair can also be expressed in continuous form as:

$$G_x(f) = 2 \int_{-\infty}^{\infty} R_x(\tau).\exp(-jk\omega_0\tau)\, d\tau \qquad (9\text{-}30)$$

and

$$R_x(\tau) = \int_{-\infty}^{\infty} G_x(f).\exp(jk\omega_0\tau)\, df. \qquad (9\text{-}31)$$

A factor of 2 is necessary for $G_x(f)$ since this is by definition a single-sided function (see equation (8-20)).

Equations (9-30) and (9-31) can be restated in real terms by consideration of the symmetrical properties of the autocorrelation function given in equation (9-17). This also implies that:

$$G_x(-f) = G_x(f) \qquad (9\text{-}32)$$

so that the power spectral density function is a real and even function of frequency. Hence, we may substitute a cosine function for the exponential term and write:

$$G_x(f) = 2 \int_{-\infty}^{\infty} R_x(\tau) \cos \omega\tau.d\tau \qquad (9\text{-}33)$$

and

$$R_x(\tau) = \int_{-\infty}^{\infty} G_x(f) \cos \omega\tau.df. \qquad (9\text{-}34)$$

These relationships express the Wiener–Khintchine Theorem, which states that the power spectral density and autocorrelation function are Fourier transforms of one another.

411

This approach to the derivation of the power spectral density has a number of useful attributes. Some of these have already been mentioned in chapter 8, where practical methods of power spectral density derivation were discussed. It is important to note that the power spectral density, being of the form $f(\omega) \cdot f^*(\omega)$, discards all the phase information contained in the time series $x(t)$. Thus, the autocorrelation will be identical for all signals having the same spectral distribution but which may, nevertheless, differ considerably from each other in the shape of their wave-forms. Consequently, when an estimate of the power spectral density function of a random signal is required, it may not be possible to obtain this by equation (8-7) since the wave-forms never repeat and it is meaningless to extend the integration period to infinity as required by the equation. For such random series the derivation of the power spectral density via the autocorrelation function may be the only accurate method possible.

9-5   CROSS-CORRELATION

The initial ideas concerning correlation given earlier refer to cross-correlation. The cross-correlation function describes the dependence of one wave-form with another. This was defined in section (9-3) for two wave-forms having a time shift τ between them and considered over an interval T, viz.

$$R_{xy}(\tau) = \lim_{T \to \infty} \frac{1}{T} \int_0^T x(t)y(t + \tau) \, dt. \tag{9-9}$$

This is similar in form to that given for autocorrelation (equation 9-12). The normalised form for the cross-correlation function was given in equation (9-10). The peak value of the normalised correlation function will be unity when the two functions are at a relative displacement which makes them identical over the period of observation, T, and less than unity at all other time displacements. As with autocorrelation the value of $R_{xy}(\tau)$ becomes zero as (τ) tends to infinity for random data having a zero mean value.

We see from equation (9-9) that the cross-correlation function is the average of the product of the wave-form $x(t)$, with a time-shifted version of the second wave-form, $y(t + \tau)$ over the period, T. If for a particular value of τ the two wave-forms, considered over the window period T are similar, then very many of the products of their corresponding values are positive, so that their sum is large. If the wave-forms are dissimilar, then a large measure of cancellation of the positive and negative products will occur and their sum will be small. As shown in equation (9-9) the average product will only approach an exact cross-correlation function as T approaches infinity. In practical terms this means that a reliable estimate will be obtained only if the correlated wave-forms contain a large number of periods of the lowest dominant frequency of interest.

A further practical point concerns distortions introduced into the wave-forms to be correlated during the pre-processing stage. Since we are seeking for similarities it is essential that any filtering introduced in each channel prior to correlation has identical frequency and phase characteristics, since otherwise any similarities will be masked by different response to the filter. This applies also to the aliasing filter and to sampling rates.

Despite their apparent mathematical similarity there are several important differences between the auto- and the cross-correlogram. The function $R_{xy}(\tau)$ is always a real-valued function which can have a positive or a negative sign. It also shows a symmetry about an ordinate at which the time-order position of the variables $x(t)$ and $y(t)$ are interchanged, thus:

$$R_{xy}(\tau) = -R_{yx}(\tau) = R^*_{xy}(\tau) \tag{9-35}$$

where * = complex conjugate.

Two further properties of the cross-correlation function are:

$$|R_{xy}(\tau)|^2 \leqslant R_x(0)R_y(0) \tag{9-36}$$

and

$$|R_{xy}(\tau)| \leqslant \tfrac{1}{2}[R_x(0) + R_y(0)]. \tag{9-37}$$

These indicate upper limits for the cross-correlation function in terms of the peak value of the autocorrelation function for $x(t)$ and $y(t)$. They can serve as useful program checks for the value of $R_{xy}(\tau)$ calculated with the digital computer since $R_x(0)$ and $R_y(0)$ are easily found as the mean square value of the functions, $x(t)$ and $y(t)$.

A typical cross-correlogram is shown in fig. 9-5. Unlike the autocorrelation function the cross-correlation function does not necessarily have a maximum value when $\tau = 0$. We also see that $R_{xy}(\tau)$ is not an even function of τ. The significance of this is that during the process of obtaining an autocorrelation, all phase information about the signal is lost, whereas with a cross-correlation the relative phases of the wave-forms are preserved.

Two important features of cross-correlation concern its scaling and wave-form recognition properties. The first is that multiplication of either wave-form by a constant, multiplies the cross-correlation function by the same constant, and, similarly, if both wave-forms are multiplied by the same constant, then the cross-correlation function will be multiplied by the square of the constant. The second is that the cross-correlation function of two sinusoidal functions of the same frequency will result in the generation of a cross-correlation function of the same frequency. This latter provides an extra-ordinarily powerful means of detection and recovery of a signal buried in extraneous noise, even when the signal is not of periodic form, providing that we have a wave-form of the same shape with which to cross-correlate it. An example of this will now be described.

We have seen that the Fourier analysis of a rectangular wave-form consists of an infinite series of odd harmonics of a sinusoidal wave-form which has a period

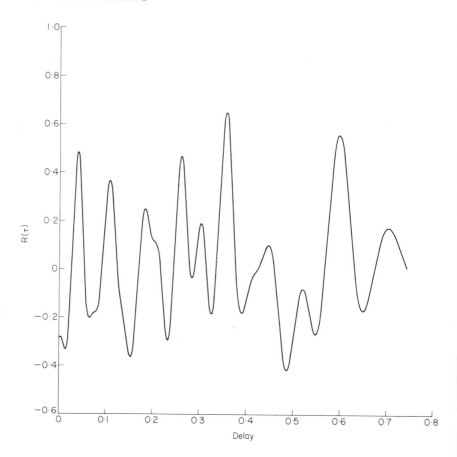

Fig. 9-5. A cross-correlogram.

equal to the fundamental period of the rectangular wave-form. The harmonic constituents will be found to have peak amplitudes proportional to the reciprocal of their harmonic number. Consequently we would expect to be able to recover the fifth harmonic (say) by cross-correlating a sinusoidal wave-form of this frequency with the square wave, and to find that the peak amplitude of the cross-correlation function is equal to one-fifth of that of the square wave. This process is illustrated in fig. 9-6 which shows the recovery of the required signal despite the presence of random noise associated with the rectangular wave-form. The process of cross-correlation may therefore be regarded as a filtering process where the frequency characteristics of the filter relative to the signal being analysed are defined by the characteristics of the referenced wave-form. This implies a relationship for the cross-correlation and the cross-power spectrum similar to that derived earlier between the autocorrelation and the power

414

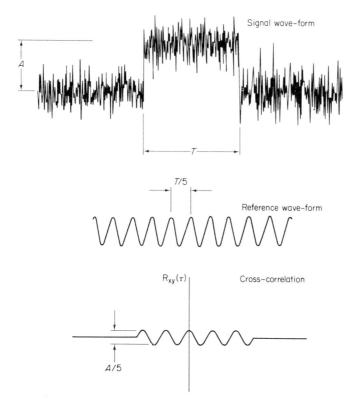

Fig. 9-6. Detection of the fifth harmonic of a square wave using cross-correlation.

spectral density. The cross-power spectral density $G_{xy}(f)$ and the cross-correlation $R_{xy}(\tau)$ can be shown to be Fourier transforms of each other, i.e.

$$G_{xy}(f) = 2 \int_{-\infty}^{\infty} R_{xy}(\tau).\exp(-jk\omega_0 t).d\tau \qquad (9\text{-}38)$$

$$R_{xy}(\tau) = \int_{-\infty}^{\infty} G_{xy}(f).\exp(jk\omega_0 t).df \qquad (9\text{-}39)$$

The complex form of equations (9-38) and (9-39) means that phase information between the two input signals, $x(t)$ and $y(t)$, is retained in the cross-spectrum as noted earlier.

We may also note that:

(a) If a constant term or harmonic term is absent from either $x(t)$ or $y(t)$ then it will be absent in the cross-correlation.

415

(*b*) The phase difference between similar harmonics is retained in the cross-correlation function. This will be seen if we consider the cross-correlation of two terms:

$$x(t) = A \sin (\omega t + a) + B$$

$$y(t) = C \sin (\omega t + c) + D \sin (\omega n t + d). \tag{9-40}$$

Here a constant term B is included in $x(t)$ but absent from $y(t)$ and the component of harmonic frequency, $\omega/(2\pi)$, present in both $x(t)$ and $y(t)$ differs in phase by $|(a - c)|$ radians. Correlation of equation (9-40) gives:

$$R_x(\tau) = \tfrac{1}{2}A^2 \cos (\omega \tau) + B^2$$

$$R_y(\tau) = \tfrac{1}{2}C^2 \cos (\omega \tau) + \tfrac{1}{2}D^2 \cos (\omega n \tau) \tag{9-41}$$

$$R_{xy}(\tau) = \tfrac{1}{2}AC \cos ((\omega \tau) + (a - c)).$$

This shows that in the case of autocorrelation all phase information is lost and in the cross-correlation case only the similarities between the signals are retained (i.e. the factors B and $D \cos (\omega n t + d)$ are lost) and the phase shift between similar harmonics retained (i.e. $(a - c)$ in the cosine term).

Other characteristics of the correlation functions will be discussed during the remainder of this chapter, which deals with the practical application of the correlation theory, described briefly above, to the measurement and analysis of signals. These applications will be described under three main headings; namely Analog methods, Digital methods, and Hybrid methods. Special importance will be attached to the process of the recovery of signals immersed in noise and to correlation methods employed for real-time working.

9-5-1   AUTO-BY-CROSS CORRELATION ANALYSIS

A powerful method of autocorrelation which makes use of the improvement in signal/noise ratio obtained with cross-correlation, is shown in fig. 9-7. This is known as auto-by-cross analysis [58] and is applicable to physical measurement situations where two transducers can be located at non-adjacent positions to

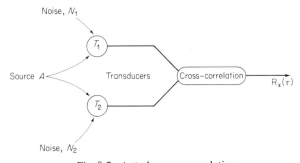

Fig. 9-7. Auto-by-cross-correlation.

detect the same signal source, A. It is assumed that a noise signal, N_1 and N_2 will be associated with each transducer output and that these two noise components will be uncorrelated with each other and with the signal source. Under these conditions it will be shown (section 9-10-1) that a cross-correlation of the two resultant signals, $A + N_1$ and $A + N_2$ will result in a correlogram substantially equivalent to the autocorrelogram of the signal source A uncontaminated with random noise. The improvement in signal/noise ratio obtained with this method is equivalent to that derived from the cross-correlation of the signal using a reference wave-form and shown in fig. 9-29.

9-6 ERRORS IN AUTO- AND CROSS-CORRELATION

A fundamental source of error is that due to finite integration time, which is limited by the length of the signal(s) to be correlated. In the case of autocorrelation of a sinusoidal signal we can write (as before):

$$R_x(\tau) = \lim_{T \to \infty} \frac{1}{2T} \int_{-T}^{T} A \sin(\omega t + \phi) A \sin[\omega(t + \tau) + \phi]\, dt$$

$$= \lim_{T \to \infty} \frac{A^2}{2T} \int_{-T}^{T} [\cos \omega\tau - \cos(2\omega t + \omega\tau + 2\phi)]\, dt$$

$$= \lim_{T \to \infty} \frac{A^2}{2T} \left[T \cos \omega\tau - \frac{1}{4\omega} [\sin(2\omega T + \omega\tau + 2\phi) \right.$$
$$\left. - \sin(2\omega T - \omega\tau - 2\phi)] \right]$$

$$= \lim_{T \to \infty} \frac{A^2}{2} \left[\cos \omega\tau - \frac{\cos \omega T}{2\,\omega T}.\sin(\omega\tau + 2\phi) \right]$$

where $R_x(\tau) = \frac{1}{2}A^2 \cos \omega\tau$, and represents the autocorrelation already given in equation (9-16), whilst $[(A^2 \cos \omega T)/4\omega T] \sin(\omega\tau + 2\phi)$ represents an error term. The peak normalised error is given by:

$$\epsilon = \frac{1}{2\omega T} \tag{9-42}$$

and is thus shown to be inversely proportional to the length of the record and the number of complete cycles included in the maximum lag period $\tau_{\max.} = T$.

An error analysis based on variance considerations will yield on expression for normalised standard error which is applicable to the more general random noise situation.

If we consider the signal to be random and band-limited to a bandwidth, B then the normalised standard error can be shown to be:

$$\epsilon \simeq \frac{1}{\sqrt{(2BT)}} \left[1 + \left(\frac{R_x(0)}{R_x(\tau)} \right)^2 \right]^{\frac{1}{2}} \tag{9-43}$$

for a lag (τ), and

$$\epsilon = \frac{1}{\sqrt{(2BT)}} \tag{9-44}$$

for zero lag. Since $R_x(\tau) \leqslant R_x(0)$ then the error will increase considerably as the correlation estimate is taken at greater and greater lags. For this and other reasons it is advisably to limit the maximum lag, to about half the length of the record.

A similar error function may be derived for cross-correlation as:

$$\epsilon = \frac{1}{\sqrt{(2BT)}} \left[1 + \frac{R_x(0)R_y(0)}{R_{xy}^2(\tau)} \right]^{\frac{1}{2}} \tag{9.45}$$

which assumes that both signals $x(t)$ and $y(t)$ have identical bandwidths, B.

When continuous signal is digitised, and the correlation estimates obtained by digital methods, then other errors are introduced. These are due to sampling and quantisation and also to the approximate methods of integration used. Consideration to these errors will be given later in section 9-8.

9-7 ANALOG METHODS

The basic method of correlation using analog techniques is to include a time delay, τ, in one of the two signals to be correlated and to find the average of the product of one signal and a time delayed version of the other, using relationship (9-9). This is shown in fig. 9-8. Note that the estimated value of $R_x(\tau)$ obtained with this method is the average of $x(t).y(t - \tau)$, which is always equal to $x(t).y(t + \tau)$ for stationary signals.

The two major problems in implementing this fundamental system using analog techniques are the provision of a time delay, τ, which will require to be varied over the time period of the record, and the need to repeatedly multiply two complete ensembles together. An analog computer is likely to be employed and since this does not generally include storage facilities special methods need to be evolved to mechanise this iterative operation. One obvious method is to set up a number of parallel circuits as shown in fig. 9-9. A separate circuit is required for each value of τ selected within the total range, T. This arrangement is frequently termed a 'real-time analyser' since it is capable of performing a complete analysis on the signal as it is generated, or read, from an external storage media. The large amount of equipment required, and the coarse values of correlation estimate obtained, due to the finite number of possible values of τ, makes this method of limited use when implemented on a general-purpose analog computer. It does, however, find wide use when mechanised on a special-purpose hybrid machine and will be referred to later in section 9-9. The

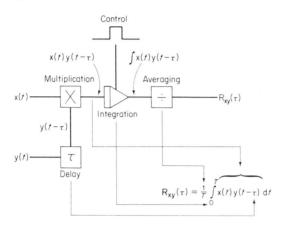

Fig. 9-8. A method of analog cross-correlation.

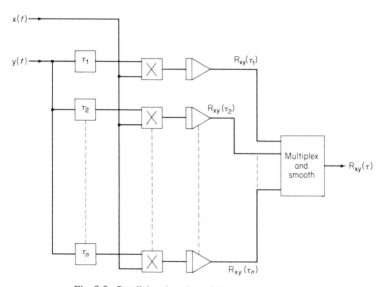

Fig. 9-9. Parallel estimation of the cross-correlogram.

most common analog method relies on temporary storage of the signal in a repeatable form, and often the manner of read-out from this store is arranged to provide the correlation delay needed for each iteration. This is known as the Sequential Correlator.

419

Other techniques for providing correlation delay are described below. A particular method, analogous to the digital shift register, is the matched filter technique, which has important applications in signal recognition, using cross-correlation of the known and unknown signal. This is described in detail in section 9-7-3.

All these methods have time as the independent variable and integration is carried out over a given time period. This need not always be the case and a number of analog methods are available in which integration over time is replaced by integration over distance resulting in considerable economy in computing time. An ingenious example of these are the optical methods developed by Piety, Ferre, Jackson and others. Two photographic transparencies are prepared from displayed time-histories of the signals to be correlated. A wide beam of light is passed through the transparencies arranged one over the other, and the light collected in a photocell. Since the beam of light is wide enough to cover the entire width of the signal, recorded on film, a complete parallel plot of the cross-correlation function can be made instantaneously for a given displacement value. By moving one film relative to the other, the cross products are obtained at increasing displacements until the entire range of displacements (corresponding to time lags) is obtained.

9-7-1   THE SEQUENTIAL CORRELATOR

The sequential correlator carries out a complete analysis for one value of τ, and then repeats the procedure taking a new value of τ, until the complete time value range of interest is analysed. The time delay can be continuously swept or stepped in discrete incremental values. The resolution of the stepped method is dependent on the incremental lag time, h and the highest frequency contained in the signal, f_N namely, $h \leqslant [1/(2f_N)]$. This may be seen to follow directly from the Fourier transform relationship of power spectra to correlation. The minimum definable resolution will be obtained if at least two average values per cycle are obtained at any given frequency.

Data storage is necessary to retain the complete continuous record values for $x(t)$ and $y(t)$, so that they can be called in repeatedly at each interation of the analysis process. A magnetic tape loop or drum replay system is often used. This is illustrated in fig. 9-10. The signal for analysis is recorded on magnetic tape and

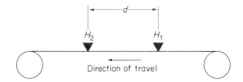

Fig. 9-10. Using a tape loop to define a correlation lag.

played back at two replay heads, H_1 and H_2. A two-track replay system is necessary for cross-correlation and it is important to ensure that the two signals have been recorded from the same head stack to minimise phase displacement between them. The signal derived from head H_2 will be delayed by comparison with that from head H_1 by:

$$\tau = d/v \text{ sec} \tag{9-46}$$

where d = distance apart of heads (in), and v = speed of the transport tape (i.p.s.). A range of delays may be obtained by altering the rotational speed of the tape transport. With this method, there is a minimum separation due to

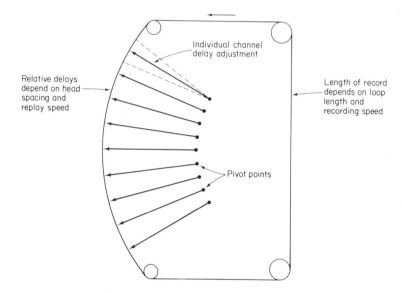

Fig. 9-11. A multiple loop delay system.

mechanical interference of the two replay heads. To overcome this, the signal applied to head H_1 may be delayed using a fixed electronic delay equal to, or greater than, the minimum delay value provided by the physical separation of the heads and the fastest transport speed.

A particularly interesting form of multiplexed delay replay system of this type is shown in fig. 9-11. This has been used to introduce a delay into each channel of information derived from a seismometer array. In order to obtain information about the direction of propagation of a seismic signal, it is necessary to obtain a series of delays, one for each channel, and each of progressively increasing value. Summation of the delayed signals will then result in a greatly augmented signal for one particular direction of propagation [2]. By proper spacing of the replay heads, it is possible to alter the direction of maximum sensitivity by varying the tape transport speed. Small adjustments to individual replay head positions can be made by mounting each head on an arm and pivoting this from the centre line of an arc formed by the tape traversal.

Variation of the delay using magnetic tape methods will, of course, also be dependent on the length of the tape loop. Whilst this is fixed in many tape loop decks, a flexible method of accommodating any length of loop is the use of a tape bin. Two capstan drives are involved. The first is engaged to pile the tape loosely in the bin until the required delay is reached. The second capstan is then engaged and both run at a constant speed. The excess tape stored within the bin will be proportional to the required delay.

A primary disadvantage of the use of magnetic tape is that its transport velocity cannot be controlled as accurately as that of a drum. Also an error can be introduced by tape stretching due to temperature, humidity, and tension variations acting on the tape. It is important to note that magnetic tape or drum methods are not real-time operations and can be extremely slow if a large number of delay points are required and/or the signal frequency is high. The autocorrelation function, using the sequential method, is obtained over a period equal to the sum of all the individual lags used, since all the information must be scanned at each incremented value of τ. Hence, the basic scan rate for an incrementally scanned system is given as:

$$\text{Rate} \leqslant \tau/T_a \text{ lags/sec} \qquad (9\text{-}47)$$

where T_a is the averaging time for the filter in seconds. As will be seen later, T_a is made equal to the record length T for minimum statistical error. The total analysis time for a given record T seconds in length is simply:

$$T_t = mT. \qquad (9\text{-}48)$$

Some reduction in analysis time is permissable if the replay speed can be increased beyond the original recording speed. Since this necessarily increases the frequency band of the signals being analysed the limitation in performance of the analog elements (particularly the multipliers) is soon reached.

9-7-2 FORMS OF DELAY

Several other forms of analog delay are in use and will be mentioned briefly below. The first of these is the rational fraction transfer function simulated using a series of integrators on the analog computer. The transfer function is given as:

$$H(j\omega) = \exp(-j\omega t) = \cos \omega t - j \sin \omega t. \qquad (9\text{-}49)$$

This represents the ideal case of a unit having no attenuation and a phase angle that varies linearly with frequency. Since this is clearly not possible in analog form an approximation must be taken to represent the delay. One solution is to expand equation (9-49) by means of a Taylor series, truncate the series at a suitable point and mechanise the resultant polynomial expression. To do this directly, however, implies that a differentiation process must be used, a process that is generally avoided in analog computation due to its propensity to introduce noise into the system. Instead a Padé approximation for the series expansion is used [3]. This defines two polynomials and obtains the approximation from their quotient as:

$$P_{nm}(x) = \frac{A_0 + A_1(x) + A_2(x)^2 + , \ldots, A_n(x)^n}{B_0 + B_1(x) + B_2(x)^2 + , \ldots, B_m(x)^m} \qquad (9\text{-}50)$$

which has some similarities with the general purpose filter derivations given in chapter 3. The coefficients A_n and B_m are specified by the series [4]:

$$A_n = 1 - \frac{n}{n+m}x + \frac{n}{n+m} \cdot \frac{n-1}{n+m-1} \cdot \frac{x^2}{2!}, \ldots,$$

$$(-1)^n \frac{n(n-1), \ldots, 3.2.1.x^n}{(n+m)(n+m-1), \ldots, (m-1)n!}$$

$$B_m = 1 + \frac{m}{n+m}x + \frac{m}{n+m} \cdot \frac{m-1}{n+m-1} \cdot \frac{x^2}{2!}, \ldots,$$

$$\frac{m(m-1), \ldots, 3.2.1.x^m}{(n+m)(n+m-1), \ldots, (n+1)m!} \qquad (9\text{-}51)$$

where n and m represent the order of the numerator and denominator polynomials respectively. This expansion is identical to the first N Taylor series terms where $N = n + m$. The order of the Padé approximation is therefore often referred to as the sum of the orders of the numerator and denominator polynomial. Examples of the fourth- and eight-order Padé approximations where $n = m$ are given as:

$$P_{2,2}(x) = \frac{12 - 6x + x^2}{12 + 6x + x^2}$$

$$P_{4,4}(x) = \frac{1680 - 840x + 180x^2 - 20x^3 + x^4}{1680 + 840x + 180x^2 + 20x^3 + x^4}. \qquad (9\text{-}52)$$

Mechanisation for a fourth-order Padé simulation is given in fig. 9-12. The use of orders beyond this can lead to operating beyond the stability margin of many analog computers. Padé circuits are 'All-Pass' in the sense of providing zero decibel attenuation at all frequencies. The phase lag produced varies with the frequency and can give quite large deviation for the high frequency components of the delayed signal. For a phase error of less than one degree the highest frequency that can be delayed can be shown [5], to be $7 \cdot 5/2\tau$ for a fourth-order circuit and $13 \cdot 5/2\tau$ for an eighth-order circuit, where τ is the value of the required delay in seconds.

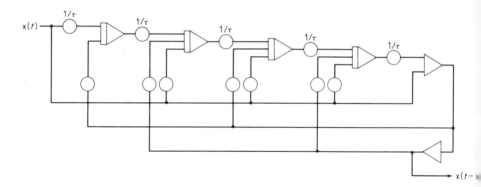

Fig. 9-12. A fourth-order Pade delay simulation.

When implemented as a lag for the computation of correlation functions a variable delay is required. For this purpose the potentiometers shown in fig. 9-12 can be replaced by multipliers and controlled simultaneously from the time incremental function, as suggested in chapter 3. This arrangement has practical limitations of about 100:1 in its range of variation and has its minimum lag value removed from zero frequency. Holst [4] has suggested an alternative form of approximation in which the variable lag is expressed as:

$$P_{nm}(x) = 1 - P_0(x) \tag{9-53}$$

where

$$P_0(x) = \frac{B_{mn}(x) - A_{nm}(x)}{B_{mn}(x)}. \tag{9-54}$$

The dynamic performance of this system is improved with a phase error considerably less than the original Padé and a lag variable down to zero value.

Where low frequencies are involved, a switched capacitor form of delay is practicable. An arrangement suggested by Stone and Dandl [6] is shown in fig. 9-13. This uses only one switch arm, thus avoiding problems of synchronisation, present when both connections to the capacitor are switched. Each capacitor shown charges to a potential $(e_s - e_0)$, given by:

$$e_c = e_s - e_0 = \frac{R_1}{R + R_1} e_i + \left[\frac{R_1}{R + R_1} - \frac{r_1}{r + r_1} \right] e_a.$$

If we let

$$\frac{R_1}{R + R_1} = \frac{r_1}{r + r_1} = A \tag{9-55}$$

then this reduces to $e_c = Ae_i$ where A represents an attenuation factor.

424

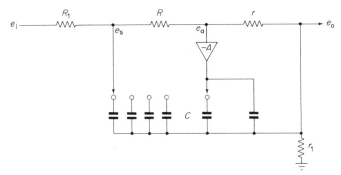

Fig. 9-13. Delay obtained by switched capacitors.

The delay is dependent on the speed of capacitor switching (S steps/sec) divided by the number of capacitors (N). A minimum of two steps per cycle of the highest input frequency of interest is necessary to preserve the shape of the input wave-form. In a practical application of this delay method, reed relays, driven from a solid-state ring counter, replace the mechanical rotary switch. This permits the number of steps per delay to be changed easily by altering the number of elements included in the ring counter.

Electronic delay lines have also been used as a method of variable delay for some correlation applications. Difficulties are experienced due to the filter characteristics of such a line and necessity to tap the line at different points to obtain different values of delay. An equivalent digital form of the analog delay line, which removes this problem, is the shift register and will be referred to later in this chapter.

9-7-3 THE MATCHED FILTER TECHNIQUE

A technique which has some similarities in its analog form with the electronic delay line is the matched filter technique [7]. This method is applicable to cross-correlation, where one of the waveforms is known apriori, e.g. in radar or communication systems, or any method of known signal identification.

It was mentioned earlier that a cross-correlation system can be considered as applying a filter to one of the signals, $x(t)$, whose characteristics are shaped by the frequency components present in the other signal, $y(t)$. This type of approach is adopted in the matched filter technique. The cross-correlation of an unknown signal, $x(t)$ with a known signal, $y(t)$, is considered as being obtainable from the output of a linear filter, whose characteristics are matched to that of $y(t)$. This can be understood if we compare the correlation integral with the convolution integral applied to linear filter theory. From equation (9-9) we can write for the cross-correlation function of $x(t)$ and $y(t)$ averaged over all time:

$$R_{xy}(\tau) = \int_{-\infty}^{\infty} x(t)y(t-\tau)\, dt.$$

425

Also from equation (7-7) the output $z(\tau)$ of a linear filter having an impulse response $h(\tau)$ to the unknown signal $x(t)$ can be expressed as:

$$z(\tau) = \int_{-\infty}^{\infty} x(t)h(\tau - t) \, dt.$$

It can be shown [8] that these two operations are in fact identical and permit the equivalence of the known signal and an impulse response function which is a time-reversed version of the known signal, i.e. $y(t - \tau) \equiv h(\tau - t)$.

Under these conditions the cross-correlation of a signal against a reference wave-form is equivalent to passing the signal through a filter matched to that wave-form. However as indicated above, the impulse response of the matched filter cannot be directly physically realisable, since $h(\tau)$ cannot have a non-zero value for negative sample values. Inversion of the negative samples or addition of

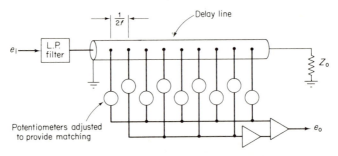

Fig. 9-14. The tapped delay line matched filter.

a delay of T seconds to the filter will permit a realisable matched filter to be obtained.

An analog solution to the problem of providing a time-reversed impulse response matching that of the reference wave-form is shown in fig. 9-14. The delay line is tapped at intervals of $1/(2f)$ sec, where f is the highest frequency contained within the reference wave-form of duration T sec, so that $2Tf$ equally spaced tapping points are required. A number of attenuators, one at each tapping point, are adjusted to provide an amplitude of signal which is proportional to the peak amplitude of the known reference waveform at that point. In order to provide smooth summation of the sample values, the input to the delay line is obtained via a low-pass filter having a cut-off frequency of f Hz. Thus, when an impulse is injected into the line the signal at each of the delay tapping points has a (sin x/x) wave-form. The peak value of these signals is adjusted by the set of attenuators and the summation of the attenuated signals will provide the time-reversed version of the reference signal. Convolution of this with the unknown signal gives the equivalent cross-correlation function. The method is applicable to the provision of multiple matched filters permitting

426

parallel correlation of several signals using the same reference wave-form and has been described previously by Lerner [9]. A digital equivalent method has been described in chapter 7 where it was termed the frequency-sampling digital filter.

9-7-4 REAL-TIME ANALOG METHODS

The use of the E.M.P. estimate of a variable has been referred to earlier (see chapters 5 and 8). This is of value when a 'running' correlation function is required, so that a continuous value of $R_{xy}(\tau)$ may be obtained. It may be used, for example, to search for a given wave-form in a lengthy recorded signal or in a real-time situation. Here a recognition of similarity is required over the immediate section of the signal being analysed and it is not a disadvantage to discount the signal available outside this analysis window.

In order to permit the continuous multiplication of the signal with a reference signal (or a time-shifted version of itself), it is necessary to introduce a time-weighted factor into the correlation process, so that a moving time 'window' is presented to the signal, causing it to be reduced to a negligible value

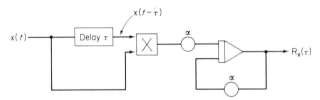

Fig. 9-15. E.M.P. autocorrelation.

after a given time. This is equivalent to a convolution process on the correlation operation. For convenience an exponential window is chosen and the E.M.P. autocorrelation function has been defined [10] as.

$$R_x(\tau) = \alpha \int_{-\infty}^{T} x(t).x(t - \tau) \exp\left[-\alpha(T - t)\right] \, dt. \qquad (9\text{-}56)$$

This is mechanised as shown in fig. 9-15 and is similar to that given for autocorrelation in fig. 9-8, except for the E.M.P. averaging circuit. The arrangement for E.M.P. cross-correlation is similar with $y(t - \tau)$ replacing $x(t - \tau)$.

A practical example of an E.M.P. real-time correlator used for seismic event selection has been described by Hutchins [11]. This carries out the cross-correlation of summed groups of signals derived from a seismic array. A trip level detector forms part of the correlator so that automatic recording of the event correlogram can be initiated when the event rises above a given threshold level. In normal operation all the array signals are undelayed. When an event is detected the E.M.P. integrator capacitor is discharged and the signal delays inserted. Thus, when the event enters the correlator for the second time, it may

427

then be recorded by suitable means. When the event amplitude drops below a specified threshold level, or after a set time period, the correlator returns to its previous condition with delays removed from the signal lines.

9-8   DIGITAL METHODS

The general problem in digital computation of correlation functions is one of speed and efficiency rather than the accuracy of estimation. A subsidiary problem is that of program size and the ability to process long data records. A discrete form of the cross-correlation equation is given by:

$$R_{xy}(\tau) = \frac{1}{N-\tau} \sum_{i=1}^{N-\tau} x_i y_{i+\tau} \quad (\tau = 0, 1, 2, \ldots, m) \tag{9-57}$$

where m is the total number of lags, and N is the number of data points for x_i and y_i. Equation (9-57) is an estimation of the integration process given by equation (9-9) for a finite time period, and shows that the summation period will decrease as the correlation lag is made larger. We would therefore expect that the reliability of this estimate to be reduced as the lag approaches that of the record length.

In many cases $Nh \gg \tau m$ and we can express equation (9-57) with little loss of accuracy as:

$$R_{xy}(\tau) = \frac{1}{N} \sum_{i=1}^{N-\tau} x_i . y_{i+\tau}. \tag{9-58}$$

It will then be necessary to truncate the correlation function as noted earlier (section 9-4), to obtain a reliable estimate. The effect of the $1/(N - \tau)$ bias to the estimate is important where the lag approaches the record length and equation (9-57) must be assumed. This will be considered later when the implementation of the indirect method is discussed.

The direct evaluation of equation (9-58) on a digital machine involves the use of regression techniques which demand large amounts of computation time. If we consider as a unit the time required to carry out one complex multiplication then approximately N^2 units will be needed for the evaluation of the correlation function. We saw in chapter 6 that convolution (which is related to correlation by a simple change in time order for one of the variables) can be obtained by carrying out two direct Fourier transforms, one inverse transform and one multiplication per data sample pair. Using the fast Fourier transform the unit time required will be approximately:

$$3N \log_2 N + N \tag{9-59}$$

which will be considerably smaller than N^2 for values of N greater than about 50. The F.F.T. has thus not only resulted in economy in calculation time but has

428

affected the way in which we carry out the calculations. Before we consider the detailed implementation of the F.F.T. to correlation estimates, other methods directly applicable to the solution of equation (9-58) will be considered. It will be apparent that the symmetry of this equation and the repetitive nature of the product terms will allow a number of grouping algorithms to be derived. Some of these exchange multiplications for additions by grouping together like multiplications [12]. Other techniques expand the cross-products into the sum of a number of squares which is often easier to implement.

One such method, using a simple 'look-up' table for the squares, will be described later. An important class of methods is based on the reduction in the number of quantisation levels that the sampled data can take. An extreme case of quantisation level reduction is the single bit or polarity-coincidence correlation defined by:

$$x_s = \text{sgn } x(t) = 1 \quad \text{for } x(t) > 0$$
$$= -1 \quad \text{for } x(t) < 0 \qquad (9\text{-}60)$$

where only the sign bit of the data sample is retained. It has been shown [13] that under these conditions very little loss of information results from the correlation of Gaussian signals having zero mean value.

9-8-1   THE HALF-SQUARE METHOD

This has been suggested by Schmid [14] as a method of avoiding the direct problem of multiplication of two quantities. Instead the cross-product is expressed as a linear combination of squares:

$$xy = \tfrac{1}{2}[(x+y)^2 - x^2 - y^2] \qquad (9\text{-}61)$$

and the correlation carried out directly from equation (9-58) expressed in the form:

$$R_{xy}(\tau) = \frac{1}{2N} \left[\sum_{i=1}^{N-\tau} (x_i + y_{i-\tau})^2 - \sum_{i=1}^{N-\tau} (x_i)^2 - \sum_{i=1}^{N-\tau} (y_{i-\tau})^2 \right]. \qquad (9\text{-}62)$$

This is an improvement on the simple device of the table 'look-up' used to determine a linearly related quantity by its storage location address.

If two numbers x and y are to be multiplied together, each resulting from, say, an 8-bit quantisation level then we can store all the possible products xy in a set of 2^{16} storage locations, each addressed by a number derived from the sum of the two numbers to be multiplied together. It is a fast method of multiplication since it requires only one machine cycle to actually derive the sum from the address. The disadvantage is, of course, the large storage required. The method given by equation (9-62) permits the three squared values to be stored in 3×2^8 locations. Storage size is traded against calculation time, involving a single-pass

429

regressive calculation to provide the stored products, and a simple retrieval routine to extract them. It is only applicable where the derived data can be located in memory storage.

The method has similar advantages to its 'quarter-squared' counterpart:

$$xy = \tfrac{1}{4}[(x+y)^2 - (x-y)^2] \qquad (9\text{-}63)$$

which is widely used for cross-multiplication using the analog computer.

9-8-2   COARSE QUANTISATION

If certain assumptions can be made concerning the signal to be correlated then considerable simplification of correlation methods is possible. These assumptions are that the signals have a stationary Gaussian probability distribution and zero

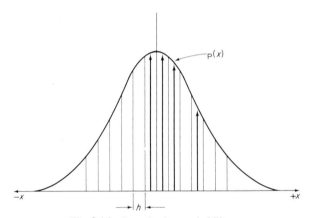

Fig. 9-16. Quantisation probability.

mean value. This permits the data samples to be expressed over fewer quantisation levels and yet still retain sufficient identity for correlation recognition. The method utilises a coarse quantisation of the signals and, under given conditions, obtains a statistical correlation error which is small and easily predictable — at least for long records. Widrow [15, 16] has described the quantisation process as one that carries out a sampling on the statistics of the input signal rather than the signal itself (see fig. 9-16). That is, we can consider the probability distribution, $p(x)$, of the signal, $x(t)$, as being sampled at amplitude intervals, h (the quantisation levels). The Nyquist sampling criterion is applicable here. Just as, in the time domain sampling, the signal is completely recoverable if the sampling rate exceeds that of twice the highest frequency component of the input signal, so in probability sampling the statistical characteristics of the signal can be recovered if the dynamic range of the signal exceeds that of two quantisation levels. When this is true we can replace the quantiser by an equivalent circuit consisting of a unit gain circuit and a white noise generator, the latter having a rectangular probability distribution (fig. 9-17). Using this equivalence we can obtain precise error information from the coarse quantised signal.

For a given class of random processes, coarse quantisation will permit considerable saving in digital equipment, particularly in the analog-to-digital conversion and storage hardware. Thus it is possible to obtain the same accuracy with 8-bit quantisation under these limited conditions as is obtainable with 13 bits under more general conditions [17]. A further improvement in accuracy is obtained by adding broad-band noise (known as 'dither') to one of the signals before quantisation. This causes the probability of the quantisation level nearest

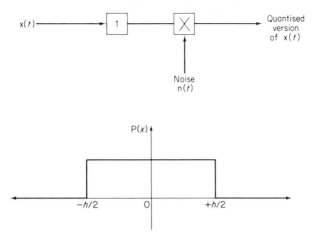

Fig. 9-17. A quantisation model.

to the signal being selected to be proportional to the closeness of the signal, to that level.

The most significant hardware simplification and reduction in computing time occurs where single-bit quantisation is employed. Either one or both signals are quantised using only two quantisation levels per signal, one level to indicate if a signal is positive and the other to indicate if the signal is negative, i.e. only the algebraic sign of the signal is preserved. Again the signals must be assumed to have a stationary Gaussian probability distribution. The correlation obtained with single-bit quantisation will be normalised.

The quantised series are produced from $x(t)$ and $y(t)$ in accordance with equation (9-60) to give two series, x_s and y_s.

A correlation function $R_{xy}^s(\tau)$ is obtained from:

$$R_{xy}^s(\tau) = \frac{1}{N} \sum_{s=1}^{N-\tau} x_s . y_{s+\tau} . \ (\tau = 0, 1, 2, \ldots, m) \qquad (9\text{-}64)$$

431

The relationship between this correlation function and that given by equation (9-10) for the complete word lengths of x_i and y_i has been derived by Van Vleck [18]. He showed that an estimate of the normalised form of the cross-correlation function $R_{xy}(\tau)$ can be obtained from:

$$R_{xy}^N(\tau) = \sin\left\{\frac{\pi}{2} \cdot R_{xy}^s(\tau)\right\} \tag{9-65}$$

A similar estimation and correction for the single-bit autocorrelation function can be obtained in the same way.

In order to maintain the same statistical accuracy obtained with complete word-length correlation it will be necessary to extend the length of the random data analysed. Weinreb [19] has shown that the variance of the autocorrelation estimate for these methods is increased by a factor of about $\frac{1}{4}\pi^2 = 2\cdot5$. This variance is proportional to record length so that in order to preserve comparable accuracy the required record length, using quantised methods, should be about $2\frac{1}{2}$ times longer than is needed for complete word-length correlation.

Note that no information is available regarding the shape of the quantised signal being correlated. The method only takes account of zero crossings so that a sinusoidal wave-form would, for example, give the same quantised correlation value as a triangular wave-form having the same period. Also, as stated earlier, only normalised correlation functions can be obtained so that the mean-square values for $x(t)$ and $y(t)$ must be calculated separately for scaling purposes. Some work has been carried out by Jespers et al. [20] which removes this limitation and permits unnormalised correlation functions to be obtained but which can also result in higher r.m.s. deviations.

9-8-3 CORRELATION USING THE FAST FOURIER TRANSFORM

It was noted earlier that correlation and convolution are related by a very simple change of variables in which only the order of one of them is changed. Thus correlation is given in discrete form as:

$$R_{xy}(\tau) = \frac{1}{N} \sum_{i=1}^{N-\tau} x_i\, y_{i+\tau} \tag{9-58}$$

whereas convolution is described by:

$$x_i^* y_i(\tau) = \frac{1}{N} \sum_{i=1}^{N-\tau} x_i\, y_{\tau-i} \tag{9-66}$$

Consequently the method for convolution described in chapter 7 is equally applicable to correlation estimation. This it will be remembered, consists of finding the direct F.F.T. of each series, multiplying the complex values together and finding the inverse transform of the product. It is an indirect method which will generally be faster than the discrete evaluation of equation (9-58). One important consequence of the multiplication of two finite Fourier transforms is· that it corresponds to convolution of periodic rather than aperiodic functions and, results in a form of repeated or 'circular' convolution which will distort the resultant product. This will be considered further later in this section.

A second indirect method which requires a similar number of complex multiplications and additions is the estimation of correlation through the power spectrum using the Wiener theorem (section 9-4-1). The ability to recover the spectrum is a useful attribute of this method. This will be illustrated for the case of autocorrelation; an essentially similar method applies to cross-correlation.

Four essential steps are necessary:

1. Transformation of a time series, x_i into a frequency series X_k, i, $k = 0, 1, \ldots, N - 1$.
2. Derivation of the power density spectrum from equation (8-15).
3. Smoothing of the power spectrum.
4. Inverse transformation of the smoothed spectrum to give the autocorrelation function (equation 9-29).

The data series x_i is assumed to be provided as N real points, where $N = 2^n$. A Fourier transform of the data is obtained using equation (6-69) as:

$$X_k(f) = \frac{1}{N} \sum_{i=0}^{N-1} x_i \exp(-j2\pi ik/N).$$

A subroutine having two arguments, N (= number of points) and X (= array of data values) would be provided for this conversion. This would input N points into a complex array X (the array is made to hold $2N + 2$ elements for reasons which are given later). Before transforming the data a number of zeros are added to each value of the real and imaginary series — one zero for each data point (e.g. if $N = 5$ then the real part of X could be: 1, 2, 3, 4, 5, 0, 0, 0, 0, 0). The transform is then evaluated using the F.F.T. and its result written back into the array overwriting the input data.

A multiplexed form of storage is used in which the real and imaginary array values are interleaved, i.e. Re X_1, Im X_1, Re X_2, Im X_2, ..., etc. The real part comprises the cosine Fourier transform:

$$X_c(f) = \sum_{i=0}^{N-1} x_i.\cos 2\pi ik/N. \tag{9-67}$$

The imaginary parts comprise the sine Fourier transform:

$$X_s(f) = \sum_{i=0}^{N-1} x_i.\sin 2\pi ik/N. \tag{9-68}$$

433

The frequency spectrum of the input time history is then computed. This is actually the power spectral density in its 'raw' state, (i.e. without any smoothing of the periodogram). The output gives a measure of power in the frequency spectrum and is given as:

$$G_x(k) = \frac{1}{N} \cdot |X_k|^2 \qquad (9\text{-}69)$$

however, X_k is complex and,

$$|X_k|^2 = [\sqrt{X_c^2 + X_s^2}]^2 = X_c^2 + X_s^2 \qquad (9\text{-}70)$$

so that:

$$G_x(k) = \frac{X_c^2 + X_s^2}{N}. \qquad (9\text{-}71)$$

These values overwrite the real part of X in the computer store. The imaginary part is made zero. The real part of X now holds the power spectral density

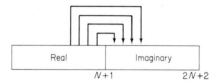

Fig. 9-18. Folding the Fourier spectrum.

which, after smoothing, may be plotted against a frequency scale of $N + 1$ points, each separated by frequency intervals $\delta f = \frac{1}{2}h$, where h is the sampling interval (i.e. a single degree of freedom). An inverse transform will provide the autocorrelation function using an essentially similar routine to that required for transformation.

Two non-obvious modifications are required to this procedure due to the properties of a finite discrete transform and the limited record length. Before carrying out the inverse transformation it is necessary to re-order the data so that the cosine series (the real part) is symmetric about the folding frequency point, $(N + 1)$, i.e. we need to fold the series as shown in fig. 9-18. The autocorrelation function is then overwritten into the real part of X and, due to the folding, is made $2N$ points in length, commencing with the zero phase shift point ($k = 0$) and repeated in inverse order after the half-way (N) point is reached. The reason for the addition of zeros and the repeating form of the autocorrelogram obtained is due to the effect of the finite length of the correlated data. When two finite Fourier transforms are multiplied together the limited length of the data series T, has an effect such that the elements which have been time-shifted past the end of the period T, re-appear at the beginning

434

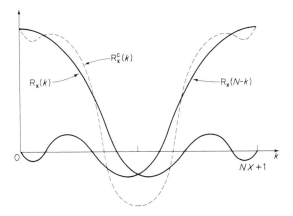

Fig. 9-19. Circular correlation.

again, (fig 9-19). This was noted earlier in connection with the convolution estimation using the same method. Thus, instead of actually computing $R_x(k)$ we obtain instead a 'circular' autocorrelation [21] defined by:

$$R^c_{x(k)} = [R_x(k) + R_x(N - k)]. \qquad (9\text{-}72)$$

The superposition of the two parts is avoided if a number of zeros are added to the original transform. The effect of this is to separate the two parts of the circular-correlation as shown in fig. 9-20.

A further source of error is present where the maximum delay becomes an appreciable fraction of the signal length being analysed. For a finite length record, equation (9-57) is applicable so that implementing equation (9-58) will impose a bias error, increasing linearly with delay value, on the correlogram,

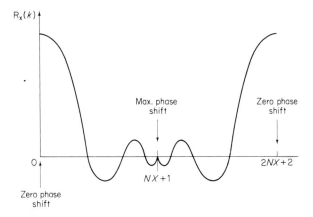

Fig. 9-20. Effect of adding zeros.

435

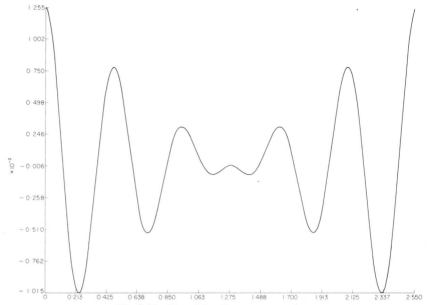

Fig. 9-21. Bias of the discrete correlogram.

commencing from the $(N - 1)$th point. If we consider, for example, the auto-correlogram of a sinusoidal wave-form, this will be given, not as a constant peak amplitude cosinusoidal wave-form, but as the product of a cosinusoid and a linearly increasing function as shown in fig. 9-21.

The resulting correlation function must therefore be multiplied by a weighting function:

$$W_k = \frac{N - k}{N} \qquad (9\text{-}73)$$

in order to correct the bias error progressively as k increases.

As a consequence of these two sources of error the procedure given earlier in this section must be modified by the addition of two further steps:

    3a. addition of N zeros to the smoothed power spectrum before step 4

    4a. multiplication of the resultant correlogram by W_k following step 4.

In the case of the equivalent convolution by finding the product of two transforms the zeros will need to be added to one of the data sequences before transformation.

9-9 HYBRID METHODS

It will be clear from the preceding treatment that the principle difficulty with analog methods of correlation analysis lies in the provision of an incremented time delay, whilst the corresponding difficulty with digital methods is the large number of complex multiplications required. Hybrid methods have been developed in an attempt to optimise both of these requirements and so provide a cheaper although not necessarily a more accurate or faster method. It has been noted earlier (chapter 3) that delays are not easy to implement on analog computer without some sacrifice in accuracy. For low frequencies and where small delays are involved. Padé approximations can be used. Tape recording and replay methods are cumbersome whilst delay lines and capacitor switching systems prove inflexible in operation. The large number of multiplications necessary in the derivation of a correlation function by digital means are time consuming although the fast Fourier transform is of value in reducing these. Accumulation of products is not generally a problem here although this can result in scaling difficulties.

The combination of digital storage and analog multiplication proves ideal for correlation calculations and several systems of this type have been described in the literature [22, 23]. When implemented on a linked hybrid machine the operational steps required can be summarised as:

1. Analog-to-digital conversion of signal and reference wave-forms.
2. Insertion of delays by digital software program loop.
3. Digital-to-analog conversion of delayed signals.
4. Analog product calculation and normalisation.
5. Averaging and scaling of the results using analog methods.
6. Display or plot of the correlogram.

The steps in this program are complicated by the need to count data samples and control the operation of both analog and digital processes. A flow diagram for a typical hybrid cross-correlation program is shown in fig. 9-22. This assumes operation using an analog computer linked to a digital machine [24]. The reference wave-form is assumed to be digitised and located within memory storage. The continuously variable unknown wave-form (considered as a real-time signal) is compared with the stored wave-form and the series of cross-multiplications carried out in intervals between the acquisition of each new data sample. A process of continuous signal identification is thus being carried out in real-time in much the same way as the continuous spectral analysis of a real-time signal considered in section 8-8. The program needs to know the period of comparison between the fixed reference wave-form and the moving signal wave-form. This can be called the 'window' period (W) and entered as an external parameter.

The selected W samples of the fixed wave-form are then accessed, and after initialising the data source (which can be an F.M. tape recorder, or analog program), the analog computer is put into the 'COMPUTE' mode and a series of W corresponding samples of the data source placed in successive core storage locations. The process of correlation begins by transferring W pairs of samples from the two storage locations via digital-to-analog converters to the analog computer. Normalisation of the data signal takes place before multiplication and integration are carried out to produce a single correlation coefficient (R_{xy}) for

437

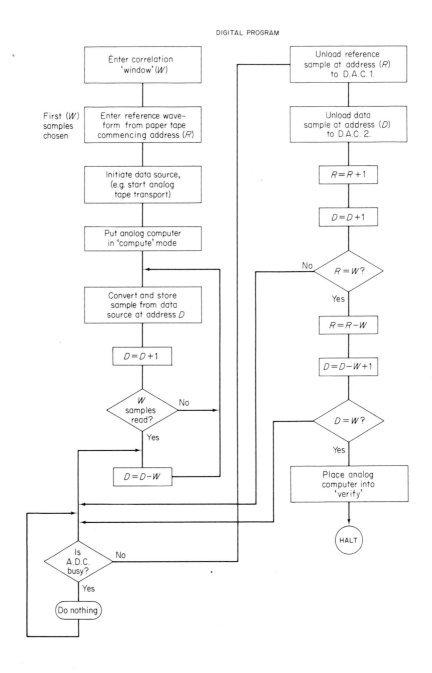

DIGITAL PROGRAM

Enter correlation 'window' (W)

First (W) samples chosen — Enter reference waveform from paper tape commencing address (R)

Initiate data source, (e.g. start analog tape transport)

Put analog computer in 'compute' mode

Convert and store sample from data source at address D

$D = D + 1$

W samples read?

No / Yes

$D = D - W$

Is A.D.C. busy?

No / Yes

Do nothing

Unload reference sample at address (R) to D.A.C. 1.

Unload data sample at address (D) to D.A.C. 2.

$R = R + 1$

$D = D + 1$

$R = W$?

No / Yes

$R = R - W$

$D = D - W + 1$

$D = W$?

Yes

Place analog computer into 'verify'

HALT

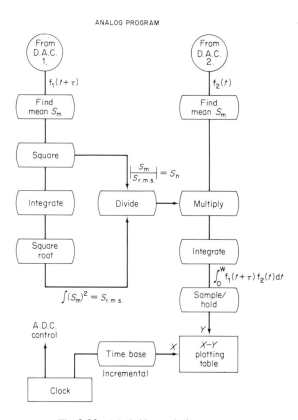

Fig. 9-22. A hybrid correlation program.

the sample set. The next set of data samples are selected by first incrementing the previous storage locations by one and carrying out the same procedure. The same set of W samples of the reference wave-form is used and the result is a second coefficient value. The procedure is continued in this way to obtain a time series of coefficient values. Return to initial data storage location is obtained automatically each time the incremented location reaches a value of W.

The presentation of pairs of samples to the digital-to-analog converters will need to be interrupted at intervals to load further data samples into the store, where the earliest used locations are then overwritten. This sampling of the data signal proceeds under the control of a separate clock signal at a rate considerably slower than the unloading of pairs of signals, so that the entire set of W pairs can be unloaded via the digital-to-analog converters between each sample of data signal input into storage. Each integrated product for W pairs of reference and data samples produces a new coefficient value, to result in a new histogram level being held in the sample/hold amplifier shown in the diagram. The instantaneous

439

value of the histogram may be plotted against an incremental time axis to form a running correlation display of the input signal. A detailed discussion of the timing problems of such a correlator has been given by Inston, Noyes and Reader [25].

A hybrid multipoint correlator which calculates a number of discrete delay points simultaneously is shown in fig. 9-23. The delay element is a digital shift register and the products obtained through the use of digital-to-analog multipliers. One signal is converted to digital form and applied as input to the shift register. The digital output from each stage of the shift register and the undelayed analog signal are simultaneously multiplied, averaged and applied as inputs to an analog multiplexor. Correlated output over a delay set by the shift

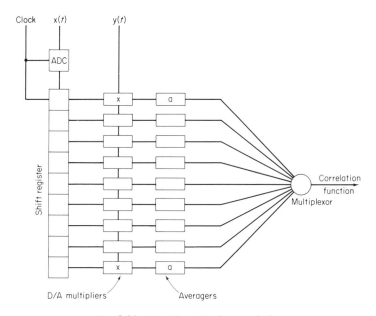

Fig. 9-23. Hybrid multipoint correlation.

register clock rate is obtained by sequentially scanning the output of the multiplexor. Due to the parallel operation of this correlator it is possible to make this operate in real time at a rate limited only by the bandwidth of the multipliers.

9-9-1   SAMPLED DATA CORRELATION

Where the analog data ensemble is obtained repetitively (e.g. from a magnetic tape loop or repeated analog computer simulation) a digitally controlled system can be implemented, which permits definition of the delay in terms of the spacing between control pulses. A schematic diagram for this method is shown in fig. 9-24 which mechanises the estimate for cross-correlation given by:

$$R_{xy} = \frac{1}{N} \sum_{i=1}^{N} x_i(t_1) y_i(t_2) \qquad (9-74)$$

where $t_2 = (t_1 + \tau)$ and τ = lag value.

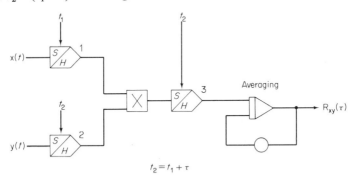

$$t_2 = t_1 + \tau$$

Fig. 9-24. Sampled data correlator.

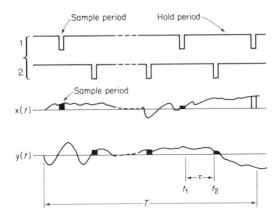

Fig. 9-25. Timing diagram for the sampled data correlator.

This differs from equation (9-57) in that x_i and y_i represent the sampling of analog signals, $x(t)$ and $y(t)$ at times t_1 and t_2 separated by a lag period τ, rather than continuous sequential sampling at a fixed sampling interval.

Continuous analog signals for $x(t)$ and $y(t)$ are applied to sample/hold amplifiers $SH1$ and $SH2$. These are placed in the sample mode for a short period δt, at times t_1 and t_2. The sampling period is considered to be small compared with the delay interval, τ (see timing diagram fig. 9-25). During the remainder of

441

this time the amplifiers are maintained in their hold mode. During this hold period the values at the output of *SH*1 and *SH*2 are multiplied together and the average value taken. This can also be an analog process although advantage can be taken of the lengthy period between sampling times to carry this operation out digitally as suggested by Bekey and Karplus [26]. During the next repetitive period sample times t_1 and t_2 are advanced one increment, (which could be δt) and the process repeated to obtain a further product value. The products are accumulated in a third sample/hold amplifier (and normalised if required) and, after N runs, an estimate for one value of the cross-correlation function (at delay $= \tau$) is obtained. The separation, τ between the position of the sampling pulses t_1 and t_2 is then advanced by a suitable increment and the entire process repeated to obtain a second cross-correlation value.

This method retains all the advantages of the analog technique, including its simplicity, and transforms the difficult problem of providing programmed delays of the analog signal into one of the generation of suitable timing pulses. The correlation will however be a fairly lengthy operation since $(NT)/\tau$ repetitions of the signal will be required to complete the correlation function. It is not, therefore, suitable for real-time operation.

Other hybrid methods include time/compression techniques, which have already been discussed in chapter 8 in connection with power spectral density estimation. The increase in speed of cross-multiplication resulting from this rapid read-out technique is equally applicable to correlation calculations.

9-9-2   DELTA MODULATION

A pulse-coded system which has some similarities to the polarity-coincidence correlation described earlier (section 9-8-2) is the use of delta modulation. In this hybrid correlator two of the basic functions required, namely that of multiplication and time delay, are carried out digitally whilst integration is implemented by analog means. Unlike the previous hybrid methods no attempt is made to quantise the signal into a discrete series of levels. Instead a simple form of pulse modulation is used in which only the rate of change of the signal is encoded.

Fig. 9-26 illustrates the principle of the modulator. The output of the modulator consists of a series of constant area pulses whose polarity is determined by the time derivative of the input signal $x(t)$, viz:

$$\frac{dx}{dt} = CP_x \qquad (9\text{-}75)$$

where C = a constant, and P_x represents a train of pulses of equal amplitude and width. Since the rate of change is taken as a coding characteristic it is not necessary for the signals to have a zero mean value so that the true covariance function is estimated. Also the superposition of a d.c. or slowly changing base to

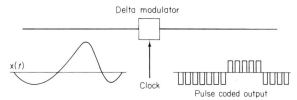

Fig. 9-26. Delta modulation.

the signal will not affect the modulated output. This has some advantages in eliminating slowly variable trends in the signal. It will be appreciated that the integration of the series of output pulses from the modulator will reconstruct the original function, so that:

$$\int CP_x \, dt = \int \frac{dx}{dt} \cdot dt = x(t). \tag{9-76}$$

The use of the modulator to effect a cross-multiplication between two signals is illustrated in fig. 9-27. A modulated form of the first input signal $x(t)$ provides

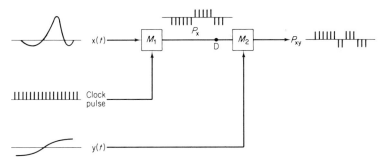

Fig. 9-27. Product evaluation using delta modulation.

the clock input for a second modulator to which is applied the second analog input, $y(t)$. The second modulator will be seen to perform the role of a sign inverter, so that its output is the analog wave-form, $y(t)$ switched in sign according to the delta impulse code P_x. Demodulation of P_{xy} by integration will recover the product signal, $xy(t)$.

When used in a delta correlator a digital shift register is interposed between the two delta modulators at D in fig. 9-27 to give a programmed delay to the pulsed form of $x(t)$. Multiplication of this delayed function by the second analog input, $y(t)$ takes place in the second modulator. In one design [27] the output of M_2 is applied to an averaging integrator which has a number of integrating capacitors, one for each stage of the shift register, and connected in turn to the

443

integrator at each sampling period. Read-out of the correlogram is obtained by sampling the value of charge stored in these capacitors at the end of the signal period.

The accuracy of the correlation estimates is not high due to the finite number of lags possible and incomplete integration during the sampling period. It has similar virtues to the polarity-coincidence correlator described earlier, in that the hardware necessary to complete the correlation can take a very simple form, so that the design is suitable for special purpose correlators which can be constructed at low cost.

9-10   RECOVERY OF SIGNALS FROM NOISE

The acquisition of data relating to a given physical phenomena is never obtained without acquiring some other (unwanted) information. This generally takes the form of random processes and noise, but could also be transients and, in some cases, periodic signals having no relevance to the subject under study. The presence of this noise element sets a limit to the attainable precision of measurement for the often small value of useful information contained within the acquired signal. Even assuming perfect data acquisition and processing equipment, certain random effects, such as the fundamental thermal changes of all matter not at absolute zero temperature, or the uncertainties caused by the quantised nature of light, electronic currents, etc., will be certain to contaminate any physical measurements we care to take.

The basic problem is thus one of improvement in the signal/noise ratio of the available information. This can be improved in a variety of ways, each applicable to a particular type of situation. In the case of radio communication, for example the problem can be couched in simple selectivity terms in which the noisy signal is passed through highly selective circuits which resonate and so generate a large output signal in the region of the frequency of interest. Filtering is of little use, however, when signal and noise occupy the same part of the frequency spectrum. Here a statistical approach to the problem can yield valuable results for those cases where the signal is (or can be caused to be) repetitive, thus permitting an averaging technique to be used. It acts on a number of ensembles containing the data so that the contribution provided by the random noise will tend to cancel out, whilst the uni-directional summation of the repetitive signal will result in an increased output. A similar result can be derived from a single ensemble by the use of phase-comparison, i.e., correlation methods. These would be used, for example, in the recognition of a known transient wave-form immersed in a noisy signal.

The choice of method is dependent on the known characteristics of a signal and, for the case of recorded information, its form of storage. In many situations a practical choice will be determined by the available equipment, but where a wider choice is possible, considerable economics in processing costs can be achieved. For example, where a real-time environment is concerned, such as a continuous analog tape recording of low frequency oceanographic data, it would be unrealistic to digitise the mass of data before attempting a signal recovery operation. Instead, an analog data reduction operation would need to precede the digitisation of the reduced data, or, alternatively, use made of a real-time hybrid system specifically designed for this purpose.

The three general processes of recovery of signals from noise, namely correlation, signal averaging and filtering will now be considered in detail both from the theoretical standpoint and in terms of their practical application to the problem of the signal/noise ratio enhancement. Correlation is the most powerful of the three and will be considered first.

9-10-1 USE OF CORRELATION

The extent to which correlation contributes to the improvement of the signal/noise ratio has been described in a notable paper by Lee *et al.* [28]. This considers the operation of a process to be carried out on discrete samples of the

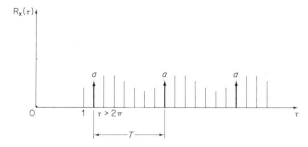

Fig. 9-28. Autocorrelation of a noisy periodic signal for large delay values.

data as given in equation (9-58), so that a statistical method for the derivation of signal/noise ratio is applicable. If we consider for example, an input to the autocorrelation process consisting of a sinusoidal signal, $x(t)$ having additional random noise, $n(t)$.

$$y(t) = x(t) + n(t) \tag{9-77}$$

then, for a displacement τ from the coherent range of noise, the sampled output of the signal would appear as fig. 9-28. For a given set of values identified at points T apart, where T is the period of the sinusoid, then the cross-product signal sample at the particular delay value τ is given as:

$$S_\tau = [x(t) + n(t)] [x(t + \tau) + n(t + \tau)]$$

$$= x(t)x(t + \tau) + n(t)x(t + \tau) + x(t)n(t + \tau) + n(t)n(t + \tau) \tag{9-78}$$

where τ is the delay value from the origin. This is, in fact, the fundamental definition of the autocorrelation process as given earlier (section 9-2). From

445

equation (9-78) the autocorrelation of a signal in the presence of noise is seen to consist of the summation of the desired autocorrelation term,

$$\frac{1}{N} \sum_0^N x(t)x(t + \tau) \, dt \tag{9-79}$$

a 'noise' autocorrelation term,

$$\frac{1}{N} \sum_0^N n(t)n(t + \tau) \, dt \tag{9-80}$$

and the sum of two cross-correlation terms,

$$\frac{1}{N} \sum_0^N x(t)n(t + \tau) \, dt + \frac{1}{N} \sum_0^N n(t)x(t + \tau) \, dt \tag{9-81}$$

where $x(t) \equiv x(ih)$ and $i = 0, 1, 2, \ldots, N$; $n(t) \equiv n(ih)$ and $h = $ the sampling interval.

The autocorrelation of a sinusoidal signal, $x(t) = A \sin \omega t$ has been derived earlier (section 9-4) as $(A^2/2) \cos \omega t$, where A is the peak value. The r.m.s. value of the autocorrelation function will be therefore $A^2/\sqrt{2}$. The variance for the desired autocorrelation term (equation (9-79)) will be the square of the r.m.s. value divided by N, i.e. $A^4_{\text{r.m.s.}}/N$ and we can write:

$$\sigma_s^2 = \frac{A^4}{2N}. \tag{9-82}$$

Taking the variance for the pure noise, (equation (9-80)) as σ^4/N the variance for the two-term equation (9-81), will be the square of twice the product of the standard deviation terms for each term: $(2A/\sqrt{2} . \sigma_n)^2$, where the standard deviation for the sinusoidal wave is, in this case, equal to the r.m.s. value of the peak, $A/\sqrt{2}$. Hence the total variance for the noisy signal will be:

$$\sigma_T^2 = \frac{1}{2N} (A^4 + 4A^2 \sigma_n^2 + 2\sigma_n^4). \tag{9-83}$$

The r.m.s. value for the autocorrelation of a sinusoidal signal without noise is given as $A^2/\sqrt{2}$ so that the signal/noise ratio of the autocorrelation of a noisy signal, $x(t) + n(t)$ is obtained from equation (9-83), as:

$$\frac{\text{signal}}{\text{noise}} = \frac{(\sqrt{N})A^2}{\sqrt{(A^4 + 4A^2\sigma_n^2 + 2\sigma_n^4)}}$$

$$= \sqrt{\left[\frac{N}{1 + 4(\sigma_n/A)^2 + 2(\sigma_n/A)^4} \right]} \tag{9-84}$$

Expressed in decibel units the signal-noise ratio for autocorrelation is:

$$R_a = 10 \log_{10} \frac{N}{(1 + 4R_i^2 + 2R_i^4)} \text{ db} \qquad (9\text{-}85)$$

where R_i is the r.m.s. value of the input noise signal ratio.

Similarly when applied to cross-correlation we have:

$$S_c = [x(t) + n(t)] [y(t + \tau)] \qquad (9\text{-}86)$$

$$= x(t)y(t + \tau) + n(t)y(t + \tau)$$

from which can be derived, as before, an expression for the signal/noise ratio namely:

$$R_c = 10 \log_{10} \frac{N}{(1 + 2R_i^2)} \text{ db}. \qquad (9\text{-}87)$$

Comparing equations (9-85) and (9-87) it is apparent that a considerable improvement is obtained if the signal can be cross-correlated with a reference wave-form as compared with an autocorrelation of the signal with a delayed replica of itself. The effect of this is shown in Fig. 9-29 where a sinusoidal signal having a considerable amount of added noise is correlated, firstly with itself and

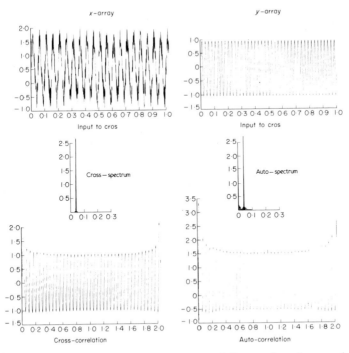

Fig. 9-29. Use of a reference wave-form to extract a periodic wave-form from a noisy signal.

447

secondly with the sinusoidal reference wave-form. Much less variation in the mean level of the cross-correlated signal is seen compared with the auto-correlated example. In both types of correlation there is some advantage in increasing the number of discrete samples. The improvement will be proportioned to the square root of the number of samples available. This is to be expected, since the effective cancellation of positive and negative peaks of the random noise will increase with the number of samples taken.

The use of cross-correlation to extract a known signal from a noisy wave-form is one example of signal detection. Here the repetitive period of the reference wave-form is more important than its shape so that a similar improvement would have been obtained with a square reference wave-form instead of the sinusoidal wave-form used, providing its period was the same. This method, when applied to the detection of sinusoidal signals, is sometimes termed phase-sensitive detection or coherent detection. An example is given in fig. 9.30 for the measurement

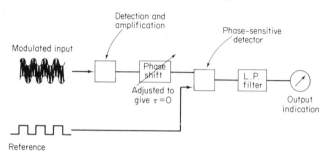

Fig. 9-30. Phase-sensitive detection.

of the peak amplitude of a high-frequency wave travelling along a wave-guide. The wave is modulated by a low-frequency signal and it is this modulation which is later extracted by means of a diode probe inserted in the wave-guide. This signal will be small in magnitude and heavily contaminated by noise so that it will be difficult to determine its peak amplitude and hence the relative peak amplitude of the high-frequency carrier. The detected signal is therefore cross-correlated with a square-wave reference wave-form of the modulation period to enhance the signal/noise ratio of the detected wave-form.

As shown in section (9-5-1) this permits only the fundamental modulation signal to be extracted having a peak value equal to half the product of the peak values of the modulated signal and the reference wave-form. Since only this peak value is desired, the output of the phase sensitive detector is passed through a low-pass filter to remove the harmonic terms (including the fundamental). The time constant of the integrating filter is adjusted manually to give a maximum output, which is then proportional to the carrier amplitude.

The noise power present at the output of such a system may be defined in terms of the noise bandwidth of the averaging filter used, which, for the case of

448

a single stage R.C. filter, is given by $B = 1/(4RC)$, so that the total noise at the output of the filter is the product of the noise bandwidth in Hertz, multiplied by the mean square value of the total noise (assuming uniform spectral density, i.e. truly random noise), which for cross-correlation is:

$$\text{total noise} = \frac{2\sigma_n^2}{4RC}. \tag{9-88}$$

and includes the effects of both sum and difference noise frequencies. Taking the mean square of the correlated peak output as $A^2/2$ then the signal/noise ratio expressed in decibel units, is:

$$R_P = 10 \log_{10} \frac{A^2 RC}{\sigma_n^2} \tag{9-89}$$

$$= 10 \log_{10} RC/(R_i)^2 \tag{9-90}$$

where R_i is again the r.m.s. value of the input noise/signal ratio per cycle. This analysis is somewhat different to that carried out previously for cross-correlation because in this case the situation is a real-time one and the filtering period, RC represents a form of averaging of the correlogram over many lag periods. ($RC \gg \tau$). The improvement in signal/noise ratio is directly related to the filtering period, and, as with the averaging methods discussed in the next section, is dependent on the permissible speed of acquisition of results. This example can be considered in the frequency domain by realising that the correlation process acts as a narrow-band filter. By correlating a noisy signal against a sine-wave reference, the output will contain only those frequency components of the signal which are very close to the frequency of the sine wave. We have, in fact, a very narrow band filter which can approach the zero frequency (Dirac-function), as we increase the averaging time and hence length of reference sinusoid wave-form.

The reference signal need not be sinusoidal since the correlation process will look for similarities between the noisy signal and *any* wave-form used for the reference signal. An example of the 'pattern recognition' ability of the cross-correlation process is the detection of radar or sonar reflection signals accompanied by noise. The ability to detect the return signal reflected by a target can be shown to be directly proportional to the product of its average power and duration [29]. Practical limitations in transmitter design control the peak value of the signal and it becomes necessary to obtain the required range by increasing the duration of the transmitted signal. Unfortunately the reduction in bandwidth which this entails also leads to a reduction in signal resolution, i.e. the ability to distinguish between echoes derived from closely-spaced objects. A solution to this resolution/range problem has been proposed by Wiener, Shannon and others. This suggests the replacement of the long narrow-band pulse by a swept frequency signal having both long duration and wide-bandwidth. This

449

technique is referred to as pulse compression and contributes enormously to the range and resolution capability of modern radar equipment. Detection of the swept frequency echo signal and the separation of this from its associated wave-form is carried out using cross-correlation of the signal with a reference wave-form derived from the transmitter output. This process compresses the energy dispersed within the long swept-frequency signal into a short well-defined pulse of consequently enhanced average power. This is shown in fig. 9-31 and illustrates the value of this technique in raising the correlated pulse to a value

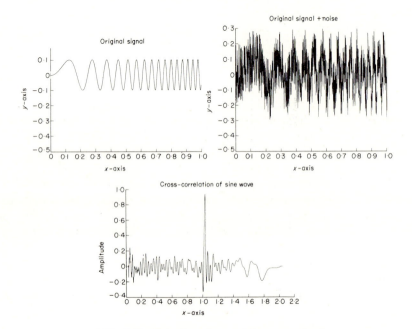

Fig. 9-31. Detection of a swept frequency signal.

well clear of the noise level. Clipping can then be carried out to give a precise location of the time position of the returned pulse.

It is interesting to consider this process as acting on a spectrally decomposed version of the received signal. This allows us to consider the signal as consisting of a series of frequencies all present at the same time and subject to a range of correlation lags. The output correlogram will consist of the summation of all the frequency terms and all the lags, i.e.

$$R_{xy}(\tau,f) = \frac{1}{NF} \sum_{f=1}^{F} \sum_{\tau=0}^{N} x_f(t)y_f(t+\tau) \, dt. \qquad (9\text{-}91)$$

Thus phase coherence will be obtained for the set of frequencies at certain specific values spaced through the total lag range. This is illustrated in fig. 9-32 which shows the condition for multiple-signal coherence of seven signals subject to seven specific values of lag. This results in a cross-correlation pulse which is short compared with the summation of the seven individual signal periods. At all other lag values expressed by equation (9-91) cancellation of the resultant values will occur.

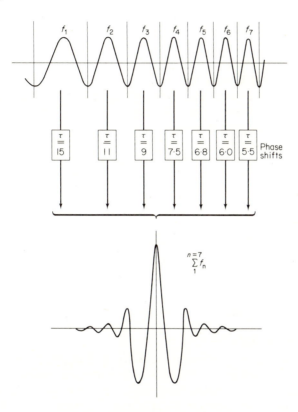

Fig. 9-32. Illustrating phase coherence.

The preceding discussion has been largely concerned with the recovery of signals from a noisy wave-form where the frequency of the signal is known *a priori*. Where the periodicity of the signal is unknown autocorrelation provides a means of detection of the periodic signals. Assuming that the extracted signal is recognisable, its period can then be measured and used to generate a local reference signal. This can then be cross-correlated with the original signal to give an improvement in signal/noise ratio greater than would have been achieved with autocorrelation alone.

451

This improvement is (from equations (9-85) and (9-87)):

$$S_{ca} = 10 \log_{10} \frac{1 + 4R_i^2 + 2R_i^4}{1 + 2R_i^2} \cdot \text{db} \qquad (9\text{-}92)$$

For very small values of input noise/signal ratio this becomes 3 db and very little advantage is derived. For large values equation (9-92) simplifies to:

$$S_{ca} \simeq 20 \log_{10} R_i \qquad (9\text{-}93)$$

and if R_i is also expressed in decibel units the improvement will be:

$$S_{ca} \simeq R_i \; \text{db}. \qquad (9\text{-}94)$$

In other words the signal/noise improvement by the use of cross-correlation, as compared with autocorrelation of large noise/signal ratios, will be such as to reduce the noise/signal ratio to unity. We can also see, from equations (9-85) and (9-87), that an improvement in integration time can be expected, since, assuming minimum improvement in signal/noise ratio (0 db) then the number of input signal samples required for autocorrelation would be:

$$N = 1 + 4R_i^2 + 2R_i^4 \; \text{samples} \qquad (9\text{-}95)$$

whereas with cross-correlation we would require:

$$N = 1 + 2R_i^2 \; \text{samples}. \qquad (9\text{-}96)$$

Another way of expressing this is to relate the improvement in signal/noise ratio with record length. We would need a larger sample of the signal to achieve the same signal/noise enhancement with autocorrelation than would be needed if a reference signal were available and cross-correlation could be carried out.

9-10-2 SIGNAL AVERAGING

An enhancement of the signal/noise ratio of repetitive signals can be carried out by making use of the redundant information inherent in the repeated signal. This is by no means a new technique and finds widespread use in long-exposure photography and is the principal method used in recording the light signals emanating from distant stars. This technique may be applied to signal time histories by summation of successive repetitions of the signal. The signal may be available in real-time form as a continuous periodic signal immersed in noise or, more generally, as a single finite ensemble of periodic or transient form, repeatedly generated from a storage device such as a magnetic tape loop or digital core storage. Successive repetitions of the wave-form are summed so that the periodic signals add coherently, whilst the random element is averaged to a small value by virtue of its incoherence. The technique is known as 'Time-Averaging', and yields an improvement in the signal/noise ratio roughly proportional to the square root of the number of repetitions of the signal. This is a useful practical rule having few exceptions and can be assumed where the wave-form of the signal is independent of the repetition rate. Ernst [30] has considered the deviation from this general assumption which can occur under certain conditions.

It is important to note, in connection with signal averaging, that the method is valueless in removing unidirectional trends or very slow changes in the signal base line. To some extent this is applicable to the case of the very low noise frequencies where the noise is inversely proportional to frequency ($1/f$ noise), although some improvement in signal/noise ratio for these very low frequencies can be obtained using the multi-sweep technique described below.

The derivation of the signal/noise ratio, R of a sampled wave-form will now be considered. The noisy signal $y(t)$ can be represented as before by:

$$y(t) = x(t) + n(t) \tag{9.77}$$

and the instantaneous sampling time by $t = hi$ where $i = 0, 1, 2, \ldots, N$. If we now consider a particular sample value, i, taken from any repetition of the record, y_k, then its value will be:

$$y_k(ih) = x_k(ih) + n_k(ih). \tag{9-97}$$

Two assumptions will now be made which are valid for most noise situations, namely:

1. That the mean value of the noise is zero.
2. That the variance of the noise, σ_n^2 is constant over the period of the record, $T = hN$.

Under these conditions the r.m.s. value of the noise will be equal to the square root of the variance and we can write the signal/noise ratio of the ith sample as:

$$R = \frac{x(ih)}{\sigma_n}. \tag{9-98}$$

After M repetitions of the signal the summated value of the ith sample will be:

$$\sum_{k=1}^{M} y_k(ih) = \sum_{k=1}^{M} x_k(ih) + \sum_{k=1}^{M} n_k(ih)$$

$$= Mx_k(ih) + \sum_{k=1}^{M} n_k(ih). \tag{9-99}$$

That is, we can expect the signal samples to add but as the variance of all the ith noise samples are considered equal, the r.m.s. values of the sum of the noise

453

samples will be equal to $M\sigma_n^2$, so that the total r.m.s. value is $(\sqrt{M})\sigma_n$ and the signal/noise ratio after M repetitions will be:

$$R_n = \frac{Mx_k}{(\sqrt{M})\sigma_n} = (\sqrt{M})R. \qquad (9\text{-}100)$$

Thus the summation of M repetitions of the noisy signal will result in a $\sqrt{M}$ improvement in the signal/noise ratio.

There are two basic ways of carrying out this signal-averaging of continuous repeating wave-forms in the time domain. The first is known as 'Multi-Sweep

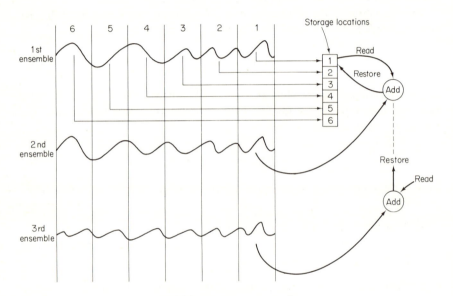

Fig. 9-33. Multi-sweep averaging.

Averaging' and is identical to the principle described above, i.e. a number of samples of the signal are taken uniformly along its time-history and stored in separate locations during the first repetition of the wave-form. Subsequent 'sweeps' across the signal selects the same number and relative time location of the samples and adds the value of these to the first set (fig. 9-33). The second method is known as 'Single Sweep Averaging'. Samples are taken after the same delay relative to the commencement of the signal with each repetition of the wave-form. These samples are summed and averaged to produce one final averaged point relative to a particular delay. The delay is then increased by one increment and the process repeated (fig. 9-34).

With multi-sweep averaging all the samples of the wave-form are simultaneously averaged so that an average of the complete wave-form is continuously available. Thus the signal/noise ratio is improved with each sweep of the wave-form. This type of averaging is equally effective in enhancing the signal/ noise ratio for a wide range of noise spectra, including low frequency $(1/f)$ noise.

Single sweep averaging is ineffective with low frequency noise since the noise component at each successive sample of the transient wave-form will be highly correlated. Consequently the mean level of the sampled wave-form will vary as the sampled delay is increased and during the slow sweep of the single point average this change will be reflected in the final averaged wave-form. It is however, fairly easy to implement using either analog or digital means and permits a long integration time per sample (since the summation for a given

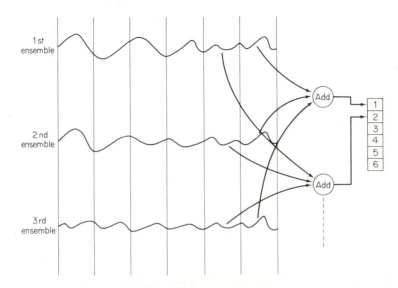

Fig. 9-34. Single-sweep averaging.

sample is carried out in parallel). In contrast the multi-averaging method requires counting of the sweeps and averaging by a different number for each sweep. This demands time-consuming logic calculations between each sample so that the process is generally fairly slow in implementation unless complex fast logic hardware is included [31].

The time resolution for multi-sweep averaging is determined by the total number of samples taken from the wave-form and is consequently limited to the analysis of fairly slow transients. Single sweep averaging can have time resolutions considerably shorter than this if long sweep times are acceptable. The maximum frequency determined by the multi-sweep averager is limited by the sampling theorem in contrast to that of the single sweep averager where the limit is determined by the width of the sampling pulse.

An analog method of single-sweep averaging uses the boxcar integrator shown in fig. 9-35. This consists of a linear gate and a passive RC integrating circuit. The gate is closed for a very short period after the same relative delay with each repetition of the input wave-form. This differs from multi-sweep averaging where integration proceeds over the entire sampling period. Since the width of the gate closure period can be made very short (less than 1 μsec) a high-time resolution can be achieved. The effective upper bandwidth of the boxcar integrator is defined as the reciprocal of this narrow gate closure period. Scanning control of gate delay relative to commencement of the repetitive wave-form permits the gated period to move slowly across the wave-form so that a single-sweep average of the signal is available at the amplifier output terminals. The input signal will need to be filtered to avoid aliasing effects so that its equivalent bandwidth is less than the upper bandwidth of the integrator.

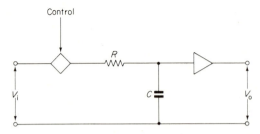

Fig. 9-35. A box-car integrator.

9-10-3   FREQUENCY DOMAIN CONSIDERATIONS

The methods discussed above consider the separation of a signal from a noisy wave-form in terms of the time domain. Often the unwanted wave-form is periodic and can be at a frequency quite close to that of the desired signal. In such cases the methods of signal averaging are still valid provided that the unwanted signal is not harmonically related to the desired signal. It will be instructive if we consider the process directly in frequency terms in order to derive some quantative information regarding signal enhancement. Viewed in this way the process will be seen to behave as a form of filtering in the frequency domain. If we consider $x(t)$ to be periodic with a period τ, then the summation of M repetitions of the wave-form is equivalent to the convolution of the input signal $y(t) = x(t) + n(t)$ with a pulse train defined as:

$$\sum_{k=1}^{M} \delta(t - k\tau)$$

where δ = a delta function,

i.e.
$$S_i = \int_{-\infty}^{\infty} y_k(ih - t) \left[\sum_{k=1}^{M} \delta(t - k\tau) \right] dt \qquad (9\text{-}101)$$

and since we are summing only those values of $y_k(ih)$ at specific sampling points τ units apart we can write:

$$S_i = \sum_{k=1}^{M} y_k(ih - k\tau). \tag{9-102}$$

This represents the output of the averaging system in the time domain. Its corresponding characteristics in the frequency domain can be derived from the translation of the impulse response of the system which is clearly:

$$h(t) = \sum_{k=1}^{M} \delta(t - k\tau) \tag{9-103}$$

and may be shown to be:

$$x_i(f) = 2(\tau) \frac{\sin \omega k\tau}{\omega k\tau}. \tag{9-104}$$

The function given by equation (9-104) is repeated over the frequency range for each discrete value of t (remembering that convolved time functions are equal to multiplied frequency functions), resulting in the comb filter shown in fig. 9-36. Repeated convolutions extend the range of k and result in narrower teeth for the comb filter. The situation is in fact identical with the result of filtering a finite length periodic signal noted earlier in section 8-4-2. The

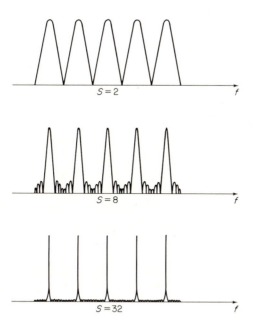

Fig. 9-36. Transfer function of a signal averaging system for increasing number of summations.

longer the signal to be filtered (in this case the sum of M repetitions) then the narrower will be the effective transfer function expressed in the frequency domain. We may also note that since the required signal, $x(t)$ is time synchronised with the repetition frequency then every frequency component of $x(t)$ coincides with the centre of one of the teeth of the comb filter. Consequently as M increases, the bandwidth of the filter is reduced and the signal/noise ratio is improved. A limit of about 60 db improvement in signal/noise ratio is set by the noise performance of the analysis equipment. As the value of M becomes greater the phase relationship between the signal and synchronising pulse becomes more important. Unless this is kept constant as M is increased the effective bandwidth of the comb teeth will widen and deteriorate the action of the signal averager. If jitter is present on the synchronising wave-form the signal/noise performance beyond a certain limit can actually be reduced with increased value of M long before the noise limit in the performance of the analyser is reached.

9-11   DETECTION OF ECHOES

The autocorrelogram is, as we have seen, an excellent detector for the periodic components of a composite signal containing a large random noise element. If the signal is of this form and the periodic components are few in number then this technique will give clear unambiguous results. However, much practical data, such as vocal signals and those deriving from periodic transient signals will have the repeating form shown in fig. 9-37 and be exceptionally rich in harmonics of the repetition frequency. Consequently although the autocorrelogram will indicate the periodic components it will not easily discriminate between their individual contributions. What happens is that the higher-frequency resonance will pass in and out of coherence many times as the time lag increases whilst the

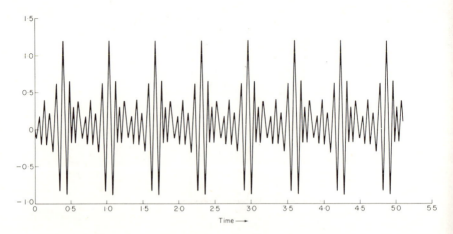

Fig. 9-37. Autocorrelogram of a harmonically rich signal.

458

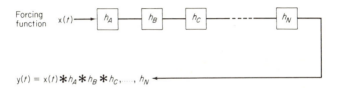

Fig. 9-38. Model for a repeated impure signal.

lower-frequency harmonic components will undergo a smaller number of co-herences. The result is a composite wave-form showing the modulation of harmonic components by each other which tends to obscure their periodicities. A Fourier transform applied to the result to determine the periodicities will bring us back to the power spectrum indicating relative frequency rather than the rate of frequency change needed to separate the harmonics.

The problem is one of identification of the pitch of a signal consisting of itself plus an echo, or rather a number of echoes [32]. The difficulty can be understood by considering an equivalent model for a harmonic generator subject to an input forcing function, $x(t)$ shown in fig. 9-38. Here $x(t)$ is applied through a series of units giving impulse response functions, $h_A, h_B, \ldots, h_N$ (which could be, for example, a series of resonant circuits). An output signal, $y(t)$ will consist of a convolution of $x(t)$ with each of the response functions:

$$y(t) = x(t)*h_A^* h_B^*, \ldots, *h_N. \qquad (9\text{-}105)$$

Transforming $y(t)$, $x(t)$, h_A etc., into the frequency domain gives:

$$Y(f) = X(f).H_A(f).H_B(f), \ldots, H_N(f) \qquad (9\text{-}106)$$

where $H_A(f)$ etc., are the transfer functions of the modulating units, i.e. $H_A(f) = F[h_A]\ldots$ etc., and $Y(f)$ and $X(f)$ are the Fourier transforms, $F[y(t)]$ and, $F[x(t)]$ respectively. To determine the effect of these separate units we could obtain the autocorrelation function:

$$R_y(T) = F[|Y(f)|^2], \qquad (9\text{-}107)$$

but this is equal to:

$$R_y(\tau) = F[|X(f)|^2.|H_A(f)|^2, \ldots, |H_N(f)|^2] \qquad (9\text{-}108)$$

and $F[|X(f)|^2]$ etc., are themselves autocorrelation functions, so we can write:

$$R_y(\tau) = R_x(\tau)*R_A(\tau)*, \ldots, *R_N(\tau). \qquad (9\text{-}109)$$

This method will clearly not affect a separation of the effects of units $A, B, \ldots, N$ since the result is a complex convolution of several correlation functions.

459

The solution is to take the logarithm of the Fourier transform of the power spectrum as suggested by Noll and others [32, 33]. Applying this to equation (9-108) gives:

$$\log |Y(f)|^2 = \log |X(f)|^2 + \log |H_A(f)|^2 + \log |H_B(f)|^2 + \ldots,$$

$$+ \log |H_N(f)|^2 \qquad (9\text{-}110)$$

which effectively separates the contribution of the different modulating units. To determine the relative frequency (pitch) of these terms a further Fourier transform is carried out on the logarithm of the power spectrum thus:

$$K_y(\tau) = F[\log |Y(f)|^2] = F[\log |X(f)|^2]$$

$$+ F[\log |H_A(f)|^2] + \ldots, + F[\log |H_N(f)|^2] \qquad (9\text{-}111)$$

where $K_y(\tau)$ is known as the Cepstrum (pronounced 'kepstrum') of the function $y(t)$. To obtain a measure of the power contained in the signal at different relative frequencies the power spectrum of this logarithmic power spectrum is taken. Since the logarithmic power spectrum is an even function of frequency we can use the cosine transform representation and write:

$$K_y(\tau) = \left[\int_0^\infty \log |Y(f)|^2 \cos(\omega\tau).d\omega \right]^2. \qquad (9\text{-}112)$$

(Note the change of domain compared with equation (6-39).)

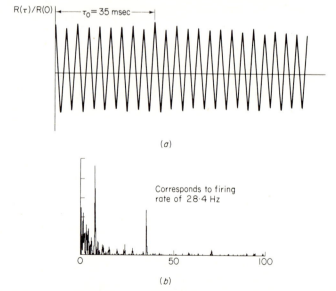

Fig. 9-39. Comparison between the autocorrelogram (a) and cepstrum analysis (b) obtained from the same repeated transient.

460

For digital computation equation (9-112) would be carried out using the F.F.T., so that the data will need to be band-limited before transformation, and smoothed after transformation and squaring. This new function has the dimensions of time like the autocorrelation function, but measures, in fact, the repetitive pitch of the harmonic contents of the signal in cycles/Hertz. Since this also gives a measure of the frequency of pitch it is sometimes referred to as quefrequency to suggest frequency in this special sense [32]. This technique has been applied successfully to determine the periodicity of an acoustic signal derived from an automobile engine [34]. In his paper Dr Thomas discusses the difficulty in making accurate estimates of firing rate from an examination of the autocorrelogram above, shown in fig. 9-39a. The result of applying cepstrum analysis to the same repeating transient signal are shown in fig. 9-39b (also reproduced from Dr Thomas's paper) and show quite clearly the signal corresponding to the engine firing rate at a position well removed from other signals which refer to a random noise element.

9-12   SYSTEMS TESTING

Mechanical systems, particulary complex cyclic mechanisms are often studied from the characteristics of the mechanical vibrations set up by the normal operation of the mechanism. Two general approaches to this study are apparent. The first involves signature analysis in which the characteristics of a vibration time-history are compared during the operational life of the mechanism to determine departure from a 'normal' signature obtained using a process of time-averaging [35].

The second uses cross-correlation techniques to measure the transfer function or mechanical impedance of the mechanism [36].

Time-averaging has been discussed in section 9-10-2 and, as applied to a group of signals obtained from a mechanism functioning normally, will enable small changes in performance to be detected long before catastrophic failures occur [37]. The technique can therefore be used to predict and diagnose malfunction in the operating mechanisms.

Comparison between the normal signature and that obtained during later operation can use statistical techniques, spectral density analysis of the difference wave-forms, or the process of cross-correlation. This latter technique is also used in the determination of system transfer function. A major advantage of correlation techniques is that they can be applied without disturbing the normal operation of the mechanical system. Earlier methods using classical Bode plotting of the amplitude and phase characteristics from a variable frequency drive generator have a number of disadvantages, not least of which is the need to carry out these measurements off-line. This is also true of impulse response methods which, although considerably faster than the Bode plot, introduce problems of implementation. The method of impulse response testing

461

is to examine the behaviour of the system in response to a single Dirac delta function applied at its input. The delayed impulse response that is produced at the output can be analysed to give the frequency amplitude and phase response of the system. Unfortunately it is rarely possible to investigate the characteristics of the system in this way. Either the pulse will result in overload conditions modifying the measurement or it will have to be reduced to such a small amplitude that the output produced will be masked by the system noise. It is also extremely difficult to maintain the energy spectrum of the impulse constant over the system bandwidth.

We saw earlier that the autocorrelation of a wide-band noise signal approximates to an impulse function and this fact can be used in an equivalent form of impulse testing. The method is shown in fig. 9-40. If a wide-band noise signal is

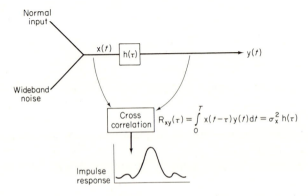

Fig. 9-40. Determination of system impulse response by cross-correlation.

applied to the system input the cross-correlation of this with the output will be proportional to the impulse response function of the system. Mathematically, if $x(t)$ is the noise input signal and $h(v)$ is the required impulse response function, then the output signal obtained is from equation (7-9):

$$y(t) = \int_{-\infty}^{\infty} h(v)x(t-v)\,dv \qquad (9\text{-}113)$$

Cross-correlating $y(t)$ with the input noise $x(t)$ gives:

$$R_{xy}(\tau) = \lim_{T \to \infty} \frac{1}{T} \int_0^T x(t)y(t+\tau)\,dt \qquad (9\text{-}114)$$

so that substituting for $y(t+\tau)$:

$$R_{xy}(\tau) = \lim_{T \to \infty} \frac{1}{T} \int_0^T x(t)\,dt \int_{-\infty}^{\infty} h(v)x(t-v+\tau)\,dv.$$

By inverting the order of integration we have:

$$R_{xy}(\tau) = \int_{-\infty}^{\infty} h(v) \, dv \, \lim_{T \to \infty} \frac{1}{T} \int_{0}^{T} x(t)x(t - v + \tau) \, dt, \qquad (9\text{-}115)$$

but the autocorrelation for $x(t)$ is:

$$R_{xx}(\tau - v) = \lim_{T \to \infty} \frac{1}{T} \int_{0}^{T} x(t)x(t - v + \tau) \, dt$$

so that equation (9-115) can be written:

$$R_{xy}(\tau) = \int_{-\infty}^{\infty} h(v)R_{xx}(\tau - v) \, dv \qquad (9\text{-}116)$$

(known as the Wiener–Lee relationship).

Thus the cross-correlation function for the system is the convolution of its impulse response function with the autocorrelation of the input wave-form to the system. Since the autocorrelation of the wide-band noise input, $x(t)$, approximates to a true Dirac impulse function multiplied by a constant, K, equivalent to its area, we can write:

$$R_{xy}(\tau) = K \int_{-\infty}^{\infty} h(v)\delta(\tau - v) \, dv \qquad (9\text{-}117)$$

which, from equation (7-6), is equal to:

$$R_{xy}(\tau) = Kh(t) \qquad (9\text{-}118)$$

where K is the r.m.s. value of the noise and $h(t)$ is the impulse response of the system.

This can also be expressed in the frequency domain by taking the Fourier transform of both sides of equation (9-116):

$$\int_{-\infty}^{\infty} R_{xy}(\tau) \exp(-j\omega\tau) \, d\tau = \int_{-\infty}^{\infty} \exp(-j\omega\tau) \, d\tau \int_{-\infty}^{\infty} h(v)R_{xx}(\tau - v) \, dv.$$
$$(9\text{-}119)$$

By definition:

$$\int_{-\infty}^{\infty} R_{xy}(\tau) \exp(-j\omega\tau) \, d\tau = G_{xy}(f) \qquad (9\text{-}120)$$

and by carrying out a similar rearrangement to that shown in equation (9-115) then equation (9-119) yields:

$$G_{xy}(f) = H_{xy}(f) \cdot G_{xx}(f) \qquad (9\text{-}121)$$

where $H_{xy}(f)$ is the transfer function for the system.

Hence the cross-spectrum obtained from the input and output of the system is equal to the product of the system transfer function and the spectrum of the input signal.

Equations (9-118) and (9-121) form extremely valuable methods of measuring the complex transfer function of a system without disturbing its normal mode of operation. In addition it can be shown [8] that such measurements are immune from extraneous noise, providing this noise is not correlated with the input noise, $x(t)$.

9-12-1   PSEUDO-RANDOM NOISE

The characteristics of the random noise required for the above method is defined by equation (9-118) and implies that it must contain equal amounts of all frequencies. This is known as 'white noise' by analogy with white light. True white noise cannot be realised since an infinite bandwidth would be needed. It is generally sufficient to specify a level power spectrum density over the frequency band of interest. Generation of a random noise signal can be obtained by utilising the random behaviour of gaseous discharges or semi-conductive materials.

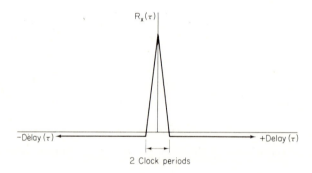

Fig. 9-41. Autocorrelation of pseudo-random noise.

These 'natural' methods of noise generation have one major disadvantage which is their unrepeatable random behaviour. It will be necessary to average the result over a very long interval to establish a small variance value. This will be the case for low-frequency measurements and result in long computational times.

A test signal which avoids these difficulties consists of a repeated pattern of random noise having a pre-determined power spectrum and amplitude probability distribution. This is known as pseudo-random noise. If the measurement time is made equal to a multiple of the pattern length then a variance of zero will be achieved resulting in short measurement times and high statistical accuracy.

Pseudo-random sequences are generated as a two-level (binary) signal using shift register techniques [38, 39]. If an analog pseudo-random signal is required then low-pass filtering of the binary sequence is carried out.

The autocorrelation function of a pseudo-random binary noise signal is triangular with a base width equal to 2 sampling periods (fig. 9-41). It thus forms a close approximation to the Dirac function required by equation (9-118) and is an ideal input noise generator for the testing method discussed above. The fact that the signal is periodic is not apparent in the correlated signal, providing the correlation window is made an integer multiple of the repeating sequence length.

Bibliography

1. WAINSTEIN, L. A. and ZUBAKOV, V. D. *Extraction of Signals from Noise.* Prentice-Hall, 1962.
2. TRUSCOTT, J. R. Approaches to seismological array processing. *I.E.R.E. Symposium,* London, May 1964.
3. PADÉ, H. Sur la représentation approchée d'une function par des fractions rationelles. *Ann. l'Ecole Normale* 9, 1-93, 1892.
4. HOLST, P. A. Padé approximations and analog computer simulation of time delays. *Simulation* 12, 6, June 1969.
5. ROGERS, A. E. and CONNOLLY, T. W. *Analog Computation in Engineering Design.* McGraw-Hill, New York, 1960.
6. STONE, R. S. and DANDL, R. A. A variable function delay for analog computers. *I.R.E. Trans.* (Elec. Comp.) EC-6, 187-9, Sept. 1957.
7. ANSTEY, N. A. and LERWILL, W. E. Correlation in real time. *Proc. Roy. Soc.,* A. 290, 430-45, 1966.
8. LEE, J. W. *Statistical Theory of Communication.* John Wiley, New York, 1960.
9. LERNER, R. M. A matched filter detection system for complicated doppler-shifted signals. *I.R.E. Trans.* (Inf. Theory) IT-6, 373-85, June 1960.
10. OTTERMAN, J. The properties and methods for computation of exponentially-mapped-past statistical variables. *I.R.E. Trans.* (Auto-Control) AC-5, 7, 2-6, Jan. 1960.
11. HUTCHINS, W. H. A real-time seismic array data analyser and its associated event selector. *J. Br. I.R.E.* 31, 5, 293-309, May 1966.
12. ROBINSON, E. A. *Multi-Channel Time Series Analysis with Digital Computer Programs.* Holden-Day Inc., San Francisco, 1967.
13. MELTON, B. S. and KARR, P. R. Polarity coincidence scheme for revealing signal coherence. *Geophysics* 23, 3, 553, 1957.
14. SCHMID, L. P. Efficient autocorrelation. *Comm. A.C.M.* 8, 115, Feb. 1965.
15. WIDROW, B. Statistical analysis of amplitude quantised sample-data systems. *A.I.E.E. Fall General Meeting, Chicago,* 555-68, Jan. 1961.
16. WIDROW, B. A study of rough amplitude quantisation by means of Nyquist sampling theory. *I.R.E. Trans.* (Circuit Theory) CT-3, 4, 266-76, Dec. 1956.
17. KORN, G. A. *Random Process Simulation and Measurements.* McGraw-Hill, New York, 1966.
18. VAN VLECK, J. H. The spectrum of clipped noise. Rep. 51, Radio Research Laboratory, Harvard University, 1943.
19. WEINREB, S. A digital spectral analysis technique and its applications to radio astronomy. *M.I.T. Res. Lab. Electronics, Tech Rep.* 412, Aug. 1963.
20. JESPERS, P., CHU, P. T. and FELTIVEIS, A. A new method for computing correlation functions. International Symposium on information theory, Brussels, Belgium, 1968, also *I.R.E. Trans.* (Inf. Theory) IT-8, 106, Sept. 1962.
21. SANDE, G. An alternative method for calculating covariance functions, Princetown Univ. Comp. Center Memorandum. Princetown, N.J. 1965.

22. CLYNES, M. E. The CAT – computer of average transients. *Med. Elec. News.* June 1962.
23. SAPER, L. M. On-line auto and cross-correlator realised with hybrid computer techniques. *Biomed. Elec. Session I.E.E.E. Convention,* 120-30, 1963.
24. BEAUCHAMP, K. G. A hybrid linkage system. *Proc. Fourth Aust. Comp. Conf.,* Adelaide, Aug. 1969.
25. INSTON, H. H., NOYES, J. G. and READER, N. L. S. A description of the A.W.R.E. hybrid data processing system with some applications. *Proc. Fourth A.I.C.A. Conf.,* Brighton, 1964.
26. BEKEY, G. A. and KARPLUS, W. J. *Hybrid Computation.* John Wiley, New York, 1968.
27. PAUL, R. J. A. Hybrid methods for function generation. *Proc. Third A.I.C.A.* Congress, Opatija, 501-12, 1961.
28. LEE, Y. W., CHEATHAM, T. P. and WIESNER, J. B. Application of correlation analysis to the detection of periodic signals in noise. *Proc. I.R.E.* **38**, 1165-71, 1950.
29. STEWART, J. L. and WESTERFIELD, E. C. A theory of active sonar detection. *Proc. I.R.E.* **47**, 872, May 1959.
30. ERNST, R. R. Sensitivity enhancement in magnetic resonance I. Analysis of the method of time averaging. *Rev. Sci. I.* **36**, 12, 1689-95, Dec. 1965.
31. CLYNES, M. and JOHN, M. Portable four-channel on-line digital average response computer CAT. *Proc. Fourth Int. Conf. Med.* Electron. N.Y.C., N.Y. July 1961.
32. BOGERT, B. P., HEALY, M. J. R. and TUKEY, J. W. The quefrequency analysis of time series for echoes, cepstrum, pseudo-autocovariance, cross-cepstrum and saphe cracking. In *Proceedings of the Symposium on Time Series Analysis,* Ed. M. Rosenblatt, John Wiley, New York, 1963.
33. NOLL, A. M. Cepstrum pitch determination. *J. Acoust. Soc. Amer.* **41**, 2, 293-309, 1967.
34. THOMAS, D. W. and WILKINS, B. R. Determination of engine firing rate from the acoustic wave-form. *Elec. Letters* **6**, 7, 193-6, 1970.
35. HOFFMAN, R. L. and FUKUNGA, K. Pattern recognition signal processing for mechanical diagnostics signature analysis. *I.E.E.E. Trans.* (Comp.) C-20, 1095-100, Sept. 1971.
36. MERCER, C. and CLARKSON, B. C. Use of cross-correlation in studying the response of lightly damped structures to random forces. *A.I.A.A. J.* Dec. 1965.
37. LAVOID, F. J. Signature analysis: Product early warning system. *Mach. Des.,* 41, 151-60, Jan. 1969.
38. GILSON, R. P. Some results of amplitude distribution experiments on shift-register generated pseudo-random noise. *I.E.E.E. Trans.* (Elec. Comp.) EC-15, 926, Dec. 1966.
39. HUTCHINSON, D. W. A new uniform pseudo-random number generator. *Comm. A.C.M.* **9**, 432-3, June 1966.

Additional references

40. STEWART, J. L. *An Introduction to Signal Theory.* McGraw-Hill, London, 1960.
41. RICE, S. O. Mathematical analysis of random noise. *Bell. Syst. Tech. J.* 23 July, 1944; 24 Jan. 1945.

42. JANIAC, K. A direct determination of correlation functions for generators of random processes. *Proc. Third A.I.C.A.* Congress, Opatija, 242-9, 1961.

43. KELLY, R. D., ENOCHSON, L. D. and RONDIVELLI, L. A. Techniques and errors in measuring cross-correlation and cross spectral density functions. *N.A.S.A. Rep.,* CR-74505, 1967.

44. ENOCHSON, L. D. and OTNES, R. K. Programming and analysis for digital time series data. Shock and vibration monograph series, SVM-3. Dept. Navy, Office of Naval Res., May 1969.

45. STOCKHAM, T. G. High-speed convolution and correlation. *A.F.I.P.S. Proc. Spring Joint Comp. Conf.* **28**, 229-33, 1966.

46. McCOWAN, D. W. Finite Fourier transform theory and its applications to the computation of convolutions, correlations and spectra, Earth Sciences Division, Teledyne Industries Inc., Virginia, U.S.A. Tech. Memo 8-66, Dec. 1966.

47. RYALL, A. and BIRTELL, J. Digital processing of array seismic recordings. U.S. Dep. Interior, Denver, Colorado, Tech. Letter, No. 2., 1962.

48. HAGAN, T. G. and TREIBER, R. Hybrid analog/digital techniques for signal processing applications. *A.F.I.P.S. Proc. Spring Joint Comp. Conf.* **28**, 379, 1966.

49. DAVENPORT, W. B. Correlator errors due to finite observation intervals. *Tech. Rep. 191,* MIT Res. Lab. of Electronics, Mar. 1951.

50. REICH, E. and SWERLING, P. The detection of a sine wave in Gaussian noise. *J. Appl. Phys.* 24, 289-96, Mar. 1953.

51. BECKER, C. L. and WAIT, J. V. Two-level correlation in an analog computer. *I.R.E. Trans.* (Elec. Comp.) EC-10, 752-8, Dec. 1961.

52. EKRE, H. Polarity coincidence detection of a weak noise source. *I.E.E.E. Trans.* (Inf. Theory) IT-9, 18-23, Jan. 1963.

53. McFADDEN, J. A. The correlation function of a sine wave plus noise after extreme clipping. *I.R.E Trans.* (Inf. Theory) IT-2, 2, 82-3, June 1956.

54. HAMPTON, R. L. A hybrid analog-digital pseudo random noise generator. *Simulation* **4**, 179-89, Mar. 1965.

55. HULL, T. E. and DOBELL, A. R. Mixed congruential random number generator for binary machines. *J. A.C.M.* **11**, 31-40, 1961.

56. WHITE, R. C. Experiments with digital computer simulations of pseudo-random noise generators. *I.E.E.E. Trans.* (Elec. Comp.) EC-15, 355-6, June 1967.

57. PETERSON, R. H. and HOFFMAN, R. L. A new technique for dynamic analysis of acoustical noise. *I.B.M. J. Res. Dev.* **9**, 205-9, May 1965.

58. FISHER, M. J. and KRAUSE, F. R. The cross-beam correlation technique. *J.* processes on hybrid computers. *Proc. Fifth A.I.C.A. Congress*, Munich 1,

59. CUENOD, M., DURLING, A. and VALISALO, P. E. Analysis of random processes on hybrid computers. *Proc. Fifth A.I.C.A. Congress*, Munich, Aug. 1968.

60. BRIGGS, P. A. N., HAMMOND, P. H., HUGHES, M. T. G. and Plumb, G. O. Correlation analysis of process dynamics. *Proc. I. Mech. Eng.* **179**, (3H), 37, 1964.

Chapter 10

TRANSIENT ANALYSIS AND THE SHOCK SPECTRUM

10-1 TRANSIENT VIBRATION OF MECHANICAL STRUCTURES

Shock phenomena are encountered in the study of system behaviour and many natural physical occurrences. Examples are found in environmental engineering, supersonic motion, earthquakes, explosions, and other spasmodic releases of energy. Shock is a transient condition and defines the transmission of energy over a period which is short compared with the natural period of the system which absorbs the shock energy. Thus the system is likely to be relatively inactive during the actual period of the applied shock but will respond in a resonant fashion immediately following the cessation of the initial transient.

A typical shock waveform derived from a mechanical structure subject to an impulse excitation is as shown in fig. 10-1. This shows a fairly rapid build-up of vibration followed by a finite decaying oscillatory wave-form having a duration dependent on the property of the system to dissipate the energy in other forms, i.e. its damping characteristic. Periodic excitation, on the other hand, will permit a build-up of the system natural or forced resonance to a peak value and will maintain this condition for as long as the periodic force is applied. A third form of excitation, that of a long but random value forcing function can also cause sustained resonances to be obtained.

Analysis of periodic or randomly excited structures using the methods developed in earlier chapters will enable some useful information to be derived concerning resonant frequencies, transmissibility, fatigue damage, etc. This chapter will consider primarily signal analysis where the applied or resultant energy is a short-lived transient defined broadly as shock and which may not be repeated, certainly not in an identical form. The treatment will be confined to mechanical shock throughout although any of the results are applicable to other forms of shock phenomena.

If we consider a short-lived shock imparted to a mechanical structure this will give rise to two problems relating to its effects and the nature of the shock itself. In the first place the acquired signal derived from transducers attached to the structure will have an unknown statistical reliability since no comparative

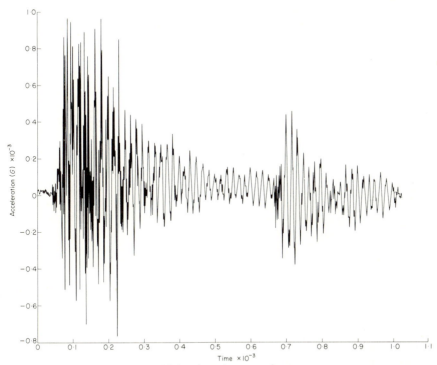

Fig. 10-1. A transient wave-form.

records will be available — every transient will be unique. Secondly the characteristic of the forcing function may not be known very accurately. Consequently the analysis of the effects of transient shock has proceeded on rather different lines to those arising from periodic or sustained random excitation. It is necessary to relate the analysis more closely to the mechanical behaviour of the structures themselves under a given set of environmental conditions. Conventional spectral analysis methods may fail, for example, by providing too much information relating to the total response of a complex structure extending over a given time period, whereas what we wish to know are those frequencies at which certain mechanical limit conditions are exceeded (e.g. fracture or collapse). A new set of definitions is needed to provide this sort of information so as to increase our understanding of the performance of structures under stress conditions. These will be developed in the following pages commencing with a study of resonant effects in ideal structures.

10-2 RESONANCES

Let us consider the behaviour of a mechanical structure subject to vibration or shock. The applied force may be transmitted from outside the structure resulting from transportation or perhaps from contact with some fixed vibration source.

469

Alternatively the structure itself may contain one or more sources of vibration. Whatever the source the structure will not be found to respond in a linear manner to vibration. Firstly it is likely to behave as a series of mechanically linked sub-structures, each responding separately to the exciting source. Secondly these sub-structures will each exhibit the phenomena of resonance, each resonating at a particular frequency or set of frequencies.

The way in which resonance affects the movement and acceleration of the sub-structure will depend on its mechanical characteristics. The mechanical engineer will want to know the precise details of the behaviour of the structure and sub-structure and, in particular, he will want to know if it is likely to exceed an allowable limit of movement or acceleration and if damage can occur due to the applied excitation force. The frequency of excitation is not in itself productive of strain or damage within the structure. It is the resonances set up by the primary excitation which will cause the excess motion. The ability of resonance to increase the likelihood of structural damage becomes apparent when it is realised that a small displacement of a large structure can be magnified several hundred times when coupled to a highly resonant and undamped sub-structure. It is for this reason that resonance is regarded as a suitable criteria for the measurement of the severity of excitation by shock and vibration forces. A study of resonance is central, therefore, to the understanding of the effects of vibration and also in devising methods of data acquisition and analysis from the vibrating mechanical structure.

Where the structure vibrates as one unit independent of the frequency range of the applied force, we regard this as a Lumped-Constant system. More usually the applied shock or vibration travels along the structure exciting various parts of it into resonance at different times. This would also be the case where the disturbance, applied at one extremity of the structure, has a high frequency component since the total mass would be unable to respond instantaneously at this rate. Such structures are referred to as Distributed-Constant systems.

10-3   ANALYSIS OF A SIMPLE MECHANICAL RESONATOR

Analysis can be carried out more readily by using a simplified model of a lumped-constant system. Extension of this model to the practical distributed case can then be made by consideration of a number of lumped-constant systems linked together in some way. A simple mechanical resonator which we can take as this basic model is the seismic mass system already referred to in chapter 2. This is shown in fig. 10-2 and consists of a mass M suspended by a spring k with its motion controlled by a form of viscous damping, c. The motion of the mass is considered to be constrained in a single degree of freedom as shown. The equation of motion for this model is:

$$Mp^2z + cpz + kz = -Mp^2x \qquad (2\text{-}5)$$

470

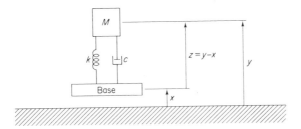

Fig. 10-2. Single degree of freedom mechanical resonator.

where z = relative displacement between base and mass = $y - x$, and p = operator d/dt.

The complete solution is given by the sum of the complementary function and the particular integral. The former represents the solution when x and its derivatives are zero and is called the Transient solution. The value of the particular integral is dependent on the form of the input acceleration, p^2x, and gives the Steady-State solution.

10-3-1 STEADY-STATE ANALYSIS

This has been derived earlier for a sinusoidal base excitation of infinite duration, $x = X \sin \omega t$ where the ratio of applied and relative displacement of the mass was shown to be (equation (2-9)):

$$\frac{Z}{X} = \frac{(\omega/\omega_n)^2}{[[(1-(\omega/\omega_n)^2]^2 + [2D\omega/\omega_n]^2]^{\frac{1}{2}}} \qquad (10\text{-}1)$$

where: $\omega_n = \sqrt{k/M}$ natural resonant frequency of the spring-mass system, and $D = c/(2\omega_n M)$, the damping factor.

The ratio Z/X defines the transmissibility of the applied displacement from base to mass and is plotted in fig. 10-3 for various damping coefficients. The magnification of base displacement is seen to be very dependent on the damping value and indicates clearly the possibility of structural damage due to enhanced displacement of the suspended mass.

It will be appreciated that the form of equation (2-5) is such that it can equally well represent the equation of motion in terms of velocity or acceleration as well as displacement. Thus equation (10-1) can also represent a transmissibility ratio if applied to resulting mass velocity Z_v/X_v or acceleration Z_a/X_a.

An approximate value of the damping factor, D can be obtained from measurements of the frequency response function, $Z/X = H(\omega)$. The width of the resonant peak at the half-power points,

$$|H(\omega)| = (\tfrac{1}{2}|H(\omega_0)|^2)^{\frac{1}{2}} = 0.707\,H(\omega_0)$$

471

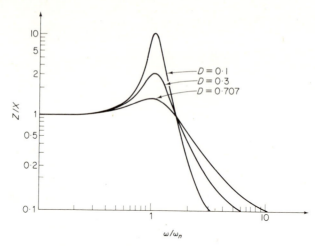

Fig. 10-3. Steady-state response of resonator.

is taken. If we let this width be $\delta\omega$, the value of D may be obtained from:

$$D = \frac{\delta\omega}{2\omega_0}.$$
(10-2)

For D values smaller than 0·1 the approximation is better than ±3% of the actual value.

10-3-2 TRANSIENT ANALYSIS

In many practical situations the maximum value of resonance shown as the positive peak in the curves of fig. 10-3 will not be reached. This is because the applied excitation to the base will last for only a short time and cannot be considered as approaching the steady-state case which, it will be remembered, is defined for an infinitely long periodic signal. The form of analysis required in this case is the transient solution.

It is assumed that an impulse has been applied to the base and that the energy contained within it has been transferred to the resonator, which is now free to execute some form of damped motion. We are therefore interested in the solution to the equation (2-5) when x and its derivatives are zero thus:

$$Mp^2y + cpy + ky = 0.$$
(10-3)

Its solution is given as:

$$y = A_1 \exp(m_1 t) + A_2 \exp(m_2 t)$$
(10-4)

where m_1 and m_2 are the roots of the auxilliary equation:

$$Mm^2 + cm + k = 0$$
(10-5)

and A_1 and A_2 are dependent on the initial displacement condition for the mass.

472

Hence, we can write for m_1 and m_2:

$$\left. \begin{array}{l} m_1 = \dfrac{-c}{2M} + \tfrac{1}{2} \sqrt{\left(\left(\dfrac{c}{M}\right)^2 - \dfrac{4k}{M}\right)} \\[4mm] m_2 = \dfrac{-c}{2M} - \tfrac{1}{2} \sqrt{\left(\left(\dfrac{c}{M}\right)^2 - \dfrac{4k}{M}\right)}. \end{array} \right\} \tag{10-6}$$

Three cases can be considered:

1. When $c^2 > 4kM$. The roots of equation (10-5) are now real and independent. On application of the transient excitation the mass will move in one direction until equilibrium is obtained. It will represent a 'stiff' movement and will be highly damped.

2. When $c^2 = 4\,kM$. The roots are now equal and negative. The movement of the mass will be more rapid and will stop just short of an oscillatory response. Any further reduction in the damping term, c will result in an overshoot and an oscillatory motion. The damping term:

$$C_c = \sqrt{(4kM)} \tag{10-7}$$

is referred to as 'Critical Damping' and the ratio c/C_c as the damping factor, D, used in equation (10-1).

3. When $c^2 < 4\,kM$. This is the more usual case and represents an oscillatory solution having complex roots, $m = r \pm js$, where:

$$r = -\,c/2M$$

and:

$$s = \sqrt{(k/M - c^2/4M^2)}.$$

The general solution is from equation (10-4):

$$\begin{aligned} y &= A_1 \exp{(r + js)t} + A_2 \exp{(r - js)t} \\ &= \exp{(rt)}[A_1 \exp{(jst)} + A_2 \exp{(-jst)}]. \end{aligned} \tag{10-8}$$

Replacing $\exp{(\pm jst)}$ by its harmonic relation gives:

$$y = K \exp{(rt)}.\cos{(st)}. \tag{10-9}$$

Thus we would expect to find an exponentially decaying cosinusoidal response to a base excitation with the rate of decay proportional to the damping coefficient. It could equally well be shown that an exponentially decaying sinusoidal response is another solution. A general transient solution can be stated as:

$$y = \exp{(rt)}[A \sin{\omega t} + B \cos{\omega t}]. \tag{10-10}$$

Here the angular frequency of oscillation $s = \omega$ is defined as:

$$\omega = \sqrt{\left(\dfrac{k}{M} - \dfrac{c^2}{4M^2}\right)}. \tag{10-11}$$

In the case of a completely undamped system ($c = 0$), then the frequency of oscillation represents the natural angular resonant frequency:

$$\omega_n = \sqrt{\left(\frac{k}{M}\right)} \qquad (10\text{-}12)$$

The three transient solutions are illustrated in fig. 10-4.

The transient solution represents the resonator response after the cessation of the input excitation and is of interest in considering the response of a structure to shock excitation providing that the simple model can be accepted. It is not directly of value for a complex structure and more powerful spectral methods will be described later in this chapter.

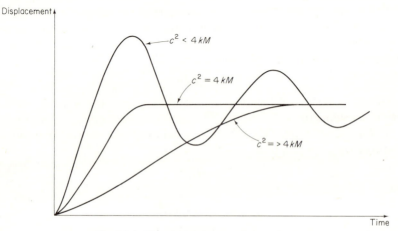

Fig. 10-4. Transient response of resonator.

In the case of periodic excitation the steady-state solution is applicable—but only if its duration is long compared with the decay response time, $1/r$. Even for a long periodic excitation transient conditions will apply at the beginning and end of the applied forcing function unless the force is applied and removed gradually. The situation is somewhat similar to the transients induced by truncation of a continuous wave-form, described in chapter 8 in connection with frequency analysis of a finite length signal.

10-4 VIBRATION AND SHOCK TESTING

In order to derive the characteristics of a mechanical structure when subjected to any applied vibration it is convenient to determine its behaviour for a number of well-defined vibration patterns. Analysis of the effect of these controlled and repeatable excitations can then provide information on the behaviour of the structure in a simplified form of the real environment.

474

One such method, used to determine the steady-state characteristic, is to vibrate the structure sinusoidally at a single frequency, which may be changed slowly with time, to cover the entire range of frequencies of interest. The extent of the vibration can be controlled in terms of displacement, velocity or acceleration. The applied excitation will be a force normal to the structure surface acting either directly by attaching the structure to the source or indirectly by means of magnetic forces. The vibrating source may be actuated by mechanical, hydraulic or electromagnetic forces.

Mechanical vibration can be obtained by attaching an eccentric mass to a rotating shaft so that the resultant out-of-balance forces are transmitted through the shaft to the mounted structure. Alternatively an eccentric crank drive can be coupled directly to the structure. In either case the frequency of excitation will be low and the method is limited to a fairly narrow range of excitation characteristics.

Hydraulic methods involve the attachment of a double-acting piston which is caused to oscillate by the admission of hydraulic fluid alternately on either side of the piston through controlling valves. The frequency of excitation can be considerably higher than with mechanical vibrators, limited only by the compliance of the hydraulic fluid. The method is a powerful one for 'on site' testing since the hydraulic vibrators can be physically small and yet provide a high oscillatory force.

Electrodynamic methods offer the most flexible source of vibratory power. The vibrator resembles a large dynamic loud-speaker having a rigid platform in place of the moving diaphragm. The structure undergoing test can be bolted to the platform and vibrated at a rate determined by the applied driving signal. The platform is attached to a driving coil, free to move within a narrow gap, across which a powerful magnetic field is maintained. Interaction of this field with an alternating current flowing in the coil creates an alternating force which vibrates the coil and table at a rate determined by the frequency of the alternating current. This type of vibrator, commonly referred to as a 'shake table', is capable of accurate control and can generate a wide range of vibration characteristics. A brief discussion of control methods is given below.

10-4-1 CONTROL AND OPERATION OF THE ELECTRODYNAMIC SHAKE TABLE

A simple form of manual control arrangement is given in fig. 10-5. The operator observes the output level indicating vibration intensity and controls the excitation amplitude to maintain the desired level of vibration at the frequency of test.

In addition to its use for determining steady-state characteristics the vibrator can be used for resonant frequency search and identification. However, manual adjustment of frequency and level is limited and time consuming and some form of swept frequency control is employed using feedback to maintain constant excitation level, fig. 10-6.

Here the level of acceleration is determined from the output of an acceleration transducer mounted on the load platform. This is used to control the excitation amplitude to the shaker power amplifier. This arrangement can result in a considerable error term being generated by noisy systems [1] and lead to less stringent test conditions being applied than are indicated by the transducer response. It can be shown that this error term is proportional to the sum of the

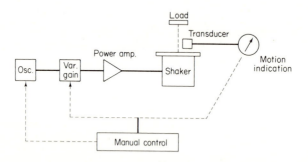

Fig. 10-5. Manual-controlled shake table.

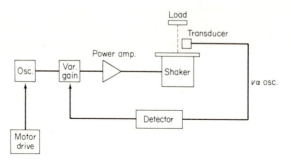

Fig. 10-6. Servo-controlled shake table.

amplitudes of the excitation frequency harmonics and can be minimised if a narrow-band filter is included in the feedback path between the accelerometer and detector as shown in fig. 10-7. The centre frequency of the filter will need to be adjusted for each new excitation frequency and in this form it is referred to as a Tracking Filter [2]. Control methods suitable for an analog tracking filter are described in chapter 3. In digital controlled systems a digital filter is used. This gives far closer control of frequency and phase than is possible with the analog tracking filter enabling dynamic control ranges of the order 70db to be obtained [3]. A major advantage arising from the use of tracking filters in sinusoidal wave-form testing is the standardisation obtained, since the results are

476

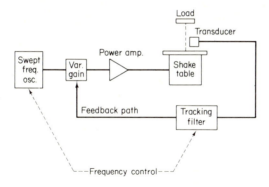

Fig. 10-7. Use of a tracking filter.

no longer dependent on differences in types of detector and servo-control. This permits results obtained on different test rigs to be usefully compared.

This form of excitation is widely used as a means of detecting resonances in structures. Selection of the sweep rate is important since this can effect the maximum response of the structure or sub-structure quite considerably. This can be seen if we consider a given sweep rate of R Hz/sec defining the time, T_B that the excitation frequency remains within the resonant bandwidth as $T_B = B/R$. The relation of T_B with the structure response time, t_r determines the maximum response obtained. Thus if:

1. $T_B \gg t_r$, steady-state conditions apply and the maximum response is approximately Q times the applied amplitude.

2. $T_B \ll t_r$; a small maximum response is obtained which is independent of Q, since little energy is transferred during the resonant period.

3. $T_B = t_r$; transient response conditions apply and a beating effect will occur which can result in a maximum response of up to Q times the applied amplitude. Under suitable conditions a continuously increasing drive frequency can be equated to a finite width impulse and used to determine the impulse response function for the system by means of cross-correlation techniques (see chapter 9). This is desirable for two reasons. Firstly it may not be possible even to approximate the required driving impulse function directly, due to the sluggish response of the controlled shake table. Secondly the application of such a rapid transient to a structure may result in damage caused by the high accelerations produced.

Random noise vibration tests are used to simulate many real environments, such as those concerned with jet noise or missile flights. A white noise generator producing essentially Gaussian noise takes the place of the variable frequency oscillator shown in fig. 10-7. The dynamics of the shake table will preclude the noise from extending over a very wide range so that the practical vibration characteristics obtained will not be truly 'white'. This can result in poor correlation between similar tests carried out on different equipment tests. The use of pseudo-random noise will be productive of more reliable results which are capable of being exactly repeated.

Where a single electrodynamic vibrator is used for resonant mode search it may give a poor indication for structures having closely adjacent resonant frequency modes. The use of several sources of excitation applied at different parts of the structure permits the excitation of a given mode and suppression of others and has been described in some detail by Taylor *et al.* [4].

10-5 IMPULSE SHOCK TESTING

The measurement of frequency response for linear systems using the shake table is referred to as the steady-state vector response method [5]. The natural frequencies, ω_n, and damping ratios, D, are determined from the steady-state values of resonances, over a range of excitation frequencies. The method is slow and although the use of a swept frequency described earlier will reduce the total time for measurement, certain errors will be introduced due to system settling time and the finite averaging period.

Transient impulse methods which were briefly introduced in the previous chapter have considerable advantages over the steady-state and slow-sweep methods because of the very short measurement times that are needed. Although theoretically a very short impuse having extremely wide bandwidth is needed it is possible to obtain reasonable results if the applied impulse is short in duration compared with the period of the system undergoing test [6]. This type of system excitation is also realistic since it is often typical of transient shocks experienced in practice.

For transient shock testing the machine used is designed to impart rapid uni-directional changes in velocity to the structure [7]. The effect can be that of a unit impulse but often methods of pulse-shaping are included using a com-pressible honeycomb impact surface so that the impact shock received by the structure follows a repeatable well-defined pattern. The characteristics of these pulses and their effects on structure resonances will be considered in section 10-6.

Methods used to generate the impact shock include free fall from towers, pneumatic guns, rocket sledges and explosive devices.

The most widely used system is the gas-gun launcher [8] which is charac-terised by its ability to provide highly reproducible results over a wide range of velocities. In one particular design of gas-gun rig [9] a round bar projectile is driven along a gun barrel by the rapid release of gas pressure. The projectile strikes a propagating bar generating a stress pulse which travels along the bar and through the test structure. A second propagating bar terminates in an oil damper and is arranged to absorb the final impact shock experienced by the structure. The terminal velocity and pulse duration for the projectile can be calculated readily from the physical characteristics of the rig and achieved under experimental conditions.

Impulse excitation suffers from a number of disadvantages including the problems of maintaining a sufficiently wide bandwidth in the generation device and difficulty in obtaining a sufficiently high energy level for the pulse. This latter is a consequence of the extremely short pulse duration and hence small area of the amplitude/duration pulse characteristic.

An alternative to this is the rapidly swept sine-wave technique introduced in the previous chapter (section 9-12). Here the change in frequency is made sufficiently rapidly that the steady-state condition is never realised at any frequency contained within the swept sine-wave and resultant effects on the structure are shown to be equivalent to a wide-band impulse. This excitation wave-form is easier to apply with conventional (i.e. non-impact) generators and permits close control over the spectral characteristics of the excitation. Other advantages were mentioned earlier in connection with cross-correlation response measurements.

Measurement of the spectral response of the structure from the impulse response or swept sine-wave response involves the calculation of the ratio of the Fourier transforms of the input excitation and the response given by:

$$H(\omega) = \frac{Y(\omega)}{X(\omega)} \qquad (10\text{-}13)$$

where all the values are complex quantities. This procedure may not be straightforward and an alternative method of deriving $H(\omega)$ using cross-correlation or the cross-spectrum has been described in earlier chapters.

It has also been shown that for a unit impulse or swept sine-wave its Fourier transform is constant for a wide range of frequencies, so that a linear relationship between the measured output and the system response can be obtained. Thus for a two degree of freedom structure we can express the relationship as:

$$Y(\omega) \propto H(\omega) = \left[1 - \left(\frac{\omega}{\omega_n}\right)^2 + j2D\left(\frac{\omega}{\omega_n}\right)\right]^{-1} \qquad (10\text{-}14)$$

where D is the damping factor.

Using this the natural frequency and damping factor can be obtained for the structure from the result of division of the Fourier transforms given in equation (10-13) (or other method of deriving $H(\omega)$). The calculation can readily be carried out using the fast Fourier transform having due regard to the precautions necessary in the division or multiplication of two transforms (see section 9-8-3).

10-6   DATA RECORDING

Recording of transducer data can be made in various ways. Analog F.M. recording has particular advantages in respect of accuracy, dynamic range and frequency range. Additional information will be necessary to define the data, e.g.

event and time markers, calibration signals and, in the case of vibration test runs, the characteristics of the excitation frequency wave-form. Where multiplexing of several transducers outputs takes place the skew error introduced must be taken into account if the results are to be time-correlated. It is preferable to avoid this type of error by recording the signals, each on a separate track of the magnetic tape or pen record. For very small signals it will be advantageous to correlate these with the excitation frequency signal (assuming this to be periodic) in order to improve the signal/noise ratio.

Problems associated with vibration test data capture include noise induced in the vibrating signal cables due to their passage through a weak magnetic field or vibration-induced changes in cable capacitance. The use of balanced pair signal leads can reduce this effect by virtue of the high common-mode rejection characteristic of the coupling amplifier.

Providing that they are properly designed then the use of head amplifiers mounted close to the transducer will improve the signal/noise ratio and reduce the effect of vibrating leads. This is, of course, assuming that attachment of the amplifier to the structure will not greatly distort the vibration results obtained.

A major problem in measuring short high-velocity transients is the increased mass loading of the structure due to the presence of the measurement device. Relativity effects can increase the apparent mass of an accelerometer to several hundred or even thousands of pounds at an acceleration force of $100\,000G$. A secondary effect is that of resonance of the transducer itself which should be high compared with the structure to which it is attached. Both of these effects can be minimised by the selection of a suitable high-G accelerometer as discussed in chapter 2.

Cable connection to transducers associated with shock testing rigs introduce relativity difficulties both on account of the increased mass and the mechanical strain experienced by rapid cable movement. Cable whip in the region of the transducer can be overcome by careful anchorage. Trailing leads, inevitably required in the shock machine, can be supported at intervals by weak anchorages which break to relieve cable strain. Stressed coil cables have also proved satisfactory in practice.

10-7 MECHANICAL LOAD-IMPEDANCE TESTING

Measurements of the effects of vibration and shock has been considered largely from the motion of the vibrating structure. There are good reasons for this since transducers can easily be attached to the structure at strategic places in order to measure the quantities of motion, whereas measurements of force, for example would require more elaborate arrangements. An alternative concept, that of loading of the main structure by the resonant effects of the sub-structure, requires the specification of these effects in terms of mechanical impedance.

Impedance testing [10] represents the determination of the dynamic characteristics of the structure in terms of a relationship between an applied excitation and its response. The two most commonly used relationships are mechanical impedance [11] and mobility [12].

Referring to fig. 10-2 mechanical impedance can be defined as:

$$Z_m = \frac{\text{force obtained at the resonator mass}}{\text{velocity given to the base}} \qquad (10\text{-}15)$$

and mobility is simply the reciprocal of this. These characteristics are measured to determine resonant frequencies (modes), transmission paths, or in order to develop a mathematical model of the structure. The method is to apply a known harmonic excitation to the structure and to measure its steady-state response at a given point. Since the response at a number of frequencies will be required to produce an impedance spectrum then the form of the excitation can be a swept sine-wave [13,14] or an impulse-response function test [4]. This latter will result in shorter testing time but requires some cross-correlation processing of the results.

Mechanical impedance can be measured at the point of excitation, when it is known as point impedance or as the ratio of the applied force at one point and velocity at another point. This latter is known as a transfer impedance. It will be appreciated that the resultant velocity will be a vector quantity having a given angle to the applied force. Consequently mechanical impedance will represent a complex quantity:

$$Z_m = \frac{F}{V} \cos\phi + j\frac{F}{V}\sin\phi \qquad (10\text{-}16)$$

having real and imaginary parts.

We will consider first the derivation of load impedance from an applied acceleration of the mass. Due to the complex nature of Z_m it is convenient to consider the applied force in complex form:

$$x_a = X_a \exp(j\omega t). \qquad (10\text{-}17)$$

The resultant acceleration is from equation (10-1):

$$Y_a = X_a \frac{(\omega/\omega_n)^2}{[[1 - (\omega/\omega_n)^2]^2 + [2D\omega/\omega_n]^2]^{\frac{1}{2}}} \qquad (10\text{-}18)$$

so that the applied force (mass x acceleration) is:

$$F = MY_a \exp(j\omega t) \qquad (10\text{-}19)$$

and the velocity obtained by integration of equation (10-17) to give:

$$V = \frac{X_a \exp(j\omega t)}{j\omega} \qquad (10\text{-}20)$$

481

Hence:

$$Z_m = j\omega M \frac{Y_a}{X_a}$$

$$= j\omega M \left[\frac{(\omega/\omega_n)^2}{[[1 - (\omega/\omega_n)^2]^2 + (2D\omega/\omega_n)^2]^{\frac{1}{2}}} \right] \quad (10\text{-}21)$$

A low impedance value means that it is fairly easy to produce motion at the point of excitation. A high impedance implies a stiff structure difficult to vibrate. At resonance a vibrating sub-structure often has a higher impedance than the main structure upon which it is mounted. It will therefore tend to vibrate the structure of lower impedance. This phenomena is referred to as loading of the main structure and can complicate the resonant behaviour of the total structure. If we consider a structure having a load impedance of Z_m vibrating at a velocity $V_m(t)$ then the effect of adding a sub-structure of load impedance, Z_s will be to alter the velocity behaviour of the total structure to:

$$V_T(t) = \frac{Z_s}{Z_s + Z_m} V_m(t) \quad (10\text{-}22)$$

where Z_s and Z_m are complex quantities.

One use of the load impedance concept is therefore to derive new time histories of motion from those derived from vibrator tests to account for the effect of adding to the structure, e.g. structural modification to an aircraft main frame. Further consideration of load impedance is given by Morrow [15] who suggests modifying the shake table feedback system shown in fig. 10-6 to include a component designed to correct for variation in load impedance of the vibrating platform. This would overcome to some extent the tendency to over-test a structural item which loads the table at certain frequencies, producing a trough in its shock-spectrum characteristics (see later).

Measurement of transfer impedance is often carried out to determine the transmission characteristics of the support upon which a vibrating structure (e.g. a machine) is placed. From the theory of electrical circuits it is well known that correct matching of impedances gives maximum transmission of energy whereas a mismatching will reduce this. In mechanical terms we need to select a supporting structure which will achieve a mismatch between the machine and its ground support. Hence the mechanical impedance for the mounting structure must be chosen to be essentially different from that of the machine in order to reduce vibration transmission.

The relevance of the analogy between the behaviour of mechanical structures and electrical networks is valuable and the use of the mobility concept rather than that of mechanical impedance is advantageous. Firestone [16] has pointed out that if we consider force as analogous to current and velocity to voltage then mobility is analogous to electrical impedance and a self-consistent conversion system is obtained. The model for a mechanical system can therefore be drawn and its elements replaced by electrical equivalent values. Mathematical analysis can then take place to determine Q values, resonances, etc., using measured mobility values.

10-8 SHOCK SPECTRA

Mechanical shock is characterised by a sudden change in motion or loading in a structure. Since this gives rise to non-periodic time histories having non-stationary characteristics a quantitative definition of the effects of shock is not easily obtained. The time-histories for shock parameters such as acceleration, velocity, or displacement generated for the duration of the shock are not very useful descriptions of the process. This is because the transient response of the structure itself affects the time-history obtained and prolongs the record to include multiple reflections of the initial shock after this has been removed.

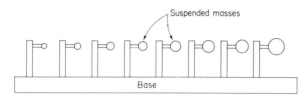

Fig. 10-8. A mechanical reed gauge.

The difficulty is one of interpretation of the effects of shock rather than its generation or measurement. In an attempt to simplify interpretation, rigid specifications of shock environment and analysis have been developed based on particular shock test machines operating under a given set of conditions. These have been of some value for limited comparative tests on the same machine. In general, however, it is fair to say that these techniques have not been very successful and that the definition of structure response to even a small number of closely defined shock excitation pulses has not proved a very accurate measure of the severity of shock. This is particularly the case where comparisons are required between the structures tested at different times and locations.

An alternative is to consider the effects of shock on a mechanical structure which can be precisely defined. The resultant structure response can then be defined reliably in terms of resonant frequencies and damping and related through these terms to the real structure. The method can be visualised in purely mechanical terms as the simulation of a general structure consisting of a large number of masses of different size, each suspended by a separate damped-spring support. The entire structure thus consists of a number of independent resonators each tuned to a different frequency and connected only by a common non-resonant base (fig. 10-8). The base is set into vibration by an applied shock pulse which in turn excites the resonators. Each will vibrate according to its resonant frequency and damping characteristics so that the extent of their individual motions may be recorded. Since any structure can be considered in terms of a group of separate mechanical units, each having its own resonant frequency, the synthesis of the results from the simulated group of resonators can broadly be considered as representative of a practical situation.

The motion of the resonators can be presented on a frequency base to form the response or shock spectrum of the applied shock pulse. Evaluation of this shock spectra is carried out by computational techniques in the form of an array of tuned analog filters or the solution of a set of equations of motion for the resonator. The latter case will be considered in the next section.

10-8-1   RESPONSE OF A RESONATOR SYSTEM

This is stated in equation (2-5) in terms of base and mass motion and may be written in terms of relative motion as:

$$p^2 z + \frac{c}{M} pz + \frac{k}{M} z = -p^2 x. \qquad (10\text{-}23)$$

But from equations (10-7) and (10-12) we can replace parameters c/M and k/M by damping and resonant frequency terms to give:

$$p^2 z + 2D\omega_n pz + \omega_n^2 z = -p^2 x \qquad (10\text{-}24)$$

where D represents the ratio of damping to critical damping factor ($= c/C_c$) and ω_n is the natural resonant frequency. This is in a convenient form for simulation using either analog or digital techniques.

10-8-2   THE SHOCK SPECTRUM

The response of a system consisting of a number of simple single degree of freedom resonators can be defined from equation (10-24) in terms of peak response against natural frequency of resonance for various damping conditions. This is the Shock Spectrum described in the response domain. This definition differs from that of the Fourier spectrum considered earlier, in that a considerable measure of data reduction takes place. Whereas with the Fourier spectrum derived from the time history of a given system parameter, the process is completely reversible so that no information is actually lost, with the shock spectrum derived from the time-history and its derivatives only the peak values are retained. Consequently the time-history cannot be recovered from the shock spectrum. The process is nevertheless valuable precisely because of its data reduction capabilities. The shock spectrum does not describe shock completely but does serve to emphasise the most important effect, that of susceptibility to damage caused when inertial loading exceeds a given level.

The shock spectrum is attributable to Biot [17] who described the effects of shock in this way in connection with the prediction of earthquake stresses. It has since been applied to the general specification of shock environment over a wide range of applications including packaging, transportation, aerospace vehicles, and industrial equipment. By limiting the consideration to a linear viscously damped single degree of freedom system, only two parameters are required to determine response to shock. These are shown in equation (10-24) as the undamped natural frequency of response (ω_n) and the critical damping fraction (D). The shock spectrum is a plot of peak response (generally acceleration $\hat{\ddot{y}}$) against ω_n. A normalised form of presentation is of more value, particularly for standardised pulse shapes [18], when the dimensionless ratios, $\hat{\ddot{y}}/\ddot{y}_{max}$ and ω_n/ω_0 are chosen. By relating the peak response to its maximum value, $\ddot{y}_{max}$ and natural resonant frequency to the fundamental frequency of the applied shock, ω_0, then the values obtained will be valid for any pulse of the same shape but different severity.

It has been noted that the response of the resonator is quite different during the period covered by the duration of the pulse from that occurring after its cessation. These two forms of the spectrum are known as the Initial and Residual Spectrum. Also since a resonator system may exhibit a higher tolerance to acceleration in one direction of displacement compared with that of the other then both positive and negative responses need to be taken. The convention used to establish polarity is to regard the direction of application of the exciting pulse to be the direction of positive response. For complete definition of shock spectra four response functions are thus required and are illustrated in fig. 10-9.

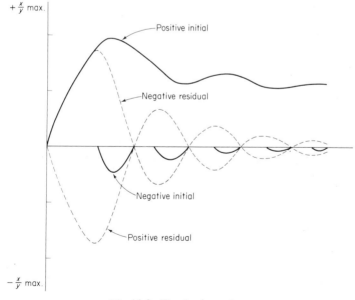

Fig. 10-9. The shock spectrum.

In practice it is often difficult to determine the exact duration of the shock so that the distinction between the initial and residual spectrum may not be clearly defined. Consequently a combination of initial and residual spectra is given and referred to as the MAXIMAX spectrum. This is defined as a plot of absolute peak response values, regardless of the time of occurrence of the peaks. It takes the form of the envelope of the maximum initial and residual spectra and is positive for simple pulse shapes. Under these conditions the maximum positive spectra is synonymous with the MAXIMAX spectra.

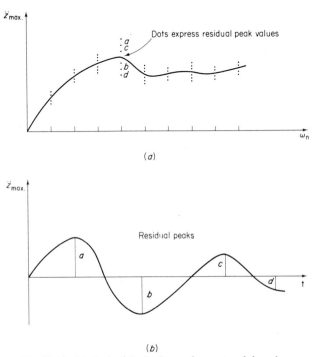

Fig. 10-10. Method of displaying peak count and damping.

For certain purposes, e.g. study of fatigue effects, the time-history of motion at each resonant frequency is required. This implies a three-dimensional display to indicate time or peak acceleration count in addition to peak acceleration value and resonant frequency. The complete time-history at each resonant frequency can be plotted to give a three-dimensional 'solid' form of display. This form has been suggested by Harris and Crede [19]. However, for many purposes the numbers of peaks at a given natural frequency of resonance exceeding a given amplitude is all that is required. A suitable computer display method for displacement shock spectra is illustrated in fig. 10-10a in which the time history at each resonant frequency is calculated and only the peak values, a, b, c, etc. shown in fig. 10-10b are stored. The maximum peak value is selected and the

486

shock spectra calculated and plotted in terms of displacement against resonant frequency. The peak values where these exceed a given minimum value relevant to a particular resonant frequency may then be plotted as the dot pattern shown, so that the peak amplitude is normalised to the maximum peak value. This type of display is indicative of the number of peaks obtained and the effects of damping.

10-8-3   SHOCK SPECTRA FOR STANDARD INPUT WAVE-FORMS
Laboratory testing of mechanical structures is carried out by subjecting the structure to a given impulse motion. The effect of this in terms of shock spectra can be determined for any input pulse shape, although we are likely to find wide variation in spectra for small differences in pulse shape. As a result the tolerances for this kind of test are best given in terms of the shock spectra characteristics rather than the pulse shape. It is necessary, therefore, to determine the type of shock spectra desirable for the test and to define from this the shape of the input pulse required. For example, shocks experienced in transportation and handling have a wide frequency range with no sharp maxima or minima. Environmental shocks associated with machinery will contain periodic oscillations and these will need to be generated by the applied excitation. A structure may have unequal sensitivity to shock in the positive and the negative direction, so that a given shock on a structure having low sensitivity to shock in one direction only could result in severe undertest. The range of requirements can be met by a comparatively small series of standardised input shapes. It will be shown later that in general these should be asymmetrical for good performance and repeatability characteristics.

The problems associated with the generation of particular pulses of motion will not be considered here. Shock test machines generate rectangular or half sine-wave pulses, but their wave-form is often a very distorted approximation to the desired shape. Pulse shaping by choice of decelerating materials in drop-test machines can produce a repeatability of ±10% [7] in the resulting shock spectra, which is adequate for many applications. Control of pulse shape with servo-controlled shake tables is easy to arrange, but the limited upper-frequency performance makes simulation of rapid changes in applied motion difficult to achieve. A further difficulty with shake tables lies in the resonance effects of the finite mass platform which become more apparent as a higher frequency performance is attempted.

The frequency characteristic of the forcing function pulse shape is relevant when considering shock spectra since the two are related by the Fourier transform. The distribution of energy contained in the pulse is given by the Fourier spectrum:

$$E(f) = \frac{1}{T} \int_{-\infty}^{\infty} x(t) \exp(-j\omega t)\, dt \qquad (10\text{-}25)$$

487

where $x(t)$ is the time history of the pulse. Solving this for a single rectangular pulse of duration T and expressing this in normalised form we have:

$$E(f) = \frac{2}{\omega T} \left| \sin \frac{\omega T}{2} \right|. \tag{10-26}$$

A similar calculation for a final peak saw-tooth gives:

$$E(f) = \frac{1}{\omega} \left| \left(1 - \frac{2}{\omega T} \sin \omega T + \frac{2}{\omega T} \sin \left(\frac{\omega T}{2}\right)\right)^{\frac{1}{2}} \right| \tag{10-27}$$

and for half sine pulse:

$$E(f) = \frac{2T}{\pi} \left| \frac{1}{1 - (\omega T/\pi)^2} \cos \frac{\omega T}{2} \right|. \tag{10-28}$$

The Fourier energy spectrum is thus seen to have a continuous form and for very low frequencies the term within the parallel brackets will tend to unity. This has two important consequences. In the first place any measuring system used for indicating the severity of shock excitation will be in error unless it exhibits an infinitely wide frequency range. Secondly where the shock duration is short compared with the periodic time of the mechanical system the severity of the shock will be determined by the area contained within the shock pulse only. The derivation of the shock spectrum from a given pulse shape is not easy to obtain analytically, although it can be derived from the Fourier spectrum (see section 10-8). Regressive or simulation computational methods are therefore

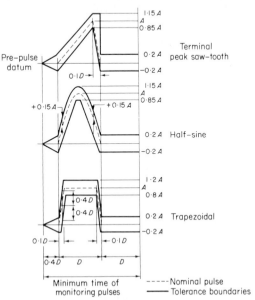

Fig. 10-11. B.S.I. pulse shape specification.

used in which the solution of the equations of motion under different conditions of frequency and damping are obtained.

A discussion of a number of pulse shapes and their relevant shock spectra will be given below. It must be stressed however, that where these pulses contain discontinuities then they may be difficult to realise in practice. The three shapes suggested by the International Electro-technical Commission and adopted by the British Standard Institution [20] are shown in fig. 10-11 and indicate the large amount of tolerance permitted in the realisation of these shapes. Blader and Ridler [21] have shown that the acceptance of these tolerances can under some circumstances result in large errors of interpretation for the shock spectra obtained so that they should be applied with care.

A number of symmetrical pulse shapes are given in fig. 10-12 together with their shock spectra, obtained under zero damping conditions. Some comments on these results are given below:

(*a*) Rectangular pulse. This is not a very desirable shape, as is apparent from the resultant shock spectra. In the case of a rectangular acceleration pulse only a positive initial spectra will be produced and the maximax spectra will show considerable variation in the peak acceleration obtained over the frequency range reaching a maximum of twice the applied acceleration.

(*b*) Half sine-wave. This also produces a poor spectra having considerable variations in peak acceleration over the frequency range. In addition a large peak in the maximax spectra is apparent at the low frequency end of the spectrum. One difficulty with this and the trapezoidal pulse described below is the lack of reproducibility of the residual spectrum. Quite small changes in pulse duration or shape may cause considerable changes in the resulting spectra.

(*c*) Terminal peak saw-tooth. A uniform undamped shock spectra is obtained which is fairly constant over the spectrum and gives the smallest response for a given peak value. The effect of the end discontinuity of the applied pulse is to induce a transient oscillation in the peak acceleration which thus simulates a condition found in the practical environment.

(*d*) Initial peak saw-tooth. A similar residual shock spectra is obtained as for the terminal peak saw-tooth, but this wave-form has the advantages of high total acceleration level and large oscillatory content [22], which makes it valuable for impact shock testing.

(*e*) Trapezoidal pulse. This gives the highest velocity change for a given peak acceleration and duration since the flat peak allows a steady build-up of movement to occur before the instantaneous value of the applied pulse begins to fall.

(*f*) Decaying sinusoid. An advantage of this wave-form is the ease by which it may be obtained using a dynamically controlled shake table. The shock spectra obtained will exhibit sharp resonant peaks whose position in the spectra can be determined by the frequency of the transient.

489

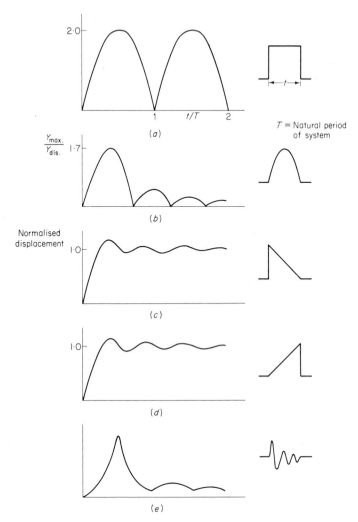

Fig. 10-12. Examples of shock spectra. (a) Rectangular pulse, (b) Half sine-wave, (c) Terminal peak saw-tooth, (d) Initial peak saw-tooth, (e) Decaying sinusoid.

It is apparent that an assymmetrical shape will give a smoother spectrum than a symmetrical pulse and will also result in a lower level of acceleration. Pendered [18] has shown that some conclusions can be drawn concerning the characteristics of the shock spectra produced for various pulse shapes. The initial part of the spectra (low-frequency end) has a linear slope dependent on the velocity change of the pulse, rather than that of the pulse shape. The end part of the spectra (high-frequency end) tends to become asymptotic to the peak value of the applied pulse, again independent of pulse shape at a time location where the pulse duration exceeds the natural half-period of resonance. No generalisation seems possible for the intermediate frequency area since this depends on all the parameters associated with the pulse.

The shock spectra shown in fig. 10-12 represent the worst possible conditions since they are based on undamped responses. Damping will have the effect of reducing the peak amplitude at intermediate frequencies during the pulse and will substantially reduce the amplitude and duration of the high frequency ripples which occur after the pulse. Reproducibility of the spectra is dependent on pulse shape and can be improved by the selection of a suitable asymmetric shape as shown above. Ripple superimposed on the pulse can be productive of unwanted

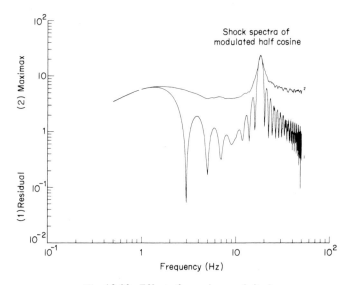

Fig. 10-13. Effect of superimposed ripple.

resonance which can invalidate the test conditions. This is a possible condition with shake table generation and its effects are shown in fig. 10-13. A large peak may be obtained in the residual spectrum which can exceed the initial peak by several orders of magnitude.

10-8-4 HIGHER-ORDER SYSTEMS

In section 10-8-2 the shock spectrum was defined in terms of the response of a single degree of freedom resonator. This represents an idealisation for a real system and has obvious limitations. For example a realistic model for a physical system can be adequately defined as a single degree of freedom system only for very low frequencies. As the frequency is increased other modes are introduced and the model must be extended accordingly. We find that the frequency response function for a complex structure, consisting of several elastically

491

interconnected masses, will exhibit a number of resonance peaks—one for each degree of freedom of the structure.

Some attempts have been made to develop multiple degree of freedom shock-spectra, notably those of Richter, Morrow, Fung and Barton [23,24,25]. The two approaches that have been attempted are:

1. To analytically study a coupled-mass system, typically of the second order.

2. To consider multiple-order systems as comprising the summation of the responses of a number of single degree of freedom systems.

In this latter case simple summation of the maxima is not permissible since they will not generally occur at the same time. The error is smallest where the resonant frequencies are close together. Where the natural frequency of vibration is high an improved accuracy is obtained by taking the algebraic sum of the peaks [24] or the square root of the sum of the squares [26].

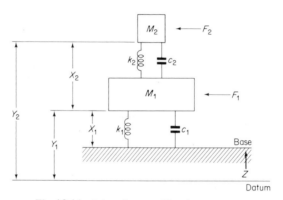

Fig. 10-14. A two-degree of freedom resonator.

The analytical approach is useful for certain limited applications, although the additional parameters required make it difficult to express a general case without resorting to multiple families of curves. Where it is possible to make some assumption as to the nature of the structure then it may be possible to determine a second order spectrum from the known separate acceleration behaviour of the individual structures. An example of this is described below for the coupled two degrees of freedom system shown in fig. 10-14. The system is considered to be initially at rest, and then excited by impulse forces F_1 or F_2 present at the two masses and originating at a force applied to the base carrying the structure. Referring to equation (2-5) the equations of motion for the two masses can be expressed as:

$$\left. \begin{array}{l} M_1 p^2 Y_1 + c_1 p X_1 + k_1 X_1 - c_2 p X_2 - k_2 X_2 = F_1 \\ M_2 p^2 Y_2 + c_2 p X_2 + k_2 X_2 = F_2 \end{array} \right\}. \qquad (10\text{-}29)$$

Substituting

$$Y_1 = X_1 + Z$$

$$Y_2 = X_2 + X_1 + Z$$

and dividing by M_1 and M_2 respectively we have:

$$\left(p^2 + p\,\frac{c_1}{M_1} + \frac{k_1}{M_1}\right) X_1 - \left(p\,\frac{c_2}{M_1} + \frac{k_2}{M_1}\right) X_2 = \frac{F_1}{M_1} - p^2 Z$$

and

$$p^2 X_1 + \left(p^2 + p\,\frac{c_2}{M_2} + \frac{k_2}{M_2}\right) X_2 = \frac{F_2}{M_2} - p^2 Z. \tag{10-30}$$

Solutions can be obtained for X_1 and X_2 and their derivatives in terms of each force separately applied, and result in a set of equations of the form:

$$X(t + h) = \exp\,(ah)X(t) + \int_0^h \exp\,[a(h - x)]u(t + x)\mathrm{d}x \tag{10-31}$$

where a is a constant, h is the displacement step length, x, u are auxiliary variables of displacement and force.

The first part of this equation represents the residual spectrum after the application of the shock and the second part gives the initial spectrum. The complete solution of equation (10-31), which gives a single value of displacement based on a previous measurement, is obtained by a regression operation particularly suited to digital calculation. Suitable methods of numerical integration can be employed and the parameters separately associated with each mass determined.

In this example we considered the case of two single degree of freedom systems each transmitting its motion directly to the other. Another situation arises when the two systems are only loosely coupled to each other but share a common excitation base (fig. 10-15). This could occur for example in mounting

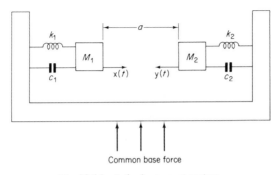

Fig. 10-15. A dual resonant system.

adjacent equipment units in a confined space and where the spacing between the units must not reduce to zero under any vibration conditions.

Assuming that the motions of the two systems are independent of each other the two displacement histories can be analysed in terms of a joint probability density thus:

$$\text{Prob } [(x + y) > a] = 1 - \int_{-\infty}^{a-x} \int_{-\infty}^{a-y} p(x, y)\mathrm{d}x\mathrm{d}y \qquad (10\text{-}32)$$

which means effectively finding the area of the free space $(a - y)$ and subtracting this from unity. An alternative approach would be to derive the cross-correlation between the two functions and to calculate the sum of the two displacements at the peak values [27].

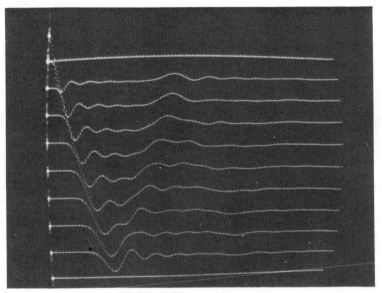

Fig. 10-16. Effects of spring surge.

Some useful information on the behaviour of coupled systems can be obtained by analog computer simulation methods. The selection of parameter values will still have to be made, but by adopting iteration techniques a large number of possibilities can be explored and the severity of the resonator motion determined over a wide range of constants. Fig. 10-16 shows the results of applying this technique to the simulation of a ten-turn coil spring. The simulation model is considered as a series of ten damped spring-mass systems each communicating its displacement to its neighbouring systems. A uni-directional displacement is assumed in the case of the end systems. The simulation model is subject to an input forcing function (in this case a step displacement transient) to determine its residual shock response under these conditions. The relative motion of each coil of the spring considered as a single resonator is displayed with respect to time and clearly shows the peak acceleration displacement travelling through the coiled spring in the form of a spring-surge phenomenon. Derivation of resonant frequency information can be obtained from these time-histories, including the effective summed response of the total spring movement in terms of this spring-surge.

10-8-5   ANALOG METHODS

The second-order equation for a damped spring-mass system is given in equation (10-24) and can be mechanised as shown in fig. 10-17 for a given value of damping and natural resonant frequency. This will be seen to be identical to a

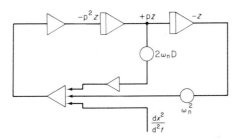

Fig. 10-17. Analog solution of a second-order equation of motion.

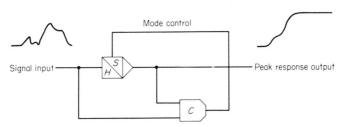

Fig. 10-18. A maximum-seeking circuit.

second order band-pass filter simulation (see chapter 3). Consequently analog shock spectrum analysis is frequently carried out in terms of a contiguous set of narrow-band filters in much the same way as that considered for Fourier analysis. In this case however only the peak response is required and can be obtained by applying the output of each filter to a maximum seeking circuit such as that shown in fig. 10-18. The comparator C is used to control the mode of a sample/hold amplifier, S/H, in order to place this in the sample mode as long as the signal is positive-going. When the signal reaches a point of inflection or becomes negative-going then the amplifier is maintained in the hold mode and the greatest possible value reached previously is held at its output. The peak response value stored can be read out by sequentially sampling the set of stored outputs at the end of the run. Alternative representations based on a comb filter have been considered and also sequential methods which rely on repetition of the input forcing function and alteration of a single filter resonant frequency with each repeat of the function.

Analog methods are particularly valuable for studying the complete time history of the shock response which can be produced in terms of simultaneous histories for displacements, accelerations and velocities associated with a multi-degree-of-freedom simulation. This is due to the parallel operation of the analog machine permitting real time simulation of complex systems.

10-8-6   DIGITAL METHODS

Digital methods fall into three categories:

1. Direct simulation of the equations of motion [28].
2. Digital filtering [29].
3. Conversion from the Fourier transform.

Solution of the second-order differential equation (10-23) is carried out on the digital computer using a numerical method such as the Runge-Kutta where the initial shock excitation, $p^2 y_i$, is described as a sampled series. Variation in step length for the analysis or sampling rate for the shock signal is necessary for computational efficiency. The former can be achieved using finite different integration methods such as the Runga–Kutta–Merson which permits step length adjustment by program during the analysis [30]. As the frequency is reduced the sampling rate should be reduced also to effect economy in computation time. Decimation below certain frequency values will achieve this but care is necessary to filter the data before decimation to avoid aliasing affects.

O'Hara [31] has described an alternative method of simulation using first forward differences to replace the process of integration in the undamped harmonic solution to equation (10-23), which can be rewritten as:

$$z = z_0 \cos \omega t + \frac{pz_0}{\omega} \sin \omega t - \frac{1}{\omega} \int_0^t p^2 x(T) \sin \omega(t - T) \, dT \qquad (10\text{-}33)$$

and

$$\frac{pz}{\omega} = -z_0 \sin \omega t + \frac{pz_0}{\omega} \cos \omega t - \frac{1}{\omega} \int_0^t p^2 x(T) \sin \omega(t - T) \, dT \qquad (10\text{-}34)$$

giving results in terms of the forward difference in relative mass velocity,

$$S_n = px_{n+1} - px_n$$

so that an iterative solution can be obtained in its simplest form as:

$$z_{n+1}\omega = z_n\omega . \cos \omega T + pz_n . \sin \omega T - \frac{S_n}{\omega T}(1 - \cos \omega T)$$

$$-S_{n-1}^2 \left(\frac{1 + \cos \omega T}{2\omega T} - \frac{\sin \omega T}{\omega^2 T^2}\right) \qquad (10\text{-}35)$$

and

$$pz_{n+1} = -z_n\omega \sin \omega T + pz_n \cos \omega T - \frac{S_n}{\omega T}(\sin \omega T)$$

$$-S_{n-1}^2 \left(\frac{1 - \cos \omega T}{\omega^2 T^2} - \frac{\sin \omega T}{2\omega T}\right) \qquad (10\text{-}36)$$

where T is the sampling interval, and S_{n-1}^2 is the second forward difference.

Greater precision can be obtained with this method by replacing the straight-line difference by a three-term parabolic equation in $px(t)$. It can be noted that in this solution for equal sample spacing, the sine and cosine terms need only be calculated once per frequency which reduces equations (10-35) and (10-36) to a simple series of product summations. Initial conditions for the equations will need to be inserted if they differ from zero in place of z_n and pz_n for the first term evaluation. The method can be extended to multi-degree of freedom spectra and is the basis of the technique referred to in section 10-8-4.

Developments in recursive digital filtering permit digital methods of shock spectrum evaluation to be carried out more cheaply and with a higher resolution than is possible by analog means. The two requirements are;

(*a*) narrow-band filtering.
(*b*) peak response detection.

Recursive digital filters for bandpass operation were discussed generally in chapter 7. A second order filter corresponding to the damped response transfer function of equation (10-1) is given by Lane [29] as:

$$y_i = Ax_i + Bx_{i-1} - Cy_{i-1} - Dy_{i-2} \qquad (10\text{-}37)$$

who shows that this formulation requires a minimum number of calculations for the shock spectrum.

Equation (10-37) is shown to be equivalent to a narrow bandpass filter having a transfer function of the form:

$$H(f) = \frac{A + B \cos \omega T}{1 + C \cos \omega T + D \cos \omega 2T} \qquad (10\text{-}38)$$

The parameters A to D are related to natural resonant frequency and damping of the equivalent mechanical resonator and will need modification for each change in analysis frequency. This representation for shock spectrum analysis has the

497

advantage that minimum error is obtained at the peak resonant value which is retained from the calculation.

Detection of peak value can be carried out by successive comparison techniques but unless each response contains a large number of response points an error in estimation will be obtained. Improved methods are area integration between each sample pair and interpolation to detect an accurate location of the peak within the sample interval.

The third method of derivation of the shock spectrum is via the Fourier transform. It will be shown later that the peak response value is proportional to the modulus of the Fourier transform for the excitation pulse scaled by its frequency of resonance. This is only applicable to the residual spectrum however so that a separate (non-Fourier) calculation is required to derive the initial spectrum.

10-9 RELATIONSHIP BETWEEN FOURIER SPECTRA AND SHOCK SPECTRA

Although the derivation of shock spectra from a time-history of motion is non-reversible it will be intuitively realised that a relationship must exist between the shock spectrum and the Fourier spectrum, since both represent a frequency domain view of the same time history. A quantitative relationship between the Fourier spectrum and the peak residual response of an undamped simple structure will be derived in this section.

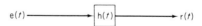

Fig. 10-19. Response of a linear system.

It is convenient to consider the structure as a linear system and derive the relationship in terms of its impulse response function. Referring to fig. 10-19 the response time-history $r(t)$ is given by:

$$r(t) = \int_0^t h(t - \tau)e(\tau)\, d\tau \qquad (10\text{-}39)$$

where τ is the incremental time variable $\leqslant t$, $h(t)$ is the system impulse response function, and $e(t)$ and $r(t)$ are the excitation input and response output (which need not be confined to displacement values).

The form of the system impulse can be obtained if we consider the equation of motion for a linear undamped system to an exciting acceleration. Thus, from equation (10-23) if $c = 0$

$$\frac{M}{k}p^2 z + z = -\frac{M}{k}p^2 x \qquad (10\text{-}40)$$

498

substituting $\omega_n^2 = k/M$ from equation (10-12)

$$\frac{p^2 z}{\omega_n^2} + z = \frac{-p^2 x}{\omega_n^2}. \tag{10-41}$$

This equation has the same form whether the initial excitation is base displacement, x, giving mass displacement, y; base velocity, px, giving mass velocity, py; or base acceleration $-p^2 x/\omega_n^2$ giving relative displacement, z.

Hence we can write a general equation of motion for any input excitation (displacement, velocity or acceleration) as:

$$\frac{p^2 r(t)}{\omega_n^2} + r(t) = e(t) \tag{10-42}$$

If we consider the excitation input, $e(t)$ to be a Dirac unit impulse where:

$$\int_{-\infty}^{\infty} e(t).dt = 1 \tag{10-43}$$

then, since its duration is vanishingly small it can be regarded as an initial condition so that we can solve for:

$$\frac{p^2 r(t)}{\omega_n^2} + r(t) = 0. \tag{10-44}$$

The general solution for this can easily be shown to be:

$$r(t) = A \sin \omega_n t + B \cos \omega_n t. \tag{10-45}$$

The initial condition for equation (10-45) when $t = 0$ will give values for the coefficients A and B as $r(t) = B = 0$ and, $pr(t) = \omega_n A$, but from equation (10-42) with $r(t) = 0$ then:

$$pr(t) = \omega_n^2 \int_0^t e(\tau)\, d\tau \tag{10-46}$$

which from the definition of the unit impulse given in equation (10-43):

$$pr(t) = \omega_n^2 = \omega_n A. \tag{10-47}$$

Therefore, $A = \omega_n$, and we can write:

$$h(t) = \omega_n \sin \omega_n t. \tag{10-48}$$

Substituting equation (10-48) in equation (10-39):

$$r(t) = \int_0^t \omega_n . \sin \omega_n (t - \tau) . e(\tau) . d\tau$$

$$= [\omega_n . \sin \omega_n t \int_0^t e(\tau) . \cos \omega_n \tau . d\tau] - [\omega_n . \cos \omega_n t \int_0^t e(\tau) . \sin \omega_n \tau . d\tau] \tag{10-49}$$

and

$$pr(t) = \omega_n^2 \cos \omega_n t \int_0^t e(\tau).\cos \omega_n \tau.d\tau$$

$$+ \omega_n^2 \sin \omega_n t \int_0^t e(\tau) \sin \omega_n \tau.d\tau. \quad (10\text{-}50)$$

Equations (10-49) and (10-50) define the response value and its derivative for an undamped resonator in terms of the input forcing function, $e(\tau)$, and the natural resonant frequency, ω_n, of the residual motion. They will now be used to obtain the response to a transient excitation of finite duration T. If we replace t by T in these equations then the initial response and velocity at the end of the applied transient can be determined. Thus, considering first the integrals in each of these equations, the Fourier transform $E(\omega)$ of the excitation time-history, $e(t)$ is:

$$E(\omega) = \int_0^T e(t).\exp(-j\omega t).dt$$

$$= \int_0^T e(t) \cos \omega t.dt - j \int_0^T e(t) \sin \omega t.dt \quad (10\text{-}51)$$

where T is the excitation pulse duration. Defining real and imaginary terms:

$$\text{Re}[E(\omega)] = \int_0^T e(t) \cos \omega t.dt$$

$$\text{Im}[E(\omega)] = \int_0^T e(t) \sin \omega t.dt. \quad (10\text{-}52)$$

Substituting in equations (10-49) and (10-50) will give the response values:

$$r(\tau, \omega_n) = \omega_n[\text{Re}[E(\omega)] \sin \omega_n \tau + \text{Im}[E(\omega)] \cos \omega_n \tau] \quad (10\text{-}53)$$

$$pr(\tau, \omega_n) = \omega_n^2[\text{Re}[E(\omega)] \cos \omega_n \tau - \text{Im}[E(\omega)] \sin \omega_n \tau]. \quad (10\text{-}54)$$

These 'Fourier' values of response function and its derivative for the period immediately following the excitation input can be substituted in equation (10-45) as initial condition constants, $A = [pr(\tau,\omega_n)]/\omega_n$ and $B = r(\tau,\omega_n)$. Thus:

$$r(t) = \frac{pr(\tau, \omega_n)}{\omega_n} \sin \omega_n t + r(\tau, \omega_n) \cos \omega t. \quad (10\text{-}55)$$

This represents a pure sinusoidal motion having a maximum response value of:

$$r_{max.} = \left[r(\tau, \omega_n)^2 + \frac{pr(\tau, \omega_n)^2}{\omega_n^2} \right]^{\frac{1}{2}}. \quad (10\text{-}56)$$

500

Substituting (10-53) and (10-54) into (10-56) the sin and cos terms cancel giving:

$$r_{\text{max.}} = \omega_n [\text{Re}^2 [E(\omega)] + \text{Im}^2 [E(\omega)]]^{\frac{1}{2}}$$

therefore:

$$r_{\text{max.}} = \omega_n |E(\omega)|. \tag{10-57}$$

Thus the peak response at a given angular frequency (ω) is the modulus of the Fourier transform of the forcing function at that particular frequency, scaled by the natural resonant frequency of the structure. This is shown in fig. 10-20 for a rectangular pulse waveform in which the normalised transform for a unit impulse is shown in (a), its modulus value $E(f)$ derived from equation (10-51) in (b) and the resulting residual shock spectrum from equation (10-57) as (c). Note

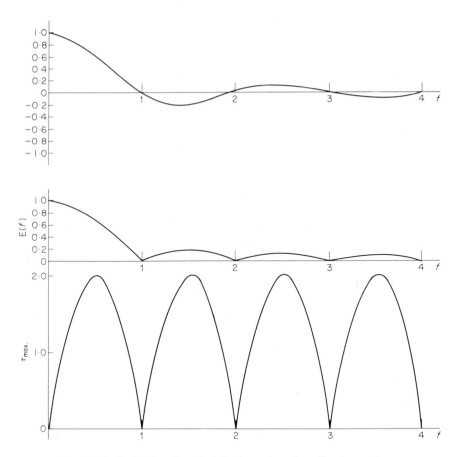

Fig. 10-20. Derivation of residual shock spectrum from Fourier spectrum.

that the peak amplitude for this is double the amplitude of the exciting force as noted earlier in section 10-6-3.

Equation (10-57) is still a general one and is applicable to the response obtained for any excitation force. Since the units in $E(\omega)$ will have the same units as the excitation $e(t)$ equation (10-57) can be written in dimensionless form:

$$\frac{r_{max.}}{e_{max.}} = \frac{\omega_n \, | \, E(\omega) \, |}{e_{max.}}. \qquad (10\text{-}58)$$

This permits the shock spectra to be expressed in terms of a single non-dimensional quantity, $\omega_n \tau$, for any given transient of duration, T, which, as noted earlier, enables the characteristic shock spectra to be compared for pulses of the same shape but different severity.

10-9-1   SHOCK SYNTHESIS

Shock testing has been described earlier in terms of the production of shock pulses having precisely defined waveshapes, such as half-sine, terminal peaks, trapezoidal, etc. In a real environment the shocks experienced by mechanical structures will generally be oscillatory in character and have frequency components distributed over a wide range of frequencies. A method of synthesis can be used in which one of the simpler forms of shock wave-form is applied to a parallel filter bank to provide the time-spreading for the frequency components which is observed in practice. The filters can be arranged to cover the range of excitation frequencies relevant to the unit under test and the resultant shock signal will be repeatable and more realistic of the natural environment.

The synthesised signal at the output of the filter bank is applied to a shaker system to excite the unit under test. This is less exacting than a swept frequency test since the amplifier/shaker system does not have to handle the full acceleration range of the shock spectrum value.

Analysis in terms of the shock spectra is carried out by determining the Fourier transform of the synthesis signal and using equation (10-57) to obtain the peak response over a range of angular frequencies.

BIBLIOGRAPHY

1. MORROW, C. T. *Shock and Vibration Engineering.* John Wiley, New York, 1963.
2. BURROW, L. R., Tracking filters standardise sinusoidal vibration tests. *Test Eng.* 2-6, April 1964.
3. MARTIN, A. and ASHLEY, C. A computer controlled digital transducer function analyser and its application in automobile testing. *Soc. Env. Eng. Symposium,* Imperial College, London, Jan. 1971.
4. TAYLOR, G. A., SKINGLE, J. and GAUKROGER, D. R. MAMA – A semi-automatic technique for exciting the principle modes of complex structures. *R.A.E. Tech. Rep.* 67211, 1967.

5. KENNEDY, C. C. and PANCU, C. D. P. Use of vectors in vibration measurement and analysis. *J. Aero, Sci.* **14**, 603-25, 1947.

6. WHITE, R. G. Use of transient excitation in the dynamic analysis of structures. *J. Roy. Aero. Soc.* **73**, 1047-60, Dec. 1969.

7. YARNOLD, J. A. L. High velocity shock machines. *Env. Eng.* **17**, 11-14, Nov. 1965.

8. KOLSKY, H. and DOUCH, L. S. Experimental studies in plastic wave propagation. R.A.R.D.E., Rep. (B), 6/61, May, 1961.

9. BILLINGTON, E. W. and BRISSENDEN, C. Dynamic stress strain curves for various plastics and fibre-reinforced plastics. *J. Phys. D. Appl. Phys.* **4**, 272-86, 1971.

10. EWINS, D. J. Some whys and wherefores of impedance testing, *Soc. Env. Eng. Symp.*, Imperial College, London. Jan. 1971.

11. PLUNKETT, R. Introduction to mechanical impedance methods for vibration problems. A.S.M.E. Colloq. on mechanical impedance methods, Dec. 1958.

12. OVERTON, J. A. and MILLS, B. The use of mobility methods for the dynamic analysis of mechanical systems. *Env. Eng.* **46**, 7-12, 1970.

13. CRONIN, D. L. Response spectra for sweeping sinusoidal excitations. *Shock and Vib. Bull,* **35**, 1966.

14. HOK, G. Response of linear resonant systems to excitation of a frequency varying linearly with time. *J. Appl. Phys.* **19**, 242-50, 1948.

15. MORROW, C. T. Application of the mechanical impedance concept to shock and vibration testing. *Noise Control* **6**, 4, July 1960.

16. FIRESTONE, F. A. A new analogy between mechanical and electrical systems. *J. Acoust. Soc. Amer.*, **4**, 249-67, 1933.

17. BIOT, M. A. Theory of elastic systems vibrating under transient impulse with an application to earthquake-proof buildings. *Proc. Nat. Acad. Sci.* **19**, 1933.

18. PENDERED, J. W. The shock spectrum, *Env. Eng.* **21**, 13-19, June 1966.

19. HARRIS, C. M. and CREDE, C. E. *Shock and Vibration Handbook.* McGraw-Hill, New York, 1961.

20. British Standard Methods for the environmental testing of electronic components and electronic equipment. BS2011 Part 2, Ed. 1967.

21. BLADER, F. B. and RIDLER, F. B. Shock spectra in transport. *Env. Eng.* **39**, 17-21, July 1969.

22. FUNG, Y. C. Shock loading and response spectra. *Amer. Soc. Mech. Eng.* Colloq. on shock and structural response, 1960.

23. RICHTER, A. P. The response of a two degree of freedom undamped system subjected to impulse loading, *Struct. Mech. Res. Lab. Rep.* University of Texas, 1960.

24. YOUNG, D., BARTON, M. V. and FUNG, Y. C. Shock spectra for non-linear spring-mass systems and their applications to design. *J. Amer. I. Aero Astr.* **1**, 1963.

25. MORROW, C. T., TROESCH, B. A. and SPENCE, H. R. Random response of two coupled resonators without loading. *J. Acoust. Soc. Amer.* **33**, 1, Jan. 1961.

26. OSTREM, F. E. and RUMERWAN, M. L. Shock and vibration transportation environmental criteria. *N.A.S.A. Rep.* CR77220, 1965.

27. HIMMELBLAU, H. and KEER, L. M. Space requirements for simple mechanical systems excited by random vibration. *J. Acoust. Soc. Amer.* **32**, 1, 76, Jan. 1960.

28. BROWN, R. D. High speed sampling and the response shock spectrum. Lecture Series, Application and methods of random data analysis, Southampton University, July 1969.
29. LANE, D. W. Digital Shock spectrum analysis by recursive filtering. *J. Sound Vib.* **33**, 2, 173-81, 1964.
30. MARTENS, H. A comparative study of digital integration methods. *Simulation* **12**, 2, Feb. 1969.
31. O'HARA, G. J. Impedance and shock spectra. *J. Acoust. Soc. Amer.* **31**, 10, 1300, Oct. 1959.

Additional references

32. CHAPMAN, C. P. A digitally controlled vibration or acoustics testing systems, Parts 1 and 2. *Prac. 1. Env. Sci.* Anaheim, California. April 1969.
33. BENDAT, J. S. Probability functions for random responses prediction of peaks, fatigue and catastrophic failure. *N.A.S.A. pub.* R-33, Washington, D.C., April 1964.
34. BENDAT, J. S., ENOCHSON, L., KLEIN, G. H. and PIERSOL, A. G. The application of statistics to flight vehicle vibration problems. *A.S.D. Tech Rep.* 61, 123, Washington D.C., 1961.
35. MORROW, C. T. The shock spectrum as a criterion of severity of shock impulses. *J. Acoust. Soc. Amer.* **29**, 5, May 1957.
36. BLADER, F. B. and RIDLER, K. D. Errors in the use of shock spectra. *Env. Eng.* **39**, 7-16, July 1969.
37. BOGART, T. F. Analog method for study of shock spectra in non-linear systems. *Shock Vib. Bull.* **33**, 6, 1966.
38. LEWIS, D. *Shock and Vibration – An Annotated Bibliography. Soc. Env. Eng.* London, 1967.
39. VIGNESS, I. Shock motions and their measurements. *Exp. Mech.* **1**, 9, Sept. 1961.

Chapter 11

NON-STATIONARY PROCESSES

11-1 INTRODUCTION

Signal analysis of time-invariant processes and systems has been considered in detail during the earlier part of this book, but little has been said concerning methods for analysis of time-varying signals, i.e. non-stationary processes. Non-stationarity is a characteristic of many physical situations such as, speech analysis, seismology, economics, aircraft noise, and the non-linear response of structures. We find, for example, that any signal derived from a process involving the Doppler effect cannot be analysed in terms of a constant frequency content since this will be found to vary with time. Similarly in the case of radar echoes from a moving target we will find that the return signal contains both stationary noise and non-stationary data and means will have to be found to differentiate between them.

Strictly speaking, real processes are always non-stationary and the concept of stationarity remains a mathematical ideal. However, very many processes change so slowly with regard to time that we can regard them as functions of time or frequency alone, the choice being a matter of convenience since the two functions are related. This is the accepted case for finite record lengths derived from supposedly 'stationary' processes, where the stationarity is assumed to extend at least over the duration of the record available for analysis.

A definition for a stationary random process has been given earlier as one in which all the statistical properties of the signal representing the process $x(t)$, do not change with time. Where these conditions are nearly realised the process is described as being 'strongly stationary'. A more usual case is where first and second moments of $x(t)$ (the mean and variance estimates) do not change with time. This is known as a 'weakly stationary' process. For practical purposes the differences between completely stationary and weakly stationary processes in the frequency domain are so slight that we can disregard this division and consider weakly stationary processes as if they were completely stationary.

505

11-1-1 TESTS FOR NON-STATIONARITY

A simple comparative test is to derive the power spectrum for one section of the signal and to compare this with the power spectrum taken from a later section of equal length. If the two power spectra are similar then the signal is considered to be stationary.

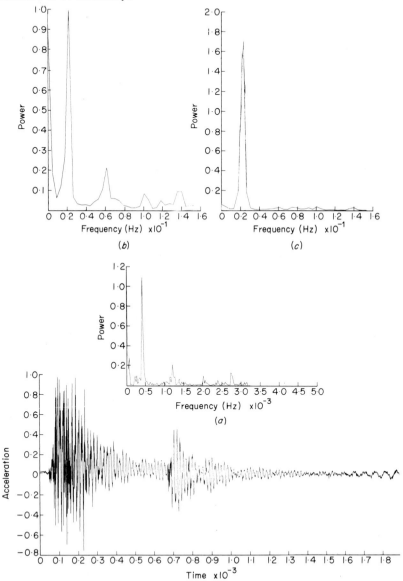

Fig. 11-1. Spectral analysis of a transient wave-form.

This test is shown applied to a non-stationary transient waveform in fig. 11-1. A power spectral density analysis taken for the complete transient (*a*) shows four pronounced spectral peaks plus a component in the region of zero frequency. Separate power spectral analysis for each half of the transient show this is a characteristic that is only present in the first half of the wave-form (*b*), and that the second half contains only the dominant resonant frequency of the signal, (*c*). Further subdivision in this way would show that the process exhibits a non-linear time variation for most of the first half of the transient and an almost stationary characteristic over the remainder of the waveform.

The method can be further developed and used to determine the optimum time/bandwidth windows required for analysis by means of the following procedure. The time-history is divided by two, each half adjusted for zero mean value and convolved with a tapering window. A smoothed power spectrum is obtained from each half for a given number of degrees of freedom. The two spectra are compared and, if different, the time history is halved again, and the process repeated; this time with double the degrees of freedom in order to

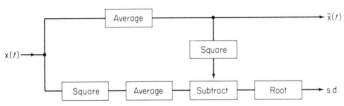

Fig. 11-2. A test for time-varying statistical moments.

maintain the BT product, (and hence statistical accuracy) constant. The optimum segmentation is obtained when an acceptable amount of agreement is obtained for the two spectra. Unless comparisons are taken throughout the length of the record, it will have to be assumed that the non-stationary trend is a linear one. Also a sufficiently long record will have to be available such that a fairly large BT product can be obtained with a small value of B in the first instance, to permit later increase of B without introducing too large a bandwidth error into the estimates.

A test for the identification of time varying statistical moments is easier to apply and is illustrated in fig. 11-2. The results of such a test are two further time-histories giving a running mean and variance (standard deviation) for the record. A quantitative assessment of the change in these input values is made to decide if the signal is non-stationary. Care will need to be taken with the amount of smoothing carried out in the averaging process, since too little smoothing can lead to local variations in mean and variance, which can indicate an apparent non-stationary characteristic. When implemented digitally the averaging can be carried out using a first-order recursive low-pass filter having the form:

$$y_i = k \cdot y_{i-1} + x_i. \tag{11-1}$$

507

Similarly using an analog filter the E.M.P. averaging integrator could be used to give:

$$\bar{x}_m = \alpha \int_{-\infty}^{T} x(t) \exp \left[-\alpha(T - t)\right] dt. \qquad (11\text{-}2)$$

The determination of k and α is best obtained on the basis of experience using a range of different values. For a single or small number of records the identification of a series for stationary analysis is determined more by good scientific judgment than statistical considerations.

The method has been extended by Priestley and Rao [1], who use a logarithmic transformation of the mean value and variance values. The transformed quantities are tested as the sum of the squares of the values obtained between given time and frequency levels against a variance obtained for the entire signal. The significance of the test can indicate whether the signal is non-stationary over a specified time and frequency band.

11-1-2 NON-STATIONARY ANALYSIS

When considering the analysis of non-stationary processes, the rate of change of the signal parameters with time is of prime importance. Generally these changes are slow enough to permit certain assumptions to be made concerning stationarity over a finite period of time, similar to the assumption that is made over the complete record length for stationary data (fig. 11-3). Thus we can apply conventional stationary methods of analysis over such a small section of the signal time-history, that the data appears to be stationary. This is the general principle behind the many forms of local, instantaneous, and evolutionary spectral analysis that have been suggested for the analysis of non-stationary data.

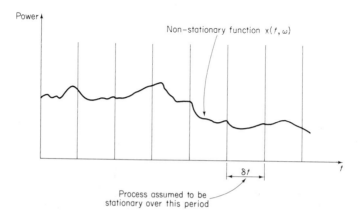

Fig. 11-3. A non-stationary process.

The ideal with this form of analysis is to determine (for example) a power/ frequency distribution defined at a unique instant in time, and to carry out this measurement at many instances during the complete time-excursion of the signal. The result gives a three dimensional 'map' of the behaviour of the process,

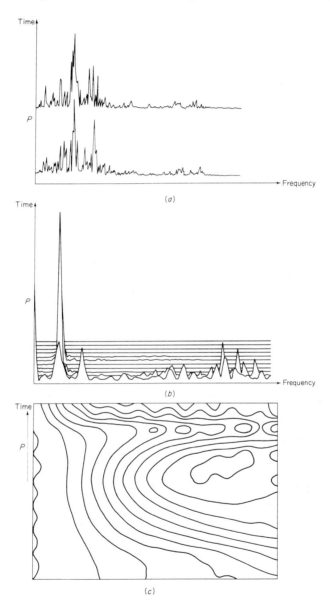

Fig. 11-4. Representations of non-stationary distributions.

509

showing power to be a function of both time and frequency. Some examples of this representation are shown in fig. 11-4. The first method shows a family of spectral density curves one for each time segment of the signal. An alternative version of this which permits a wider dynamic range of display is the peak-perspective method shown in (b). Here the range of excursion for the peaks for any given segment is allowed to overlap any earlier variations from previous segments so that the information from these earlier segments is effectively removed from the overlap periods. This method is fairly easy to program and takes little program storage. It does however suffer from the disadvantage of discarding a certain fraction of the information. A third method, that of contour display, is economical in presentation but presents a fairly formidable software problem, requiring considerable program storage and retention of the total data field (unlike the preceding methods which need only store the data for one segment at a time).

The analysis of a time-varying signal presents a number of new problems. Apart from the difficulties inherent in the finite nature of the signal obtained, which were discussed previously in connection with stationary signals, we have the added difficulty of precise determination of the spectrum at a precise point in time. This situation was described earlier in chapter 8 and referred to as an uncertainty principle, where the two requirements are found to be mutually exclusive. It has been pointed out by Daniels [2] that a direct comparison exists between this principle and Heisenberg's principle of uncertainty in quantum mechanics relating to the position of a charged particle in space and the frequency of its wave function. If its energy is exactly known and hence its wave function, then its precise location in space will be impossible to obtain. Similarly if its location is known precisely then its wave function covers a very broad spectrum and its energy is ill-defined. It must be emphasised however that this difficulty is a purely practical one and effectively prohibits the realisation of filters having both short impulse response times and narrow bandwidths. Analytical methods will still enable the estimation of power at precise time/frequency locations as described later.

In the practical case we have other difficulties to solve after accepting a suitable filter design compromise. We saw earlier that in considering power spectral estimates for a stationary process two types of errors are introduced. These are a bias error caused by the finite bandwidth of the analysis filter and a random dynamic error resulting from the finite product of the bandwidth and record length. These error relationships are:

(a) Bias error $\propto B$,
(b) Dynamic error $\propto (BT)^{-1/2}$,

and can both be reduced to a small value if the record length, T, is taken to be very long. This permits a small value of B to be chosen to reduce bias error

510

whilst maintaining the product BT large enough to result in a small dynamic error.

Where the signal is time-varying then the power spectrum for the non-stationary process will include a third type of error. This expresses the difference between the spectral density at a given time t_1 and that estimated as an average value over a time period $t_1 - t_2 = T'$. This will give a bias error proportional the value of T' chosen, and is referred to as the time-interval bias error [3]. Ideally we would like to make T' as short as possible to reduce the time interval bias error, and at the same time, make the analysis bandwidth narrow to reduce the bandwidth bias error. However, for this short section of the signal $T' \ll T$ the product BT' will be small and the dynamic error proportionately large. Hence we cannot solve the problem simply by taking longer signal records.

Other problems associated with non-stationary signals, which render analysis difficult, are:

1. We can no longer average quantities over the entire record length, since the spectral properties are no longer independent of the period of time over which they are taken. It is possible to average over a much shorter period of time, short enough for stationarity to be assumed but this average value is not completely relevant unless taken over a three-dimensional volume. The computation of meaningful averages for time-dependent data is indeed difficult. To obtain statistically reliable estimates very large numbers of records will need to be taken. In some cases this may not be possible since the process is unrepeatable (e.g. a seismic record) or the cost of repeating the record prohibitive (e.g. a satellite launch).

It is necessary therefore, to devise methods which will deal with the single record case also.

2. The concept of ergodicity must necessarily be rejected since a single sample record may no longer be assumed to be representative of a set of similar records taken from the same source. It is possible to consider a form of ergodicity in which the average of the process is considered repeatable, relative to a particular time instant, an assumption which is relevant to many physical environments.

3. As we shall see later, there are problems in interpreting the power spectrum from the autocorrelation, using the Wiener–Khintchine relationship since this is no longer directly applicable. Forms of this relationship have been developed by various workers but all lead to a double integral form of the non-stationary equations which have no direct physical meaning.

4. The same difficulty is apparent if we obtain the spectral density of the signal directly from the Fourier transform since the frequency series is not an orthogonal process, i.e. the increments of frequency are not correlated with each other. Any attempt to modify the transform to take this into account will again lead to a two dimensional function of frequency.

All these difficulties will be referred to again during this chapter in connection with practical methods of non-stationary analysis.

Where a number of non-stationary records M can be obtained from the same process then ensemble averages can perform the same function as record time averaging from an ergodic process. The improvement in signal/noise ratio can be considerable and is much used in the E.E.G. analysis of small noisy signals [4,5]. Some examples and a quantitative mathematical analysis has been given by Bendat [6] who derives a ratio for the record length required in the case of an unknown distribution to that of a Gaussian distribution, where the improvement is given by $\sqrt{M}$. Derivation of a time-varying spectrum from ensemble averaging of a series of autocorrelation functions taken from repetitions of the measurement has also been carried out [7]. This enables the Wiener–Khintchine method to be applied in the spectrum evaluation without incurring the difficulties of a double integral form.

Developments in the area of non-stationary analysis have been to a large extent focused on theoretical analysis and the need to develop and test suitable mathematical models. This is in contrast to the progress in stationary random analysis which has proceeded by alternate empirical methods and theoretical development. Consequently it is necessary to consider some of the mathematical concepts proposed for non-stationary analysis before describing practical methods which, it must be emphasised, lack generality in their application.

11-2 MATHEMATICAL CONCEPTS

The most important of these is the generalised harmonic analysis of Wiener [8] in which he defines a more general type of Fourier integral given as:

$$x(t) = \int_{-\infty}^{\infty} \exp{(j\omega t)}.\, dx(\omega). \qquad (11\text{-}3)$$

This defines a time-history $x(t)$ in terms of a sine–cosine expansion and a function of frequency $x(\omega)$, which is uniquely determined by $x(t)$. It can be shown that this equation can represent a wide range of functions including both deterministic and random signals. The values of the increment in $x(\omega)$ are necessarily larger than those in (ω) so that the rate of change of $x(\omega)$ need not be finite and the increment in $x(\omega)$ not differentiable, which would be the case if very rapid changes in $x(\omega)$ were experienced. Stationary processes are considered as a special case of the general Fourier integral, and in this case $x(\omega)$ is differentiable, (see fig. 11-5) and we have:

$$\frac{dx(\omega)}{d(\omega)} = X(\omega). \qquad (11\text{-}4)$$

This gives the frequency series for the Fourier integral included in equation (11-3) which then reduces to the inverse Fourier transform for a continuous

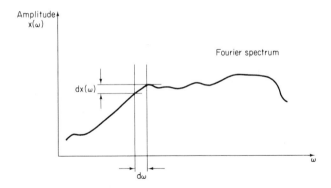

Fig. 11-5. Power spectra defined in terms of generalised harmonic analysis.

value of $x(t)$ given by equation (6-37). The power spectrum was defined as an average of the power obtained over the record length for a given range of frequencies and can be written from equation (8-12), in terms of a small frequency increment, $d\omega$ as:

$$\overline{dP(\omega)} = E\{|dx(\omega)|^2\} \qquad (11\text{-}5)$$

so that the spectral density function can be defined as:

$$G(\omega) = \frac{\overline{dP(\omega)}}{d\omega} \qquad (11\text{-}6)$$

for a specific angular frequency, (ω).

We saw in chapter 6 that if $x(t)$ has a periodic form then we can represent this by a complex Fourier series:

$$x(t) = \sum_{n=-\infty}^{\infty} A_n \exp(j2\pi nt/k). \qquad (11\text{-}7)$$

If $x(t)$ is finite and has a non-periodic form, we can describe this under certain conditions by the Fourier integral:

$$x(t) = \int_{-\infty}^{\infty} X(\omega)\exp(j\omega t)\, d(\omega) \qquad (11\text{-}8)$$

or its discrete value version:

$$x_i = \sum_{n=-N/2}^{N/2} X_n \exp(j2\pi in/N). \quad (i = 1, 2, 3, \ldots, N) \qquad (11\text{-}9)$$

It can be shown [9] that all of these forms are special cases of equation (11-3) and that representation in this general form permits the development of a method of analysis for non-stationary processes. This has been termed Evolutionary Spectral Density Analysis and will be considered in detail later in this chapter.

The advantage of the general representation of equation (11-3) is that it can be used to represent all functions of form, $x(t)$, including periodic functions which would give rise to a Dirac series, and, as will be shown later, can be adapted to represent the non-stationary case.

The inclusion of exp (jωt) in equation (11-3) implies that $x(\omega)$ is orthogonal so that the equation can only be used in this form applied to stationary data. This is understood when we consider the decomposition of exp(jωt) into its constituent sine and cosine terms. Unless the corresponding product term $x(\omega)$ is itself orthogonal, which implies that it too can be decomposed into a set of sine or cosine terms, then their cross-multiplication will not permit the selection of individual frequency terms. We lose, in effect the filtering property of the orthogonal process and so fail to distinguish between $\omega_1 = \omega_2$ and $\omega_1 \neq \omega_2$.

Where the process $x(t)$ is non-stationary then this can no longer be represented by means of equation (11-3). Some classes of non-stationary data, can, however, be represented by an equivalent form:

$$x(t) = \int_{-\infty}^{\infty} \exp (j\omega t).\ dZ(\omega) \qquad (11\text{-}10)$$

where increments in $Z(\omega)$ are correlated and hence no longer orthogonal in form [9]. To define the energy distribution, using this relationship, we have to take into account, not only the mean-squared value of each increment $dZ(\omega)$, but the correlation between them. This leads to a two-dimensional function of frequency which is not capable of physical interpretation:

$$f(\omega,v)\ d\omega\ dv = \overline{dZ(\omega).\ dZ^*(v)} \qquad (11\text{-}11)$$

where * implies the complex conjugate. A similar difficulty is encountered if we attempt to obtain the power-spectra for non-stationary processes from the correlation coefficient. The non-stationary cross-correlation function is defined as:

$$R_{xy}(t_1, t_2) = E[x(t_1)y(t_2)] \qquad (11\text{-}12)$$

for two time-constants t_1 and t_2. Note that here the correlation is dependent on the absolute values of t_1 and t_2 and not on the lag, $(t_1 - t_2)$, which is the case for stationary data. The cross-spectral density function can be derived from equation (11-12), using a double Fourier transform, viz.

$$G_{xy}(\omega_1, \omega_2) = \int\!\!\int_{-\infty}^{\infty} R_{xy}(t_1, t_2) \exp [j(\omega_1 t_1 - \omega_2 t_2)]\ dt_1\ dt_2 \qquad (11\text{-}13)$$

which has an inverse relationship:

$$R_{xy}(t_1, t_2) = \int\!\!\int_{-\infty}^{\infty} G_{xy}(\omega_1, \omega_2) \exp [j(\omega_1 t_1 - \omega_2 t_2)]\ d\omega_1.\ d\omega_2. \qquad (11\text{-}14)$$

These expressions and their derivants [11, 12] are of value in the analytical evaluation of time-dependent processes, but since they have no direct physical meaning are difficult to apply to the analysis of real data. Some work on the application of this method to noise analysis has been carried out by Roberts [13] who indicates the difficulties in determining the spectrum in this way for all but the simplest cases.

The autocorrelation for a non-stationary process is given as:

$$R(t, \tau) = E[x(t), x(t + \tau)] \tag{11-15}$$

which differs from the stationary case in that the instantaneous time of correlation t must be included since the function will no longer be independent of t. Consequently the power spectral density from the correlation coefficient must be stated as:

$$G(t, f) = \int_0^\infty R(t, \tau) \exp(2\pi f \tau) \, d\tau \tag{11-16}$$

[14] which yields a different spectrum for each value of t. Where the non-stationary process is repeatable then a whole ensemble of signals $x(t)$ can be obtained and an estimate of the power-spectrum determined from averaging a series of estimates of the form:

$$\frac{1}{M} \sum_{i=1}^{M} G_i(t, f) \tag{11-17}$$

to give a spectral density averaged over a period T viz.

$$G(t, f) = \frac{1}{T} \int_{t-T}^{t} \frac{1}{M} \sum_{i=1}^{M} G_i(t, f) \, dt. \tag{11-18}$$

In a practical case it is convenient to average over each individual record before summation giving:

$$G(t, f) = \frac{1}{TM} \sum_{i=1}^{M} \int_{t-T}^{t} G_i(t, f) \, dt. \tag{11-19}$$

The choice of averaging interval, T, will depend on resolution requirements and degree of non-stationarity over the record length. For large ensembles, (value of M) then T can be less than the record length giving the effect of small amounts of smoothing and high resolution in the time domain.

11-2-1   THE INSTANTANEOUS SPECTRA

An early attempt to define a time-varying spectra in terms of a Fourier transform was made by Page [15]. He recognised that the time variation is invariably a slow one in terms of the constituent signal frequencies and defined what he called an Instantaneous Power-Spectra as a differential of time. This spectra depends only on the past history of the signal and not upon its future values, which are implied in the classical definition of the Fourier transform.

A power term is proposed as:

$$P_T(\omega) = \left| \int_0^T x(t) \exp(-j\omega t)\, dt \right|^2 \qquad (11\text{-}20)$$

which is identical to the unnormalised form of Shuster's periodogram referred to in chapter 8. This is used to define a spectrum of $x(t)$ as:

$$G(\omega) = \lim_{T \to \infty} \overline{\{P_T(\omega)\}} \qquad (11\text{-}21)$$

The instantaneous spectrum, $P_t(\omega)$ defined over a period T is written as:

$$P_t(\omega) = \frac{d\,\{p_t(\omega)\}}{dt} \qquad (11\text{-}22)$$

where:

$$p_t(\omega) = \int_0^T P_t(\omega)\, dt. \qquad (11\text{-}23)$$

This type of derivation can be considered in terms of a simple difference equation in which, at a particular instant in time, the spectra can be represented as the power difference between two spectra obtained at times T and $(T + \delta T)$, i.e.

$$P_t(\omega) = \frac{1}{T} E \left\{ \left| \int_0^{T+\delta T} x(t) \exp(-j\omega t)\, dt \right|^2 \right.$$

$$\left. - \left| \int_0^T x(t) \exp(-j\omega t)\, dt \right|^2 \right\} \qquad (11\text{-}24)$$

which has also been noted by Priestley [9]. It represents an attempt to reduce the time interval bias error by taking the rate of change of the spectral density at a given frequency. The difference between the two spectra given by equation (11-24) lacks a means of physical interpretation and so has limited applicability. The methods of short-term spectra discussed in the following pages are more useful and enable time-varying spectra to be obtained for localised energy distribution which can then be compared to stationary spectra derived from a similar process.

11-3   SHORT-TERM SPECTRA

The methods of non-stationary spectral analysis most productive of practical results can be grouped into three main classes. These are, the evaluation of short term spectra, evolutionary spectra, and complex spectral demodulation. The latter two techniques are somewhat similar in concept.

Short-term spectral analysis rests on the assumption that a long non-stationary record can be divided into a number of short segments, each of which can be considered as having stationary characteristics within the segment length. Segmentation can be carried out in either the time or frequency domain. The selection of optimum segment size is dependent on the rate of change of non-stationarity and the minimum length of data necessary to provide a reasonable estimate of spectra. These contiguous segments are analysed and the results pieced together, using some form of three-dimensional display (time, frequency, power) to determine the time-dependence of the spectra. An excellent example of this form of display is given in a paper by Singleton and Poulter [16] in which they describe an application of short-term spectral analysis.

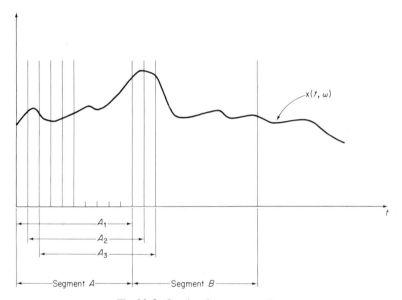

Fig. 11-6. Overlapping segmentation.

A simple practical technique for determining time-dependence has been suggested by Granger and Hataneka [17]. This involves the calculation of spectra for a number of overlapping time segments. A spectra for the first segment of length S is obtained and the commencing point for the next segment of similar length, taken from point $0 \cdot 1\ S$ in advance of the first commencing point (see fig. 11-6). Spectra for similar overlapping segments are taken and their difference then gives a measure of the time-dependence of the data.

The derivation of a general expression for the energy distribution in the time and frequency domain has been given by Ackroyd [18] and is applicable to short-term spectral analysis. He assumes an ideal filter having unity gain between narrow frequency bands, $(f_1 + \delta f)$ and f_1 also $-f_1$ and $-(f_1 + \delta f)$, and having

517

zero gain elsewhere. A voltage signal $x(t)$ is considered to be applied through this filter to a load resistance having unit value so that a current $i(t)$ flows only when $x(t)$ contains frequency components within the two ranges given above. This current is given by:

$$i(t) = \int_{f_1}^{f_1+\delta f} X(f) \exp (j\omega t)\, df + \int_{-f_1-\delta f}^{-f_1} X(f) \exp (j\omega t)\, df \quad (11\text{-}25)$$

where $X(f)$ is the Fourier transform of $x(t)$.

We can express this in terms of the real component only as:

$$i(t) = 2\,R_e \left[\int_{f_1}^{f_1+\delta f} X(f) \exp (j\omega t)\, df \right] \quad (11\text{-}26)$$

so that the power provided at the filter terminals is given as:

$$p(t) = x(t).\, i(t). \quad (11\text{-}27)$$

The energy contained within a narrow frequency band $(f_1 + \delta f)$ for a narrow time interval, $(t + \delta t)$ is therefore defined as:

$$W = \tfrac{1}{2} \int_{t_1}^{t_1+\delta t} x(t).i(t).\, dt \quad (11\text{-}28)$$

since equation (11-27) defines twice the power contained in the real part of the filter pass-band.

The energy density at t_1 and f_1 is given by:

$$e(t_1, f_1) = \lim_{\substack{\delta t \to 0 \\ \delta f \to 0}} \tfrac{1}{2} W(\delta t, \delta f). \quad (11\text{-}29)$$

Hence substituting equations (11-26) and (11-28) into (11-29), a general expression for the energy density at a specific time and frequency can be written as:

$$e(t, f) = x(t).\,\mathrm{Re}[X(f) \exp (j\omega t)]. \quad (11\text{-}30)$$

If a known expression for $x(t)$ is available then equation (11-30) can be applied analytically to derive a three-dimensional representation of the energy density of $x(t)$ in the time-frequency domains. Direct evaluation of equation (11-30) is confined to cases where the form of $x(t)$ is known, since in the analytical case the time variable can be fixed whilst a complete spectrum is taken. Other definitions for short-term spectra have been shown to be special cases of equation (11-23), notably those of Page [15] and Fano [19].

A practical application of Fano's method derives a 'running' average for the frequency content of the continuous signal taken at specific frequencies. An exponential weighting function is chosen to give prominence to the most recent part of the time history. The result was described in chapter 5 as an exponentially-mapped-past (E.M.P.) variable.

The E.M.P. power spectrum can be derived from the Fourier transform for $x(t)$ multiplied by the exponential weighting function [20]:

$$X(f) = \alpha \int_{-\infty}^{t_1} x(t) \cdot \exp\left[-\alpha(t_1-t)\right] \cdot \exp(-j\omega t) \cdot dt \qquad (11\text{-}31)$$

where t_1 relates to an arbitrary time at which the estimate is taken, and α is a constant determining the period of averaging.

To retain the same time variable we write:

$$X(f) = \alpha \exp(-j\omega t_1) \int_{-\infty}^{t_1} x(t) \cdot \exp\left[-\alpha(t_1 - t)\right] \cdot \exp\left[-j\omega(t_1 - t)\right] \cdot dt. \qquad (11\text{-}32)$$

Expanding $\exp(-j\omega t_1)$ and $\exp\left[-j\omega(t_1-t)\right]$ gives:

$$X(f) = \alpha \exp(-j\omega t_1)\left[\int_{-\infty}^{t_1} x(t) \cdot \exp(-\alpha(t_1 - t)) \cdot \cos \omega(t_1 - t) \cdot dt\right.$$

$$\left. -j \int_{-\infty}^{t_1} x(t) \cdot \exp\left[-\alpha(t_1 - t)\right] \cdot \sin \omega(t_1 - t) \cdot dt\right]. \qquad (11\text{-}33)$$

The E.M.P. power-spectrum is obtained by taking the product $X(f) \cdot X^*(f)$ as:

$$G(f) = \alpha^2 \left[\int_{-\infty}^{t_1} x(t) \cdot \exp\left[-\alpha(t_1 - t)\right] \cdot \cos(t_1 - t) \cdot dt\right]^2$$

$$+ \alpha^2 \left[\int_{-\infty}^{t_1} x(t) \cdot \exp\left[-\alpha(t_1 - t)\right] \cdot \sin(t_1 - t) \cdot dt\right]^2. \qquad (11\text{-}34)$$

11-3-1 ANALOG METHODS

A contiguous band analyser incorporating an E.M.P. averaging circuit is shown in fig. 11-7. The signal to be analysed is applied simultaneously to a number of narrow-band filters, each tuned to a different frequency f_1, f_2, f_3 etc. An output from each filter is squared and averaged over the segment period. Averaging for the filter output is carried out using an E.M.P. Integrator. This introduces an exponential time-weighting which can be compensated for in a scaling of results. (In certain circumstances this may be valuable in providing a truer approximation to stationarity over the segment period.) Results are scaled to give an output proportional to the power contained in the signal over the filter bandwidth at the filter centre frequency. The outputs from each averaging integrator are displayed each on a separate channel of a strip chart recorder (see fig. 11-8). The power contained in the random input signal (or energy in a periodic signal) taken over a limited number of frequency points, is displayed in terms of its value and time of occurrence. Stationary signals will be recognised as giving a similar proportional output on each channel at the same instant in time. Non-stationary signals, on the other hand, will show the peak outputs displayed along the time-axes and permit the rate of change of the time-dependent process to be determined by the slope of a line connecting similar events.

519

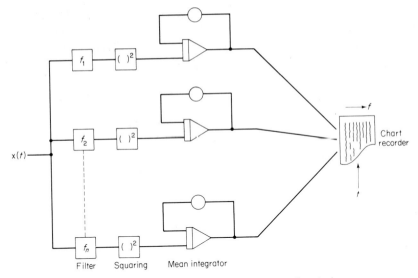

Fig. 11-7. Analog method of short-term spectral analysis.

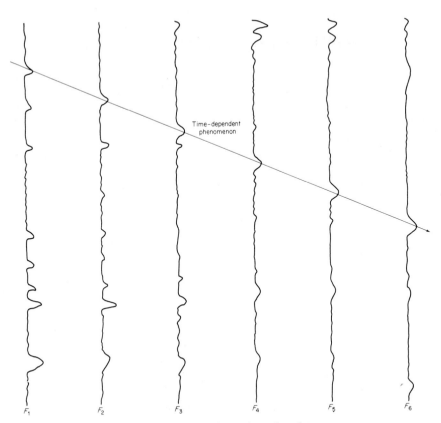

Fig. 11-8. Display of time-dependent data.

An equivalent estimation was described in section 6-8-1 to obtain the E.M.P. Fourier amplitude spectrum. Squaring and addition of the E.M.P. sinspec and cospec functions will obtain a power spectrum estimation for a continuous signal averaged exponentially over a period $\simeq \alpha/5$ at one particular frequency. If the signal is capable of being repeated (e.g. from magnetic loop storage) then a frequency/time power indication can be obtained by repeated analysis at incremented values of the centre frequency. These methods have been used to detect Doppler echoes obtained from a radar pulse reflected from a moving target, when the signals received are heavily obscured by stationary noise. They provide, however, fairly crude estimates of non-stationarity and more valuable results can be obtained by the adaption of hybrid methods described below.

11-3-2   HYBRID METHOD

A hybrid method carried out in the time domain is shown in fig. 11-9 and represents a faster version of the contiguous filter bandwidth method. This uses the programmable segmentation facilities available within the digital computer core memory. Referring to fig. 11-9 the random input signal is converted to digital form and successive samples placed in sequential locations within memory storage. This read-in sequence need only be carried out once, and at a rate determined by the highest frequency contained within the signal and the computer read-in loop instruction time.

The stored data is divided into contiguous segments over which stationarity can be assumed. Serial read-out of this stored data can take place at a fast rate determined by the computer loop instruction time. This will translate the segmented signal into a higher frequency ensemble giving the advantages in

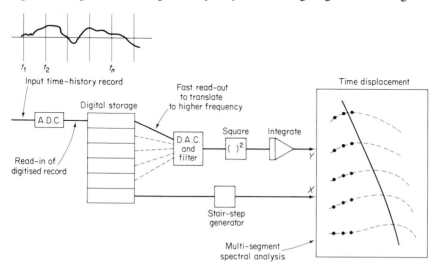

Fig. 11-9. Hybrid method of short-term spectral analysis.

521

spectral analysis procedures described in chapter 6. The samples from the first segment are applied through a digital-to-analog converter to a narrow-band analog filter tuned to the lowest frequency of interest, f_1. The output is squared and integrated to provide a power spectral density point for curve 1 (segment 1). It is convenient to read subsequent segments in turn via this same filter setting to obtain similar points for curves 2 to M. The next read-out (again of segment 1) takes place following increase of the filter centre frequency to $f_1 + \delta f$ (by heterodyne or direct method). This provides the second power spectral density point for curve 1. Subsequent read-outs provide the remaining M of the second-power spectral density points on all the curves. The filter centre frequency is then incremented to $f_1 + 2\delta f$ and the process repeated. This basic method can be modified to accept continuous data by dropping off the oldest sample with each new sample read in (see chapter 6).

11-3-3   DIGITAL METHODS

As with the evaluation of time-invariant spectral density three techniques are available for short-term spectral analysis:

1. Computation via the Fourier transform of a correlation function.
2. The periodogram approach.
3. Use of digital filters.

The derivation via the clipped autocorrelation function is valuable for Gaussian processes and is discussed in section 11-6. Otherwise the calculation via the correlation function for each segment, assuming limited stationarity over the segment length, is as described earlier, giving one spectral density estimate averaged over the analysis bandwidth for each time segment. Mathematically this is the preferred method for reasons given in chapter 8.

The calculation of the spectrum from the squared value of the modulus of the Fourier transform is applicable for method 2. The variance is reduced if the segments, each containing a power of two data points, are analysed in pairs [21]. The fast Fourier transform for each pair of segments is obtained separately and added before the modulus squared value is taken. The procedure is repeated for each pair of segments and the results plotted in the frequency domain giving one representation for each pair of segments taken.

Digital filters can be used for short-term spectra and are particularly applicable to the evolutionary and complex demodulation methods to be described. The use of narrow-band recursive filters represents a fast method when applied to multi-channel analysis so that it is practicable to obtain an averaged version of the spectral density from a number of time-coincident signals in a real-time situation. The resolution can be poor with short-term evaluation since the number of data points for each segment is likely to limit the number of filter weights that can be used.

11-4   EVOLUTIONARY SPECTRA

The calculation of short-term spectra implies replacing the process of averaging over the entire record by averaging over local segments of the record. This leads to a loss in resolution over the time domain which further reduces the accuracy of the process for spectral estimation. Priestley [22] has suggested an alternative method of estimation which avoids this difficulty. He defined a particular form of non-stationary spectral function which approaches the ideal of single point spectral density and is therefore closely related to the well-known stationary methods. Because of this relationship it enables physically meaningful values to be obtained from the estimation. This method is known as evolutionary spectral density estimation and can, under certain conditions, be derived from a single record. As noted earlier this requirement is a necessary pre-requisite for the analysis of certain types of random data.

A generalised form of the Fourier integral was defined in section 11-2 (equation (11-3)). This can be represented further by an even more general expression:

$$x(t) = \int_{-\infty}^{\infty} F(t, \omega)\, dx(\omega) \qquad (11\text{-}35)$$

where $F(t,\omega)$ is an unspecified function of time and frequency and $x(\omega)$ is, as before, an orthogonal process. In equation (11-3), $F(t,\omega)$ was chosen to be $\exp(j\omega t)$. This is a natural choice where $x(t)$ represents a stationary process since it can be shown that this permits the decomposition of the process into a series of sine and cosine waves of different frequencies, and therefore allows a spectral function to be estimated which is capable of simple physical interpretation, (equations (11-5), (11-6)).

Parzen [23], has pointed out that there are an infinite number of possibilities for the series $F(t,\omega)$ and that, even in the stationary case, $\exp(j\omega t)$ is not the only possible choice, although it is probably the one most amenable to calculation. The difficulty when choosing a function for a non-stationary analysis is that it must permit $x(\omega)$ to remain orthogonal in order to retain a physical interpretation for the spectrum. (We saw earlier in section (11-2) that if $x(\omega)$ departs from orthogonality then a double frequency form of the spectrum results.) For this reason a choice of $F(t,\omega) = \exp(j\omega t)$ is inadmissable in the non-stationary case. It is necessary to choose a function which, whilst remaining non-stationary for each realisation of $x(\omega)$, retains some form of frequency representation.

To avoid this difficulty Priestley introduces an oscillatory function:

$$F(t, \omega) = A(t, \omega) \exp(j\omega t) \qquad (11\text{-}36)$$

523

into equation (11-3) giving:

$$x(t) = \int_{-\infty}^{\infty} A(t, \omega) \exp{(j\omega t)} \, dx(\omega) \tag{11-37}$$

as a representation of a time-history $x(t)$ having an orthogonal increment $dx(\omega)$. The time history $x(t)$ is considered to be an oscillatory process where for each fixed value of ω the Fourier transform of the modulating function $A(t,\omega)$ has an absolute maximum at the origin. Thus $F(t,\omega)$ can be considered as representing a series of sine and cosine waves modulated in amplitude by a time-dependent function $A(t,\omega)$. The idea of a periodic frequency term is thus retained providing that the modulating function varies slowly with time. In this respect it accepts similar limitations to those imposed by Page in his instantaneous spectra representation. Referring to equation (11-37) the spectrum can therefore be represented as:

$$E[x^2(t)] = \int_{-\infty}^{\infty} |A(t, \omega)^2| \, dF(\omega) \tag{11-38}$$

where $dF(\omega) = E(|dx(\omega)|^2)$ which represents the average power over the record length for each frequency (shown as $d\overline{P(\omega)}$ in equation (11-5)).

We can also write:

$$E[x^2(t)] = \int_{-\infty}^{\infty} dF(t, \omega) \tag{11-39}$$

so that:

$$dF(t, \omega) = |A(t, \omega)|^2 \, dF(\omega) \tag{11-40}$$

describes the energy distribution of the process in the time and frequency domains. If we now replace the power-spectrum $dF(\omega)$ by $G(\omega).d(\omega)$ (equation (11-6)) and similarly the corresponding spectrum for a non-stationary signal; $dF(t,\omega) = G(t,\omega) \, d\omega$, into equation (11-40) we have:

$$G(t, \omega) = |A(t, \omega)|^2 G(\omega) \tag{11-41}$$

which is the evolutionary spectral density defined in a region centred on (t,ω). Choice of $A(t,\omega)$ is confined to functions which change slowly with time and thus have a Fourier transform which is largest near to zero frequency. Given a particular value of $A(t,\omega)$, it is possible to obtain exactly the corresponding evolutionary spectrum in terms of a covariance function of $x(t)$ using a relationship similar to that of Wiener–Khintchine in conventional stationary analysis. However, estimation of the non-stationary spectrum in terms of a value smoothed in both time and frequency directions can be made without the need for definition of this value.

11-4-1 ESTIMATION OF EVOLUTIONARY SPECTRAL DENSITY

If we assume that our non-stationary signal has the form described by equation (11-37) then it is possible to process this in terms of time and frequency windows. That is, we can now consider the effects of linear transformations purely in terms of spectral frequencies as we can with a stationary process. It is not necessary to define $A(t,\omega) \exp(j\omega t)$ for a particular process if a correct estimating procedure is adopted. An optimum choice for $A(t,\omega)$ can be made based upon the segmentation selected for the non-stationary signal (i.e. the fraction of the record over which the process is assumed to be stationary).

Before describing this estimating procedure it is necessary to introduce the two segmentation terms, T_F and T_X, shown in fig. 11-10. T_F is defined as the maximum time interval over which $A(t,\omega)$ is constant and is referred to as the characteristic width of $A(t,\omega) \exp(j\omega t)$. T_X describes a segment of the record (total length T) and is the maximum time interval over which the process can be

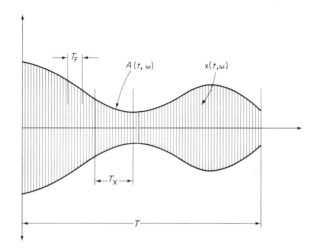

Fig. 11-10. Segmentation in terms of evolutionary spectral analysis.

considered to be stationary. It is equal to the largest possible value of T_F, since beyond this point the process can no longer be considered to be orthogonal. T_X can be described as the characteristic width of the process $x(t)$. It can be shown that the maximum resolution in the time domain can be obtained when $T_F = T_X$.

Given a non-stationary signal $x(t,\omega)$ of finite length T, an estimation for the time varying spectra $f(t,\omega)$ can be obtained at each instant of time over the period $t = 0$, to $t = T$, by using the following procedure. The signal $x(t)$ is passed through a filter centred on ω, and having a frequency response function, $H(\omega)$ corresponding to an impulse response function in the time domain of $h(t)$. The output of the filter, $y(t)$ will be a convolution integral of the form:

$$y(t) = \int_{-\infty}^{\infty} h(\tau).x(t-\tau) \exp\left[-j\,\omega(t-\tau)\right] d\tau \qquad (11\text{-}42)$$

525

the window representing the filter is assumed to have a width in the time domain of T_F, defined previously. The filter response function $H(\omega)$ is normalised to give unity power over $\omega = \pm \infty$, so that

$$\int_{-\infty}^{\infty} |H(\omega)|^2 \, d\omega = 1. \tag{11-43}$$

As with the periodogram (and for the same reasons) we need to smooth the resultant filtered signal in the frequency domain. The output of the filter, $y(t)$ is squared and convolved with a weighting function, $W_T(\tau)$, of characteristic width, T_G, to give a value for the evolutionary spectrum centered on a time, t, and an angular frequency ω, viz.

$$\hat{f}(t, \omega) = \int_{-\infty}^{\infty} W_T(\tau) \, | \, y^2(t - \tau) \, | \, d\tau. \tag{11-44}$$

The value of ω_0 may then be varied to obtain an estimate for the complete spectrum at this time. The procedure is shown in fig. 11-11 and will be seen to

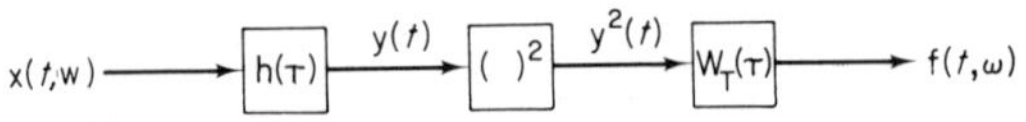

Fig. 11-11. Evolutionary spectral evaluation.

take the general form of estimation for stationary signals with the addition of time domain filtering. The form of window for $h(\tau)$ and $W_T(\tau)$ (both expressed in the time domain) can be obtained from any of the standard windows used for spectral analysis and described in chapter 8.

Having determined the form for $h(\tau)$ and $W_T(\tau)$ the parameters describing the filter widths, (T_F and T_G), are obtained so as to minimise the relative mean-square error of $f(t,\omega)$. A method of selection for these parameters, together with a detailed discussion of the design relationships for evolutionary spectral response is given in [24]. An empirical determination of $|A(t,\omega)|^2$ based on a curve-fitting to known spectral results for an equivalent stationary case is described by Hammond [12].

This procedure for non-stationary analysis has proved effective for smooth and slow changes with time. Determination of optimum window parameters and detailed analysis in time and frequency can be demanding in computer time, and for many purposes, the complex demodulation procedures described in the next section would be preferred. Both methods give rise to a large error when the rate of change of the non-stationary trend becomes comparable with the rate of change or response time of the signal source.

526

11-5   COMPLEX DEMODULATION

This is a method of power spectral density analysis which has been developed to enable the behaviour of a time series within a relatively narrow frequency band to be studied. In this sense it is a method of short-term analysis by contiguous frequency bands, but differs in the way the spectrum is obtained. Apart from its use in non-stationary spectrum analysis it has been used for ensemble averaging and prediction.

The method of complex demodulation was first suggested by Granger and Hataneka [17] for the analysis of discrete economic time series and proved of considerable value before the rediscovery of the fast Fourier transform algorithm. He derived a similar representation to that of Priestley for a non-stationary series, namely:

$$x_t = \int_{-\pi}^{\pi} \exp\left(j\omega t\right) a(t, \omega) \, dz(\omega) \qquad (11\text{-}45)$$

where $a(t,\omega)$ is a function changing slowly with time. He shows that if this series is multiplied by a low-pass filter function having a bandwidth, $2B$, then the resultant complex product may be used to obtain the average amplitude and phase over this frequency band. This use of complex demodulation permits the transfer of the frequency band of analysis into the place originally occupied by the low-pass filter bandwidth which commences at zero frequency, so that the characteristics of the contiguous bands can be obtained by using a single low-pass filter and the variance, and hence power density, determined for each frequency band.

The method, as described by Granger, Godfrey [17, 25] and others is a digital one. It assumes that a real valued series x_i is available or can be derived from sampled continuous data. The procedure reduces this to a number of complex frequency series of shorter length associated with various short ranges of angular frequency (ω), from which the amplitude and phase over this frequency band at a single value of time or for the entire record or ensemble can be derived.

Given the real series $x_i (i = 1, 2, \ldots, N)$ the products:

$$x_i \sin \omega_0 t$$
$$x_i \cos \omega_0 t \qquad (11\text{-}46)$$

are derived from a given frequency value number. These two series are then subject to a low-pass (digital) filter to obtain:

$$y_i' = F(x_i \sin \omega_0 t)$$
$$y_i'' = F(x_i \cos \omega_0 t) \qquad (11\text{-}47)$$

527

so that:

$$2[(y_i')^2 + (y_i'')^2]^{\frac{1}{2}} = A \qquad (11\text{-}48)$$

which represents an estimate of the amplitude of the complex series at time t, and could be termed the instantaneous spectrum of the signal at this point in time. An estimate of the phase angle at time t, is given as:

$$\text{Arctan}\left(\frac{y_i'}{y_i''}\right) = \theta \qquad (11\text{-}49)$$

which is also an 'instantaneous' quantity.

In its practical implementation on a digital machine the difficulties evolve round the optimum design of the digital filter to minimise leakage effects and the speed of carrying out the computational process.

An alternative representation is to define the complex demodulation as:

$$y_i(t, \omega_0) = F[x_i \exp (j\omega_0 t)] \qquad (11\text{-}50)$$

This implies that to obtain the spectrum we will need to take the square of the sum of $x_i \sin \omega_0 t$ and $x_i \cos \omega_0 t$ rather than the sum of squares of $x_i \sin \omega_0 t$ and $x_i \cos \omega_0 t$. This will obtain the same values in this case due to the action of the low-pass filter as shown below.

If $x_i = A \sin (\omega_0 t + \theta)$ then:

$$x_i \exp (j\omega_0 t) = A \sin (\omega_0 t + \theta) (\cos \omega_0 t - j \sin \omega_0 t)$$

$$= \tfrac{1}{2}A \sin \theta + \tfrac{1}{2} \sin (2\omega_0 t + \theta) - \tfrac{1}{2}Aj \cos \theta - \tfrac{1}{2}Aj \cos (2\omega_0 t + \theta). \qquad (11\text{-}51)$$

Subjecting equation (11-51) to a low-pass filter operation:

$$y_i = F[x_i \exp (j\omega t)] \qquad (11\text{-}52)$$

and obtaining the modulus gives:

$$2(y_i^2)^{\frac{1}{2}} = A$$

as required. (This may be compared with the derivation given earlier.) Hence we can regard equations (11-47) and (11-50) as giving similar results. The time average of the modulus squared value of $y_i(t,\omega_0)$ is taken as an estimate for the spectral density over the band $(\omega_0 \pm B)$ at a particular time, t, where B is the bandwidth of the low-pass filter. This is obtained over the time period of the signal for a range of centre frequencies, ω_0. The method has been used particularly to determine trends in economic series and some interesting examples are given by Godfrey in his paper.

An analog version of this technique has been known for some considerable time and was described in section 8-8-2, where equations (11-47) were called sinspec(f) and cospec(f) respectively. Since the digital implementation of this method implies $2N^2$ complex multiplications for each frequency band it can be expensive of digital computer time unless extensive decimation is allowed. Consequently the equivalent analog method has much to recommend it.

11-5-1 A HYBRID EXAMPLE

In equation (11-45) the function $a(t,\omega)$ is assumed to change slowly with time although its actual form is not specifically stated. This will introduce an error dependent on the time interval taken for the analysis.

Where the form of $a(t,\omega)$ is known this error can be minimised by optimum choice of the time-averaging band. This is the case for a Doppler-shifted signal and in the example which follows the analysis of the non-stationary characteristic of the noise radiated from a supersonic aircraft will be considered. Here the noise, as intercepted at a fixed point along the aircraft flyover path, is subject to a number of modifications such as Doppler frequency shift, inverse square law of transmission and atmospheric attenuation all of which result in a highly non-stationary signal pattern. These causes of non-stationarity are highly deterministic so that a form of $a(t,\omega)$ can be estimated and used, either to limit the number of analyses required in the digital case, or as a modulation characteristic for determination of the sinspec and cospec function in the analog case. The correction for Doppler shift in this way for the constant velocity case has been described by Mangiarotty and Turner (26), using a digital computer. A hybrid method is described below which also permits correction for change in aircraft velocity (measured by ground observation) and inclusion of the other factors mentioned above.

For reasons which will become apparent later a fixed-bandwidth heterodyne technique of power spectral analysis is used. Third-octave proportional bandwidth methods, traditionally used for noise analysis, present a difficulty in frequency correction since it would be necessary to vary the bandwidth of analysis to preserve the correct third-octave relationship at the corrected frequency.

Complex modulation of the signal is achieved first by multiplying the signal with the output of a local oscillator having frequency f_a. The resultant signal is then passed through a low-pass filter of bandwidth B so as to reject all noise signals outside the range: $f_a - B < f < f_a + B$. It should be noted that the filter centre frequency is now equal to the heterodyne frequency thus permitting change under program control by altering the local oscillator frequency. The practical implications of this are somewhat more complicated since simple heterodyning leaves an ambiguity of phase, (see chapter 8). This is removed by heterodyning the signal with the sine and cosine function of the controlling frequency, low-pass filtering the products, and squaring and adding the results.

A flow diagram for the method is shown in fig. 11-12. This is mechanised using an iterative analog computer linked to a digital machine [27]. The noise

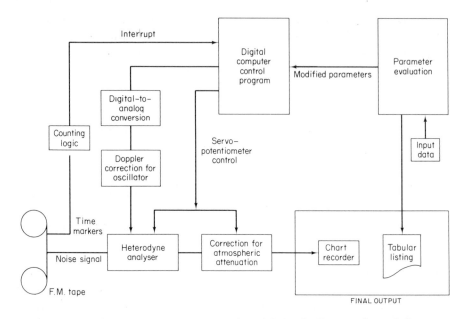

Fig. 11-12. Hybrid method of complex demodulation for flyover noise analysis.

signal, recorded at ground level on analog magnetic tape, forms a loop containing three data tracks; the recorded noise, a clock track synchronised with the shutter of the flight path recording camera, and a start/end marker track which denotes the portion of the loop to be used. The program is initiated from the start marker when the analog section of the program proceeds to count a lagged version of the frame markers. The initial lag is required to take account of the transit time for the noise signal from the aircraft. The analyser is put into operation after N_1 counts of the same markers, lagged by a time T_1 where: $N_1 + T_1$ = initial range of aircraft/velocity of sound and commences with the frequency determining potentiometers set to values corresponding to the lowest analysis frequency. As each subsequent lagged frame marker is read this is routed to the digital computer as an interrupt flag, causing a new value of equivalent aircraft velocity to be output to the digital-to-analog converter in order to control the servo-potentiometer and hence the heterodyne analysis frequency.

The aircraft range is obtained by integrating the velocity values given and is used to modify the noise signal and, correct for inverse square-law transmission characteristic. This gives an amplitude modulated form to the signal which is then applied as input to a heterodyne spectral analyser (see section 8-8-2). The analysis centre frequency is controlled from the voltage controlled oscillator shown. Doppler correction is achieved by amplitude modulation of this control voltage to the heterodyne oscillator by means of the function, $(a + v)/a$ where a = velocity of sound and v = velocity of the aircraft. The output of the analyser thus represents the level of noise available at a fixed point in relation to the moving aircraft following correction for the inverse law of attenuation, as a function of time, and for the Doppler frequency shift at a given frequency corresponding to the heterodyne frequency.

This results in a mean-square value of the noise level, corrected over the time interval of the signal, and at a specific frequency corresponding to the un-modulated input through the heterodyne oscillator. Scaling correction is made for atmospheric attenuation (assumed to be a function of frequency only for a given aircraft run) and for the analysis bandwidth. The results are plotted as a family of curves, giving noise power against time for a number of analysis frequency values (see fig. 11-13).

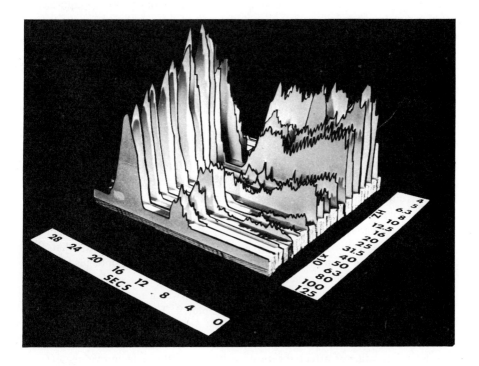

Fig. 11-13. Three-dimensional example of noise analysis record.

11-6 CORRELATION

Autocorrelation for a non-stationary random process $x(t)$ calculated at two times t_1 and t_2 may be defined as:

$$R_x(t_1, t_2) = E[x(t_1).x(t_2)] \qquad (11\text{-}53)$$

531

If we now replace the variables t_1 and t_2, by their interval difference, $\tau = (t_2 - t_1)$ and relate the times to a particular instant $t = \frac{1}{2}(t_1 + t_2)$ then equation (11-53) can be written:

$$R_x(t, \tau) = E[x(t - \tfrac{1}{2}\tau)x(t + \tfrac{1}{2}\tau)] \qquad (11\text{-}54)$$

Silverman [28] and Piersol [3] have noted that this form of the correlation function can lead to a limited form of 'local stationarity' for the non-stationary process.

Thus if the non-stationary signal has the form:

$$x(t, \tau) = A(t)x(t) \qquad (11\text{-}55)$$

where $A(t)$ is a slowly varying deterministic function and $x(t)$ is an equivalent stationary signal we can express the relationship of the correlation functions for both sides from equation (11-55) as:

$$R_x(t, \tau) = E[A(t - \tfrac{1}{2}\tau) \cdot A(t + \tfrac{1}{2}\tau)]R_x(\tau). \qquad (11\text{-}56)$$

For small delays (i.e. in the region of the correlation peak):

$$R_x(t, \tau) = A(t)R_x(\tau). \qquad (11\text{-}57)$$

Similarly for a small frequency variation of the function $A(t)$ the power spectrum is:

$$G_x(t, f) = A(t)G_x(f) \qquad (11\text{-}58)$$

which is similar to that obtained by Priestley (equation (11-41)).

Thus over a limited delay around the peak of the correlation and, if a deterministic variation in the non-stationary signal is known, then satisfactory results can be achieved. It has been demonstrated by Piersol [3] that the assumptions made in deriving equations (11-57) and (11-58) can be justified for some important classes of practical data.

The use of a clipped autocorrelation function, which requires only the zero crossings of the signal, was introduced in chapter 9. This can also be applied to the non-stationary case. Using a similar model to that given by equation (11-55) and regarding $A(t)$ as an amplitude modulating function for $x(t)$ we can see that its inclusion will not affect the zero crossings for the data, providing $A(t)$ is positive at all times. This enables a method of correlation to be determined for a non-stationary process based on the autocorrelation function for the clipped function:

$$x^s(t) = 1 \text{ for } x(t) \geqslant 0$$
$$= 0 \text{ for } x(t) < 0. \qquad (11\text{-}59)$$

This is related to $R_x(\tau)$ by:

$$R_x(\tau) = \sin\left[\tfrac{1}{2}\pi R_x^s(\tau)\right]. \qquad (11\text{-}60)$$

Since the zero crossings for clipped non-stationary data using this model will be indistinguishable from an equivalent stationary signal we can regard equation (11-60) as representing the autocorrelation required, i.e.

$$R_x(t, \tau) = \sin \left[\tfrac{1}{2}\pi R_x^s(\tau)\right]. \tag{11-61}$$

If this estimate is used for the derivation of spectral density then it will be necessary to modify the output by a variance factor for each value of the time interval t, to correct for change in stationarity over this interval. Since the analysis of non-stationary signals can be time-consuming there are obvious advantages in using this method if the data characteristics permit this type of simplified estimate. As pointed out in section 9-8-2 a longer length of data series is required to obtain the same statistical accuracy if the full dynamic range of the signal is to be utilised.

BIBLIOGRAPHY

1. PRIESTLEY, M. B. and RAO, T. S. A test for non-stationarity of time series. *J. Roy. Stat. Soc.*, B, **31**, 1, 140-99, 1969.
2. DANIELS, H. E. Contribution to the discussion of evolutionary spectra and non-stationary processes. *J. Roy. Stat. Soc.*, B, **27**, 234, 1965.
3. PIERSOL, A. G. Power spectra measurements for spacecraft vibration data. *J. Spacecraft* **4**, 12, 1613-17, Dec. 1967.
4. BARLOW, J. S. An electronic method for detecting evoked responses of the brain and for reproducing their average waveforms. *Electroenceph. Clin. Neurophysiol.* **9**, 340-3, 1957.
5. SIEBERT, W. M. *Processing Neuroelectric Data.* Tech. Press, M.I.T., Cambridge, Mass. Research Mono, 351, 1959.
6. BENDAT, J. S. Mathematical analysis of average response value for non-stationary data. *I.E.E.E. Trans.* (Bio. Med. Eng.) BME-11, 72-81, July 1964.
7. MARK, W. D. Spectral analysis of the convolution and filtering of non-stationary random processes. *J. Sound. Vib.* **11**, 1, 19-64, 1970.
8. WIENER, N. Generalised harmonic analysis. *Acta. Maths. Stockh.* **55**, 117, 1930.
9. PRIESTLEY, M. B. Power spectral analysis of non-stationary random processes. *J. Sound. Vib.* **6**, 86, 1967.
10. CRAMER, H. On some classes of non-stationary processes. Proc. Fourth, Berkeley Symposium on mathematics, statistics and probabilities, California, 2, 57, 1960.
11. BENDAT, J. S. and PIERSOL, A. G. *Random Data: Analysis and Measurement Procedures.* John Wiley, New York, 1971.
12. HAMMOND, J. K. On the response of single and multi-degree of freedom systems to non-stationary random excitations. *J. Sound. Vib.* **7**, 3, 393, 1968.
13. ROBERTS, J. B. On the harmonic analysis of evolutionary random vibrations. *J. Sound. Vib.* **2**, 3, 393, 1965.
14. ZHELEZNOV, N. A. Some problems of the spectral correlation theory of non-stationary signals. *Radioteknika I. Electronika*, **4**, 3, 359-79, 1959.
15. PAGE, C. G. Instantaneous power spectra. *J. Appl. Phys.* **23**, 1, Jan. 1952.

16. SINGLETON, R. C. and POULTER, T. C. Spectral analysis of the male killer whale. *I.E.E.E. Trans.* (Audio Acoust.) AU-15, 104-13, 1967.
17. GRANGER, C. W. J. and HATANEKA, M. *Spectral Analysis of Economic Time Series.* Princeton University Press, 1964.
18. ACKROYD, M. H. Instantaneous and time-varying spectra—an introduction. *Radio Electron. Eng.* **39**, 3, 145-52, 1970.
19. FANO, R. M. Short-term autocorrelation functions and power spectra. *J. Acoust. Soc. Amer.* 22, 546-50, 1950.
20. OTTERMAN, J. The properties and methods for computation of exponentially mapped past statistical variables. *I.R.E. Trans.* (Auto Control) AC-5, 7. 11-17, Jan. 1960.
21. ENOCHSON, L. D. and OTNES, R. K. Programming and analysis for digital time series data. The Shock and Vibration Information Center, U.S. Dept of Defence, Washington, D.C., 1968.
22. PRIESTLEY, M. B. Evolutionary spectra and non-stationary processes. *J. Roy. Stat. Soc.* B, **27**, 204, 1965.
23. PARZEN, E. Statistical reference on time-series and hilbert space methods. *Stanford University Tech. Rep.* 23, 1959.
24. PRIESTLEY, M. B. Design relations for non-stationary processes, *J. Roy. Stat. Soc.* B, **29**, 570, 1967.
25. GODFREY, M. D. Prediction for non-stationary stochastic processes. In *Advanced Seminar in Spectral Analysis of Time Series.* Ed. B. Harris, John Wiley, New York, 1967.
26. MANGIAROTTY, R. A. and TURNER, B. A. Wave relation and doppler effect correction for motion of a source, observer and the surrounding medium. *J. Sound. Vib.* **6**, 1, 110-16, 1967.
27. BEAUCHAMP, K. G., THOMASSON, P. G. and WILLIAMSON, M. E. A hybrid computer solution of a non-stationary process. Symposium C.A.D. in Eng., University of Waterloo, May 1971.
28. SILVERMAN, R. A. Locally stationary random processes, *I.R.E. Trans.* (Inf. Theory) IT-3, Sept. 1957.

Additional References
29. THRALL, G. P. and BENDAT, J. S. Mean and mean square measurements of non-stationary random processes. N.A.S.A. CR226, Washington, D.C., May 1965.
30. PRIESTLEY, M. B. Time-dependent spectral analysis and its applications to prediction and control. Lecture series, Applications and methods of random data analysis, Southampton University, July 1969.
31. DREXLER, J. and KROPAX, O. Some problems of analysing non-stationary random strain processes in mechanical systems. Lecture series, Applications and methods of random data analysis, Southampton University, July 1969.
32. ABDRABBO, N. A. and PRIESTLEY, M. B. On the prediction of non-stationary processes. *J. Roy. Stat. Soc.,* B, **29**, 570, 1967.
33. LIN, Y. K. Non-stationary response of continuous structures to random loading. *J. Acoust. Soc. Amer.,* **35**, 227, 1963.
34. LEVIN, M. J. Instantaneous spectra and ambiguity functions. *I.E.E.E. Trans.* (Inf. Theory) IT-10, 95-7, 1964.
35. RIHACZEK, A. W. Signal energy distribution in time and frequency, *I.E.E.E. Trans.* (Inf. Theory) IT-14, 369-74, 1968.

36. HINICH, M. Estimation of spectra after hard clipping of Gaussian processes. *Technometr.* **9**, 391, Aug. 1967.
37. MARK, W. D. Spectral analysis of the convolution and filtering of non-stationary stochastic processes. *J. Sound Vib.* **11**, 19, 1970.
38. PRIESTLEY, M. B. Some notes on the physical interpretation of spectra of non-stationary stochastic processes. *J. Sound Vib.* **17**, 1, 51-4, 1971.

INDEX

AUTHOR INDEX